Birkhäuser

Trends in Mathematics

Trends in Mathematics is a series devoted to the publication of volumes arising from conferences and lecture series focusing on a particular topic from any area of mathematics. Its aim is to make current developments available to the community as rapidly as possible without compromise to quality and to archive these for reference.

Proposals for volumes can be submitted using the Online Book Project Submission Form at our website www.birkhauser-science.com.

Material submitted for publication must be screened and prepared as follows:

All contributions should undergo a reviewing process similar to that carried out by journals and be checked for correct use of language which, as a rule, is English. Articles without proofs, or which do not contain any significantly new results, should be rejected. High quality survey papers, however, are welcome.

We expect the organizers to deliver manuscripts in a form that is essentially ready for direct reproduction. Any version of $\TeX$ is acceptable, but the entire collection of files must be in one particular dialect of $\TeX$ and unified according to simple instructions available from Birkhäuser.

Furthermore, in order to guarantee the timely appearance of the proceedings it is essential that the final version of the entire material be submitted no later than one year after the conference.

More information about this series at http://www.springer.com/series/4961

José Luis Cisneros-Molina • Dũng Tráng Lê
Mutsuo Oka • Jawad Snoussi
Editors

Singularities in Geometry, Topology, Foliations and Dynamics

A Celebration of the 60th Birthday of José Seade, Merida, Mexico, December 2014

Birkhäuser

Editors
José Luis Cisneros-Molina
Unidad Cuernavaca
Instituto de Matemáticas, UNAM
Cuernavaca, Morelos, Mexico

Dũng Tráng Lê
Centre de mathématiques et informatique
Université d'Aix-Marseille
Marseille, France

Mutsuo Oka
Tokyo University of Science
Tokyo, Japan

Jawad Snoussi
Unidad Cuernavaca
Instituto de Matemáticas, UNAM
Cuernavaca, Morelos, Mexico

ISSN 2297-0215 ISSN 2297-024X (electronic)
Trends in Mathematics
ISBN 978-3-319-39338-4 ISBN 978-3-319-39339-1 (eBook)
DOI 10.1007/978-3-319-39339-1

Library of Congress Control Number: 2016963255

Mathematics Subject Classification (2010): 32Sxx, 58Kxx, 14B05, 14E15, 14H20, 14J17, 14P25, 37F75, 57R45

Printed on acid-free paper

This book is published under the trade name Birkhäuser, www.birkhauser-science.com
The registered company is Springer International Publishing AG
The registered company address is: Gewerbestrasse 11, 6330 Cham, Switzerland

Contents

Contents

Preface

The workshop "Singularities in geometry, topology, foliations and dynamics" took place in Mérida, Mexico, from December 8 to 19, 2014. It was a celebration of José Seade's 60th birthday. This meeting was preceded by a two week long school, held at the Institute of Mathematics of Universidad Nacional Autónoma de México (UNAM), in Cuernavaca, Mexico.

The workshop was held in a historical building of the Universidad Autónoma de Yucatán, located in downtown in Mérida. It was supported and financed by various entities of UNAM (Instituto de Matemáticas, Posgrado de Matemáticas, Dirección General de Asuntos del Personal Académico), as well as by the Consejo Nacional de Ciencia y Tecnología (CONACyT) and the Abdus Salam International Centre for Theoretical Physics (ICTP). The main organizing institution was the Institute of Mathematics of UNAM.

During the two weeks of the workshop, a total of forty-four plenary talks were presented, as well as ten poster presentations. There were a total of 121 participants coming from 14 different countries, a list of which appears below. The themes in singularity theory discussed at this meeting include the topology of singularities and characteristic classes, resolutions of spaces and of foliations, contact structures, Milnor fibrations, metric and bi-Lipschitz behaviour, equisingularity, moduli of spaces and foliations, among others.

José Seade, also known to his colleagues as Pepe Seade, was originally trained as an algebraic topologist at the University of Oxford, where he wrote his Ph.D. thesis under the direction of Brian Steer and Nigel Hitchin. Since his very first publications he showed his interest into singularities. Over a period of 35 years of productive research, Pepe's work in singularity theory has explored a variety of subthemes: vector fields, characteristic classes, mappings and foliations, Milnor fibrations, contact structures, and the topology of local singularities. Even then, his strong dedication to the field of singularities has not prevented him from working in other fields, such as Kleinian groups and dynamical systems, where his research has also had an unmistakable impact. Since 1981, Pepe has published 63 research papers as well as 4 books. Two of his books have been awarded the Ferran Sunyer i Balaguer prize: one a book on the topology of singularities and the other a book on complex Kleinian groups.

Pepe has also played an important role in integrating the Mexican mathematical community into a variety of important international mathematical networks. This is due in large part to his abilities in organizing international meetings and facilitating the formation of research groups, as well as his readiness to help young people obtain financial support or make scientific contacts abroad. These activities – in Mexico, America, and worldwide – have helped make Mexico an international center for singularity theory.

These are just some of the reasons explaining why so many mathematicians from all over the world attended the workshop.

This volume consists of 13 original research articles – submitted by some of the participants of the workshop – covering various aspects of singularity theory. At least one co-author of each paper was present at the conference or took part in its preparation.

The scientific committee of the workshop consisted of Roberto Callejas-Bedregal (Universidade Federal da Paraíba, Brazil), José Luis Cisneros-Molina (UNAM, Cuernavaca, Mexico), Javier Fernández de Bobadilla (Instituto de Ciencias Matemáticas, Spain; Institute for Advanced Study, Princeton, USA), Xavier Gomez-Mont (CIMAT, Mexico), Renato Iturriaga (CIMAT, Mexico), Anatoly Libgober (University of Illinois at Chicago, USA), David Massey (Northeastern University, USA), Mutsuo Oka (Tokyo University of Science, Japan), Anne Pichon (Institut de Mathématiques de Luminy, France), Marcelo Saia (ICMC, USP, São Carlos, Brazil), Jawad Snoussi (UNAM, Cuernavaca, Mexico), Mark Spivakovsky (Institut de Mathématiques de Toulouse, France), Alberto Verjovsky (UNAM, Cuernavaca, Mexico).

The organizing committee in Mexico consisted of Vanessa Alderete (UNAM, Cuernavaca), Waldemar Barrera (UADY, Mérida), Omegar Calvo (CIMAT, Guanajuato), José Luis Cisneros-Molina (UNAM, Cuernavaca), Jesús Muciño (UNAM, Morelia), Juan Pablo Navarrete (UADY, Mérida), Ramón Peniche Mena (UADY, Mérida), Jawad Snoussi (UNAM, Cuernavaca), Manuel Alejandro Ucan Puc (UNAM, Cuernavaca).

The editorial committee of this volume comprises José Luis Cisneros-Molina (Instituto de Matemáticas UNAM, Unidad Cuernavaca, Mexico), Mutsuo Oka (Tokyo University of Science, Japan), Dũng Tráng Lê (Université Aix-Marseille, France) and Jawad Snoussi (Instituto de Matemáticas UNAM, Unidad Cuernavaca, Mexico)

The members of the editorial committee are grateful to all the referees who did a fantastic job in reviewing all the submitted papers, sometimes proposing interesting and useful modifications. We would like also to thank the entire team of the Trends in Mathematics series for their wonderful work.

Acknowledgments. We are grateful to the Consejo Nacional de Ciencia y Tecnología (CONACyT) for his financial support to the meeting through the grant CONACyT 224652. We also acknowledge the support of CONACyT-CNRS-LAISLA. The first editor was supported by the grants UNAM-DGAPA-PAPIIT IN106614 and CONACyT 253506. The fourth editor was supported by the grant UNAM-DGAPA-PAPIIT 107614.

List of Participants

Alanis Lilia CIMAT, Guanajuato, Mexico	Participant
Alderete Acosta Vanessa Instituto de Matemáticas, Unidad Cuernavaca, Universidad Nacional Autónoma de México	Organizing Committee
Alvarez Patiño Raul Mexico	Participant
Ardila Ardila Jonny Federal University of Rio de Janeiro, Brazil	Participant
Aroca José Manuel Universidad de Valladolid, Spain	Speaker
Aroca Bisquert Fuensanta Instituto de Matemáticas, Unidad Cuernavaca, Universidad Nacional Autónoma de México	Speaker
Arroyo Camacho Aubin Instituto de Matemáticas, Unidad Cuernavaca, Universidad Nacional Autónoma de México	Participant
Artal Enrique University of Zaragoza, Spain	Speaker
Barbosa Grazielle Feliciani Universidade Federal de São Carlos, Brazil	Participant
Barranco Mendoza Gonzalo Instituto de Matemáticas, Unidad Cuernavaca, Universidad Nacional Autónoma de México	Participant
Barrera Waldemar Universidad Autónoma de Yucatán, Mexico	Organizing Committee
Birbrair Lev Universidade Federal do Ceara, Brazil	Speaker
Ble Gamaliel Universidad Juárez Autónoma de Tabasco, Mexico	Participant

Boileau Michel Université Aix-Marseille, France	Speaker
Brambila-Paz Leticia CIMAT, Guanajuato, Mexico	Participant
Brasselet Jean-Paul Université Aix-Marseille, France	Speaker
Cabrera Ocañas Carlos Instituto de Matemáticas, Unidad Cuernavaca, Universidad Nacional Autónoma de México	Participant
Cadavid Aguilar Natalia CINVESTAV, Mexico	Participant
Callejas-Bedregal Roberto Universidad Federal de Paraiba, Brazil	Scientific Committee/ Speaker
Calvo José Omegar CIMAT, Guanajuato, Mexico	Organizing Committee/ Speaker
Camacho Caldrón Angelito Instituto de Matemáticas, Unidad Cuernavaca, Universidad Nacional Autónoma de México	Participant
Cano Felipe Universidad de Valladolid, Spain	Speaker
Cano Cordero Ángel Instituto de Matemáticas, Unidad Cuernavaca, Universidad Nacional Autónoma de México	Speaker
Casamiglia Gabriel Universidade Federal Fluminense, Brazil	Participant
Castellanos Víctor Universidad Juárez Autónoma de Tabasco, Mexico	Participant
Chen Ying ICMC, Universidade de São Paulo, Brazil	Participant

Cisneros Molina José Luis Instituto de Matemáticas, Unidad Cuernavaca, Universidad Nacional Autónoma de México	Organizing & Scientific Committee/ Speaker
Corrêa Maurício Universidade Federal de Viçosa, Brazil	Participant
de la Peña José Antonio CIMAT, Guanajuato, Mexico	Speaker
de la Rosa Castillo Miguel Ángel Universidad Autónoma de Zacatecas, Mexico	Participant
Dos Santos Raimundo Araujo Instituto de Ciencias Matematicas y Computacion, Universidad de Sau Paulo, Sau Carlos, Brazil	Speaker
Ebeling Wolfgang Leibniz University, Hannover, Germany	Speaker
Elizondo Javier Instituto de Matemáticas, Universidad Nacional Autónoma de México	Participant
Fernández Duque Miguel Universidad de Valladolid, Spain	Participant
Ferrarotti Massimo University of Pisa, Italy	Participant
Ghassab Ali Sharif University of Thechnology, Iran	Participant
Giles Arturo CIMAT, Guanajuato, Mexico	Participant
Gómez Morales Mirna University of Warwick, England	Participant
Gómez-Mont Xavier CIMAT, Guanajuato, Mexico	Scientific Committee/ Speaker
González Urquiza Adriana Instituto de Matemáticas, Unidad Cuernavaca, Universidad Nacional Autónoma de México	Participant

González-Dorrego María del Rosario — Participant
Universidad Autónoma de Madrid, Spain

Goryunov Victor — Speaker
University of Liverpool, England

Greuel Gert-Martin — Speaker
University of Kaiserslautern, Germany

Grulha Jr. Nivaldo — Participant
Instituto de Ciências Matemáticas y Computação, Universidad de Saõ Paulo, Saõ Carlos, Brazil

Guadarrama Miguel Ángel — Participant
Instituto de Matemáticas, Universidad Nacional Autónoma de México

Guillot Santiago Adolfo — Participant
Instituto de Matemáticas, Unidad Cuernavaca, Universidad Nacional Autónoma de México

Guzmán Durán Carlos Rodrigo — Participant
CIMAT, Guanajuato, Mexico

Hefez Abramo — Speaker
Universidade Federal Fluminense, Niteroi, Brazil

Hernández Orozco José Antonio — Participant
Centro de Ciencias Matemáticas, UNAM, Mexico

Ilyashenko Yuli — Speaker
Cornell University, USA

Inuma Zamora Francisco Miguel — Participant
Universidade Federal do Rio de Janeiro, Brazil

Jaurez Rosas Jessica Angélica — Participant
Instituto de Matemáticas, Universidad Nacional Autónoma de México

Jurado Liliana — Participant
Universidade Federal do Rio de Janeiro, Brazil

Kanarek Blando Herbert Universidad de Guanajuato, Mexico	Participant
Kashani Seyed Mohammad Tarbiat Modares University, Iran	Poster
László Tamás Alfred Renyi Institute of Mathematics, Hungarian Academy of Sciences, Hungary	Participant
Lê Dũng Tráng Université Aix-Marseille, France	Speaker
Libgober Anatoly University of Illinois at Chicago, USA	Scientific Committee/ Speaker
López de Medrano Lucía Instituto de Matemáticas, Unidad Cuernavaca, Universidad Nacional Autónoma de México	Participant
López de Medrano Santiago Instituto de Matemáticas, Universidad Nacional Autónoma de México	Speaker
Luengo Ignacio Universidad Complutense de Madrid, Spain	Speaker
Magaña Cáceres Julio César Centro de Ciencias Matemáticas, UNAM, Mexico	Participant
Martínez León Victor Universidad Federal do Rio de Janeiro, Brazil	Participant
Martins Luciana Universidade Estadual Paulista, Brazil	Participant
Martins Rafaella Universidade de São Paulo, Brazil	Participant
Mattei Jean-Francois Université Paul Sabatier, France	Speaker
Meersseman Laurent Université de Rennes, France	Speaker

Menegon Neto Aurelio Universidad Federal de Paraíba, Brazil	Speaker
Miranda Aldicio José Universidade Federal de Uberlândia, Brazil	Poster
Mond David University of Warwick, England	Speaker
Mukhtar Muzammil Abdus Salam School of Mathematical Sciences, GC University, Pakistan	Participant
Navarrete Juan Pablo Universidad Autónoma de Yucatán, Mexico	Organizing Committee
Navarro Ana Claudia Universidade de Sao Paulo, Brazil	Poster
Neumann Walter Columbia University, USA	Speaker
Nguyên Thi Bích Thuy ICMC, Universidade de São Paulo, Brazil	Poster
Novacoski Josnei Antonio Université Paul Sabatier, France	Participant
Nuño Ballesteros Juan José Universidad de Valencia, Spain	Speaker
Oka Mutsuo Tokyo University of Science, Japan	Scientific Committee/ Speaker
Olmedo Rodrigues João Hélder Universidade Federal Fluminense, Brazil	Participant
Ortiz Adriana Instituto de Matemáticas, Universidad Nacional Autónoma de México	Participant
Parusinski Adam Université de Nice Sophia Antipolis, France	Speaker
Pe Pereira María Instituto de Ciencias Matemáticas, Madrid, Spain	Speaker

Peniche Mena Ramón Universidad Autónoma de Yucatán, Mexico	Organizing Committee
Pereira Miriam Universidade de Paraíba, Brazil	Participant
Pérez Esteva Salvador Instituto de Matemáticas, Unidad Cuernavaca, Universidad Nacional Autónoma de México	Participant
Pérez Gavilán Torres Jacinta University of Cologne, Germany	Participant
Pham Thuý Huong Technische Universität Kaiserslautern, Germany	Participant
Pichon Anne Université Aix-Marseille, France	Scientific Committee/ Speaker
Plenat Camille Université Aix-Marseille, France	Participant
Pontigo Herrera Jessie Diana Instituto de Matemáticas, Universidad Nacional Autónoma de México	Participant
Popescu-Pampu Patrick Université Lille 1, France	Speaker
Rodríguez Guzmán Diego IMPA, Brazil	Participant
Romano Velázquez Faustino Agustín Instituto de Matemáticas, Unidad Cuernavaca, Universidad Nacional Autónoma de México	Participant
Roque Márquez Christopher Jonatan CINVESTAV, Mexico	Participant
Rosales González Ernesto Instituto de Matemáticas, Universidad Nacional Autónoma de México	Participant

Ruas Maria Aparecida Universidade de São Paulo, Brazil	Speaker
Rudolph Lee Clark University, USA	Speaker
Saavedra Calderón Víctor Manuel Instituto de Matemáticas, Unidad Cuernavaca, Universidad Nacional Autónoma de México	Participant
Sad Paulo IMPA, Brazil	Speaker
Saia Marcelo José Universidade de São Paulo, Brazil	Scientific Committee
San Saturnino Jean-Christophe Institut de Mathématiques de Toulouse, France	Participant
Saravia Molina Nancy Edith Pontificia Universidad Católica del Perú	Participant
Scardua Bruno César Universidade Federal do Rio de Janeiro, Brazil	Speaker
Schürmann Jörg University of Münster, Germany	Speaker
Seade Kuri José Antonio Instituto de Matemáticas, Universidad Nacional Autónoma de México	Participant
Serjiescu Vlad Université Joseph Fourier, France	Participant
Siersma Dirk Universiteit Utrecht, The Netherlands	Speaker
Sigurdsson Baldur Central European University, Hungary	Participant
Snoussi Jawad Instituto de Matemáticas, Unidad Cuernavaca, Universidad Nacional Autónoma de México	Organizing & Scientific Committee

Soares Marcio Universidade Federal de Minas Gerais, Brazil	Speaker
Souza Taciana Oliveira Universidade Federal de Uberlândia, Brazil	Poster
Spivakovski Mark Université Paul Sabatier, Toulouse, France	Scientific Committee
Teissier Bernard Institut Mathématique de Jussieu, France	Speaker
Tosun Meral Department of Mathematics, Galatasaray University, Turkey	Speaker
Trejo Abad Sofía Universidade de São Paulo, Brazil	Participant
Trotman David Université Aix-Marseille, France	Speaker
Ucan Puc Manuel Alejandro Instituto de Matemáticas, Unidad Cuernavaca, Universidad Nacional Autónoma de México	Organizing Committee
Uribe Vargas Ricardo Université de Bourgogne, France	Speaker
Vega Efraín Instituto de Matemáticas, Universidad Nacional Autónoma de México	Participant
Verjovsky Alberto Instituto de Matemáticas, Unidad Cuernavaca, Universidad Nacional Autónoma de México	Scientific Committee/ Speaker
Zúñiga Ávila Haremy Yazmín Instituto de Matemáticas, Unidad Cuernavaca, Universidad Nacional Autónoma de México	Participant

Extending the Action of Schottky Groups on the Complex Anti-de Sitter Space to the Projective Space

Vanessa Alderete, Carlos Cabrera, Angel Cano and Mayra Méndez

This paper is dedicated to Pepe Seade in celebration of his 60th Birthday Anniversary.

Abstract. In this article we show that if a complex Schottky group, acting on the complex anti-de Sitter space, acts on the corresponding projective space as a Schottky group, then the space has signature (k, k). As a consequence, we are able to show the existence of complex Schottky groups, acting on $\mathbb{P}^n_{\mathbb{C}}$, such that the complement of whose Kulkarni's limit set is not the largest open set on which the group acts properly and discontinuously. This is the starting point towards the understanding of the notion of the role of limit sets in the higher-dimensional setting.

Mathematics Subject Classification (2000). Primary 37F30, 32M05, 32M15; Secondary 30F40, 20H10, 57M60.

Keywords. Schottky groups in higher dimensions, limit sets, complex hyperbolic spaces.

Introduction

Classical Schottky groups in $\mathrm{PSL}(2,\mathbb{C})$ play a key role in both complex geometry and holomorphic dynamics. On one hand, Koebe's Retrosection Theorem says that every compact Riemann surface can be obtained as the quotient of an open set in the Riemann sphere which is invariant under the action of a Schottky group. On the other hand, the limit sets of Schottky groups have a rich and fascinating geometry and dynamics, which has inspired much of the current knowledge we have about fractal sets and 1-dimensional holomorphic dynamics. In this article

Supported by grants of the PAPIIT's projects IN106814, IN108214 and IN102515 and CONACYT's project 164447.

we study the behavior of complex Schottky groups in $\mathrm{PU}(k,l)$ acting on $\mathbb{P}^{k+l-1}_{\mathbb{C}}$. More precisely we show:

Theorem 0.1. *If a purely loxodromic free discrete subgroup of* $\mathrm{PU}(k,l)$ *acts as a complex Schottky group on* $\mathbb{P}^{k+l-1}_{\mathbb{C}}$*, then* $k=l$*. Moreover in this case,*

1. *the group* Γ *acts as a complex Schottky group on the complex anti-de Sitter space;*
2. *the limit set* $\Lambda_{PA}(\Gamma)$ *is contained in the complex anti-de Sitter space and homeomorphic to the product* $\mathcal{C}\times\mathbb{P}^{k-1}_{\mathbb{C}}$*, where* $\mathcal{C}$ *is the triadic Cantor set.*

The limit set $\Lambda_{PA}(\Gamma)$ will be defined in Theorem 1.9. As a partial reciprocal of the previous theorem we have:

Theorem 0.2. *Let* $\Gamma\subset\mathrm{PU}(k,k)$ *be a group acting as a complex Schottky group on the complex anti-de Sitter space. If* Γ *is generated by* $\gamma_1,\dots,\gamma_n$*, then there is* $N\in\mathbb{N}$ *such that* $\Gamma_N=\langle\langle\gamma_1^N,\dots,\gamma_n^N\rangle\rangle$ *acts as a complex Schottky group on* $\mathbb{P}^{2k-1}_{\mathbb{C}}$.

The paper is organized as follows: in Section 1, we review some general facts and introduce the notation used along the text. In Section 2, we answer a question by J. Parker showing that no complex Schottky group is hyperbolic. In Section 3, we provide a lemma that helps us to describe the dynamics of compact sets. Finally, in Section 4, we provide a proof of the main results of this article.

1. Preliminaries

1.1. Projective Geometry

The complex projective space $\mathbb{P}^n_{\mathbb{C}}$ is defined as:

$$\mathbb{P}^n_{\mathbb{C}}=(\mathbb{C}^{n+1}\setminus\{0\})/\mathbb{C}^*,$$

where $\mathbb{C}^*$ acts by the usual scalar multiplication. This is a compact connected complex n-dimensional manifold equipped with the Fubini-Study metric d_n.

If $[\ ]:\mathbb{C}^{n+1}\setminus\{0\}\to\mathbb{P}^n_{\mathbb{C}}$ is the quotient map, then a non-empty set $H\subset\mathbb{P}^n_{\mathbb{C}}$ is said to be a projective subspace of dimension k if there is a $\mathbb{C}$-linear subspace $\widetilde{H}$ of dimension $k+1$ such that $[\widetilde{H}\setminus\{0\}]=H$. In this article, $e_1,\dots,e_{n+1}$ will denote the standard basis for $\mathbb{C}^{n+1}$.

Given a set of points S in $\mathbb{P}^n_{\mathbb{C}}$, we define:

$$\mathrm{Span}(S)=\bigcap\{P\subset\mathbb{P}^n_{\mathbb{C}}\mid P \text{ is a projective subspace containing } S\}.$$

Clearly, $\mathrm{Span}(S)$ is a projective subspace of $\mathbb{P}^n_{\mathbb{C}}$.

1.2. Projective and Pseudo-projective Transformations

Every linear isomorphism of $\mathbb{C}^{n+1}$ defines a holomorphic automorphism of $\mathbb{P}^n_{\mathbb{C}}$. Also, it is well known that every holomorphic automorphism of $\mathbb{P}^n_{\mathbb{C}}$ arises in this way. The group of projective automorphisms of $\mathbb{P}^n_{\mathbb{C}}$ is defined by:

$$\mathrm{PSL}(n+1,\mathbb{C}):=\mathrm{SL}(n+1,\mathbb{C})/\mathbb{C}^*,$$

where $\mathbb{C}^*$ acts by the usual scalar multiplication. Then $\mathrm{PSL}(n+1,\mathbb{C})$ is a Lie group whose elements are called projective transformations. We denote by $[[\ ]]: \mathrm{SL}(n+1,\mathbb{C}) \to \mathrm{PSL}(n+1,\mathbb{C})$ the quotient map. Given $\gamma \in \mathrm{PSL}(n+1,\mathbb{C})$, we say that $\widetilde{\gamma} \in \mathrm{SL}(n+1,\mathbb{C})$ is a lift of γ if $[[\widetilde{\gamma}]] = \gamma$.

1.3. Complex Anti-de Sitter Space and its Isometries

Let us start by constructing the complex anti-de Sitter space. To do that consider the following Hermitian matrix:

$$H = \begin{pmatrix} & & & & & & 1 \\ & & & & & \iddots & \\ & & & & 1 & & \\ & & & \mathrm{Id}_{l-k} & & & \\ & & 1 & & & & \\ & \iddots & & & & & \\ 1 & & & & & & \end{pmatrix}$$

where Id_{l-k} denotes the identity matrix of size $(l-k)\times(l-k)$, and the off-diagonal blocks in the upper right and the lower left are of size $k \times k$. We set

$$U(k,l) = \{g \in \mathrm{GL}(k+l,\mathbb{C}) : gHg^* = H\}$$

and denote by $\langle,\rangle : \mathbb{C}^{n+1} \to \mathbb{C}$ the Hermitian form induced by H. Clearly, $\langle,\rangle$ has signature (k,l), and $U(k,l)$ is the group preserving $\langle,\rangle$, see [11]. The respective projectivization $\mathrm{PU}(k,l)$ preserves the set

$$\mathbb{H}^{k,l}_{\mathbb{C}} = \{[w] \in \mathbb{P}^n_{\mathbb{C}} \mid \langle w,w\rangle < 0\},$$

which is the pseudo-unitary complex ball. We call the boundary, denoted by $\partial\mathbb{H}^{k,l}_{\mathbb{C}}$, *the complex anti-de Sitter space*. In the rest of the article we will be interested in studying those subgroups of $\mathrm{PSL}(n+1,\mathbb{C})$ preserving the pseudo-unitary complex ball.

Given a projective subspace $P \subset \mathbb{P}^n_{\mathbb{C}}$ we define

$$P^{\perp} = [\{w \in \mathbb{C}^{n+1} \mid \langle w,v\rangle = 0 \text{ for all } [v] \in P\} \setminus \{0\}].$$

An important tool in this work is the following result, see [6, 11].

Theorem 1.1 (Cartan Decomposition). *For every $\gamma \in \mathrm{PU}(k,l)$ there are elements $k_1, k_2 \in \mathrm{PU}(n+1)\cap\mathrm{PU}(k,l)$ and a unique $\mu(\gamma) \in \mathrm{PU}(k,l)$, such that $\gamma = k_1\mu(\gamma)k_2$ and $\mu(\gamma)$ have a lift in $\mathrm{SL}(n+1,\mathbb{C})$ given by*

$$\begin{pmatrix} e^{\lambda_1(\gamma)} & & & & & & & & \\ & \ddots & & & & & & & \\ & & e^{\lambda_k(\gamma)} & & & & & & \\ & & & 1 & & & & & \\ & & & & \ddots & & & & \\ & & & & & 1 & & & \\ & & & & & & e^{-\lambda_k(\gamma)} & & \\ & & & & & & & \ddots & \\ & & & & & & & & e^{-\lambda_1(\gamma)} \end{pmatrix}, \tag{1.1}$$

where $\lambda_1(\gamma) \geq \lambda_2(\gamma) \geq \ldots \geq \lambda_k(\gamma) \geq 0$.

1.4. Pseudo-projective Transformations

The space of linear transformations from $\mathbb{C}^{n+1}$ to $\mathbb{C}^{n+1}$, denoted by $M(n+1,\mathbb{C})$, is a linear complex space of dimension $(n+1)^2$. Note that $\mathrm{GL}(n+1,\mathbb{C})$ is an open dense set of $M(n+1,\mathbb{C})$. Hence $\mathrm{PSL}(n+1,\mathbb{C})$ is an open dense set in $\mathrm{QP}(n+1,\mathbb{C}) = (M(n+1,\mathbb{C}) \setminus \{0\})/\mathbb{C}^*$; the latter is called the space of pseudo-projective maps. Let $\widetilde{M} : \mathbb{C}^{n+1} \to \mathbb{C}^{n+1}$ be a non-zero linear transformation and $\mathrm{Ker}(\widetilde{M})$ be its kernel. We denote by $\mathrm{Ker}([[\widetilde{M}]])$ the respective projectivization. Then $\widetilde{M}$ induces a well defined map $[[\widetilde{M}]] : \mathbb{P}^n_{\mathbb{C}} \setminus \mathrm{Ker}([[\widetilde{M}]]) \to \mathbb{P}^n_{\mathbb{C}}$ given by

$$[[\widetilde{M}]]([v]) = [\widetilde{M}(v)] .$$

The following fact shows that we can find sequences in $\mathrm{QP}(n+1,\mathbb{C})$ such that the convergence as a sequence of points in a projective space coincides with the convergence as a sequence of functions.

Proposition 1.2 (See [4]). *Let* $(\gamma_m) \subset \mathrm{PSL}(n+1,\mathbb{C})$ *be a sequence of distinct elements, then*

1. *there are a subsequence* $(\tau_m) \subset (\gamma_m)$ *and* $\tau_0 \in M(n+1,\mathbb{C}) \setminus \{0\}$ *such that* $\tau_m \xrightarrow[m\to\infty]{} \tau_0$ *as points in* $\mathrm{QP}(n+1,\mathbb{C})$*;*
2. *if* (τ_m) *is the sequence given by the previous part of this lemma, then* $\tau_m \xrightarrow[m\to\infty]{} \tau_0$*, as functions, uniformly on compact sets of* $\mathbb{P}^n_{\mathbb{C}} \setminus \mathrm{Ker}(\tau_0)$.

We need the following lemmas. Further details and the proof of the next one can be found in [3].

Lemma 1.3. *Let* $(\gamma_m), (\tau_m) \subset \mathrm{PSL}(n+1,\mathbb{C})$ *be sequences such that* $\gamma_m \xrightarrow[m\to\infty]{} \gamma_0$ *and* $\tau_m \xrightarrow[m\to\infty]{} \tau_0$. *If* $\mathrm{Im}(\tau) \cap \mathrm{Ker}(\gamma) \neq \emptyset$*, then*

$$\gamma_m \tau_m \xrightarrow[m\to\infty]{} \gamma_0 \tau_0 .$$

For the proof and details of the next lemma, see [4].

Lemma 1.4. *Let $\gamma \in \mathrm{PU}(1,n)$ be a loxodromic element with attracting fixed point $a \in \partial\mathbb{H}^n_{\mathbb{C}}$ and repelling fixed point $r \in \partial\mathbb{H}^n_{\mathbb{C}}$, then $\gamma^m \xrightarrow[m\to\infty]{} a$ uniformly on compact sets of $\mathbb{P}^n_{\mathbb{C}} \setminus r^{\perp}$.*

1.5. The Grassmanians

Let $0 \leq k < n$, we define the Grassmanian $\mathrm{Gr}(k,n)$ as the space of all k-dimensional projective subspaces of $\mathbb{P}^n_{\mathbb{C}}$ endowed with the Hausdorff topology. One has that $\mathrm{Gr}(k,n)$ is a compact, connected complex manifold of dimension $k(n-k)$. A method to realize the Grassmanian $\mathrm{Gr}(k,n)$ as a subvariety of the projective space of the $(k+1)$-th exterior power of $\mathbb{C}^{n+1}$, in symbols $\mathbb{P}(\bigwedge^{k+1}\mathbb{C}^{n+1})$, is done by the so called Plücker embedding which is given by

$$\begin{aligned}\iota : \mathrm{Gr}(k,n) &\to \mathbb{P}(\bigwedge^{k+1}\mathbb{C}^{n+1})\\ \iota(V) &\mapsto [v_1 \wedge \cdots \wedge v_{k+1}],\end{aligned}$$

where $\mathrm{Span}(\{v_1,\cdots,v_{k+1}\}) = V$. We can induce an action of $\mathrm{PSL}(n+1,\mathbb{C})$ on $\mathrm{Gr}(k,n)$ and $\mathbb{P}(\bigwedge^{k+1}\mathbb{C}^{n+1})$ as follows:

Let $[[T]] \in \mathrm{PSL}(n+1,\mathbb{C})$, take $W = \mathrm{Span}(\{w_1,\ldots,w_{k+1}\}) \in \mathrm{Gr}(k+1,n+1)$ and a point $w = [w_1 \wedge \cdots \wedge w_{k+1}] \in \mathbb{P}(\bigwedge^{k+1}\mathbb{C}^{n+1})$. Now set

$$T(W) = \mathrm{Span}([[T]](w_1),\ldots,[[T]](w_{k+1}))$$

and

$$\bigwedge^{k+1} T(w) = [T(w_1) \wedge \cdots \wedge T(w_{k+1})],$$

then we have the following commutative diagram:

$$\begin{array}{ccc} \mathrm{Gr}(k,n) & \xrightarrow{\quad T \quad} & \mathrm{Gr}(k,n) \\ \downarrow \iota & & \downarrow \iota \\ \mathbb{P}(\bigwedge^{k+1}\mathbb{C}^{n+1}) & \xrightarrow{\bigwedge^{k+1} T} & \mathbb{P}(\bigwedge^{k+1}\mathbb{C}^{n+1}). \end{array} \tag{1.2}$$

1.6. The Kulkarni Limit Set

When we look at the action of a group on a general topological space, there is no natural notion of a limit set. A nice starting point is the so-called Kulkarni limit set (see [7]).

Definition 1.5. Let $\Gamma \subset \mathrm{PSL}(n+1,\mathbb{C})$ be a subgroup. We define

1. the set $\Lambda(\Gamma)$ as the closure of the set of cluster points of Γz, where z runs over $\mathbb{P}^n_{\mathbb{C}}$;
2. the set $L_2(\Gamma)$ as the closure of cluster points of ΓK, where K runs over all the compact sets in $\mathbb{P}^n_{\mathbb{C}} \setminus \Lambda(\Gamma)$;
3. the *Kulkarni limit set* of Γ as:

$$\Lambda_{\mathrm{Kul}}(\Gamma) = \Lambda(\Gamma) \cup L_2(\Gamma);$$

4. the *Kulkarni region of discontinuity* of Γ as:
$$\Omega_{\text{Kul}}(\Gamma) = \mathbb{P}^n_{\mathbb{C}} \setminus \Lambda_{\text{Kul}}(\Gamma).$$

For a more detailed discussion on this topic in the 2-dimensional setting see [3]. The Kulkarni limit set has the following properties (see [3, 4, 7]).

Proposition 1.6. *Let Γ be a complex Kleinian group. Then:*

1. *The sets $\Lambda_{\text{Kul}}(\Gamma)$, $\Lambda(\Gamma)$, $L_2(\Gamma)$ are Γ-invariant and closed.*
2. *The group Γ acts properly discontinuously on $\Omega_{\text{Kul}}(\Gamma)$.*
3. *Let $\mathcal{C} \subset \mathbb{P}^n_{\mathbb{C}}$ be a closed Γ-invariant set such that for every compact set $K \subset \mathbb{P}^n_{\mathbb{C}} \setminus \mathcal{C}$, the set of cluster points of ΓK is contained in $\Lambda(\Gamma) \cap \mathcal{C}$, then $\Lambda_{\text{Kul}}(\Gamma) \subset \mathcal{C}$.*
4. *The equicontinuity set of Γ is contained in $\Omega_{\text{Kul}}(\Gamma)$.*

1.7. Complex Schottky Groups

Recall the classification of projective transformations (see [2]):

Definition 1.7. Let $\gamma \in \text{PSL}(n+1, \mathbb{C})$, then γ is said to be

1. *loxodromic* if γ has a lift $\widetilde{\gamma} \in \text{SL}(n+1, \mathbb{C})$ such that $\widetilde{\gamma}$ has at least one eigenvalue outside the unit circle;
2. *elliptic* if γ has a lift $\widetilde{\gamma} \in \text{SL}(n+1, \mathbb{C})$ such that $\widetilde{\gamma}$ is diagonalizable and all of its eigenvalues are in the unit circle;
3. *parabolic* if γ has a lift $\widetilde{\gamma} \in \text{SL}(n+1, \mathbb{C})$ such that $\widetilde{\gamma}$ is non-diagonalizable and all of its eigenvalues are in the unit circle.

Complex Schottky groups are defined as follows, compare with definitions in [5, 8, 9, 12, 13].

Definition 1.8 (See [1]). Let $\Gamma \subset \text{PSL}(n+1, \mathbb{C})$; we say that Γ is a *complex Schottky group* acting on $\mathbb{P}^n_{\mathbb{C}}$ with g generators if

1. there are $2g$, for $g \geq 2$, open sets $R_1, \ldots, R_g$, $S_1, \ldots, S_g$ satisfying the following properties:
   (a) each of these open sets is the interior of its closure,
   (b) the closures of the $2g$ open sets are pairwise disjoint;
2. the group has a generating set $\{\gamma_1, \ldots, \gamma_g\}$ with the property $\gamma_j(R_j) = \mathbb{P}^n_{\mathbb{C}} \setminus \overline{S_j}$ for each j.

Examples of complex Schottky groups were constructed by A. Guillot, C. Frances, M. Mendez, M. W. Nori, J. Seade and A. Verjovsky, see [5, 8, 9, 12–16]. A standard result is the following.

Theorem 1.9 (See [1]). *Let $\Gamma \subset \text{PSL}(n+1, \mathbb{C})$ be a complex Schottky group with g generators, then Γ is a purely loxodromic free group with g generators. If $D = \bigcap_{j=1}^{g} \mathbb{P}^n_{\mathbb{C}} \setminus (R_j \cup S_j)$, then $\Omega_\Gamma = \Gamma D$ is a Γ-invariant open set where Γ acts properly discontinuously. Moreover, Ω_Γ has compact quotient and the limit set $\Lambda_{\text{PA}}(\Gamma) = \mathbb{P}^n_{\mathbb{C}} \setminus \Omega_\Gamma$ is disconnected.*

In Section 4 we give an example of a group Γ for which $\Lambda_{\text{Kul}}(\Gamma) \neq \Lambda_{\text{PA}}(\Gamma)$.

2. Complex Schottky Groups Cannot be Hyperbolic Ones

Now we address the following question by J. Parker: If Γ is a purely loxodromic free discrete subgroup of $\mathrm{PU}(1,n)$, is it true that Γ is a complex Schottky group? The answer is no, and this is the content of the following.

Theorem 2.1. *Let* $\Gamma \subset \mathrm{PSL}(n+1,\mathbb{C})$ *be a complex Schottky group acting on* $\mathbb{P}^n_{\mathbb{C}}$*, then* Γ *can not be conjugate to a subgroup of* $\mathrm{PU}(1,n)$.

Proof. On the contrary, let us assume that there is a complex Schottky group $\Gamma \subset \mathrm{PSL}(n+1,\mathbb{C})$ acting on $\mathbb{P}^n_{\mathbb{C}}$ which is conjugate to a subgroup of $\mathrm{PU}(1,n)$. Let γ be a generator of Γ; by Theorem 1.9, we know that γ is loxodromic. Let $\widetilde{\gamma} \in \mathrm{SL}(n+1,\mathbb{C})$ be a lift of γ. It is a well-known fact, see [11], that after conjugating by a projective transformation, we have:

$$\widetilde{\gamma} = \begin{pmatrix} re^{2\pi i\phi} & & & & \\ & e^{2\pi i\phi_1} & & & \\ & & \ddots & & \\ & & & e^{2\pi i\phi_{n-1}} & \\ & & & & r^{-1}e^{-2\pi i\phi} \end{pmatrix}$$

where $\Pi_{j=1}^{n-1} e^{2\pi i\phi_j} = 1$. If R and S are the sets associated to γ, as in the definition of complex Schottky group, then:

Claim 1. We have $\{[e_1],[e_{n+1}]\} \subset R \cup S$. Let us assume that $[e_1] \notin R \cup S$. In this case $[e_1] \in \mathbb{P}^n_{\mathbb{C}} \setminus S = \gamma\overline{R}$, then $[e_1] = \gamma^{-2}([e_1]) \in \gamma\overline{R} \subset R$, which is a contradiction. Similarly, one can show that $[e_{n+1}] \in R \cup S$.

In the following we will assume that $[e_1] \in R$ and $[e_{n+1}] \in S$.

Claim 2. It is verified that $P = \langle\langle [e_2], \ldots, [e_n] \rangle\rangle \subset \Lambda_\Gamma$. It is clear that given any point x in P, there is a sequence $n_m \in \mathbb{Z}$ of distinct numbers such that $\gamma^{n_m} x \xrightarrow[m\to\infty]{} x$, thus x cannot belong to any region where the group Γ acts properly discontinuously. In particular $x \in \Lambda_\Gamma$.

Observe that since $[e_1]$ and $[e_{n+1}]$ have infinite isotropy group, we can deduce that $\{[e_1],[e_{n+1}]\} \subset \Lambda_\Gamma$. The following is a reminiscent of the Lambda Lemma due to J.P. Navarrete, see [10].

Claim 3. Either $\ell_1 = \langle\langle [e_1],[e_2] \rangle\rangle \subset \Lambda_\Gamma$ or $\ell_2 = \langle\langle [e_2],[e_{n+1}] \rangle\rangle \subset \Lambda_\Gamma$. Let us assume that $r < 1$ and $\ell_2 \not\subset \Lambda_\Gamma$, then there is $q \in \Omega_\Gamma \cap \ell_2$. Define $\ell = \langle\langle q, r \rangle\rangle$, where $r \in \ell_1 \setminus \{[e_1],[e_2]\}$, then $\gamma^m \ell \xrightarrow[m\to\infty]{} \ell_1$. Thus given any $z \in \ell_1$ there is a sequence $(z_m) \subset \ell$ such that $z_m \xrightarrow[m\to\infty]{} z_0 \in \ell$ and $\gamma z_m \xrightarrow[m\to\infty]{} z$, in virtue of Lemma 1.4, it is clear that $z_0 = q$, therefore $z \in \Lambda_\Gamma$, which completes this claim.

Through similar arguments one can show:

Claim 4. There is $\tau : P \to P$ conjugate to an element in $SU(n-1)$, such that given $z \in P$ either $\langle\langle z, [e_1] \rangle\rangle \subset \Lambda_\Gamma$ or $\langle\langle \tau z, [e_{n+1}] \rangle\rangle \subset \Lambda_\Gamma$.

After all these claims we conclude the proof of the Theorem. Since $[e_1]$ and $[e_{n+1}]$ lie in different connected components of Λ_Γ we deduce that either $[e_1]^\perp \subset \Lambda_\Gamma$ or $[e_{n+1}]^\perp \subset \Lambda_\Gamma$. But Γ is a free group thus Γ can not be elementary. Then $[e_1]^\perp \subset \Lambda_\Gamma \cup [e_{n+1}]^\perp \subset \Lambda_\Gamma$, which is a contradiction. Hence Γ is not a complex Schottky group. □

3. The λ-lemma

In [5], C. Frances studied Lorentzian Kleinian groups in several dimensions. On her Ph.D thesis (see [9]), M. Mendez developed Frances ideas and techniques. In particular, she proved Lemma 3.1 and applied it to the study of complex orthogonal Kleinian groups in dimension three. The techniques in [9] allow us to compute, in a very precise way, the accumulation points of orbits of compact sets of divergent sequences of $\mathrm{PU}(k,l)$. In this section we present these techniques which will be very useful in the study of complex Schottky groups in higher dimensions.

Recall that a divergent sequence (g_m) in a topological space X is a sequence leaving every compact set in X. If $(g_m) \subset \Gamma \subset \mathrm{PU}(k,l)$ is a divergent sequence and x is a point in $\mathbb{P}_{\mathbb{C}}^{k+l-1}$, we define $\mathcal{D}_{(g_m)}(x)$ as the set of all the accumulation points of sequences of the form $(g_m(x_m))$, where $(x_m) \subset \mathbb{P}_{\mathbb{C}}^{k+l-1}$ is a sequence converging to x.

The following key lemmas, proved by M. Méndez (see [Men15]), will help to determine the sets $\mathcal{D}(g_m)(x)$.

Lemma 3.1. *Let $(u_m), (\widehat{u}_m), (g_m) \subset \mathrm{PU}(k,l)$ be sequences and $U \subset \mathbb{P}_{\mathbb{C}}^{k+l-1}$ a non-empty open set. If $u = \lim_{m\to\infty} u_m$ and $\widehat{u} = \lim_{m\to\infty} \widehat{u}_m$, then*

$$\mathcal{D}_{(u_m g_m \widehat{u}_m)}(U) = u\big(\mathcal{D}_{(g_m)}\widehat{u}(U)\big). \tag{3.1}$$

Lemma 3.2. *Let us consider the closed polydisc $B_\epsilon(z) = I_\epsilon(z_1) \times \cdots \times I_\epsilon(z_{k+l})$, where $I_\epsilon(y)$ denotes the closed disc of $\mathbb{C}$ of center y and radius ϵ. If $(g_m) \subset U(k,l)$ is a divergent sequence and $[g_m(B_\epsilon(z))] \xrightarrow[m\to\infty]{} B_\epsilon^\infty([z])$ in the Hausdorff topology, then*

$$\mathcal{D}_{([[g_m]])}([z]) = \bigcap_{\epsilon>0} B_\epsilon^\infty([z]). \tag{3.2}$$

Now we specify the ways on which a sequence diverges using Cartan Decomposition given in Theorem 1.1.

Definition 3.3. Let $(\gamma_m) \subset U(k,l)$ be a divergent sequence. We will say that (γ_m) tends *simply to infinity* if

1. the compact factors in the Cartan Decomposition of γ_m converge in $U(k+l) \cap U(k,l)$;

2. there are s natural numbers $n_1, \ldots, n_s \in \mathbb{N}$ such that $\sum_{j=1}^{s} n_j = k$, and corresponding sequences $(\alpha_{1m}), (\alpha_{2m}), \ldots, (\alpha_{sm}) \subset \mathbb{R}$, and block matrices $(D_{1m}) \subset \mathrm{GL}(n_1, \mathbb{R}), \ldots, (D_{sm}) \subset \mathrm{GL}(n_s, \mathbb{R})$ with $\det(D_{im}) = 1$ satisfying

$$\mu(A_m) = \begin{pmatrix} e^{\alpha_{1m}} D_{1m} & & & & & & & & \\ & \ddots & & & & & & & \\ & & e^{\alpha_{sm}} D_{sm} & & & & & & \\ & & & 1 & & & & & \\ & & & & \ddots & & & & \\ & & & & & 1 & & & \\ & & & & & & e^{-\alpha_{sm}} D_{sm}^{-1} & & \\ & & & & & & & \ddots & \\ & & & & & & & & e^{-\alpha_{1m}} D_{1m}^{-1} \end{pmatrix},$$

where the differences $\alpha_{im} - \alpha_{jm} \xrightarrow[m\to\infty]{} \infty$, for $i > j$, and the blocks D_{im} converge to some $D_i \in \mathrm{GL}(n_i, \mathbb{R})$ as $m \to \infty$. Moreover, we will say that (γ_m) tends *strongly to infinity* if $\alpha_{sm} \xrightarrow[m\to\infty]{} \infty$.

In order to generalize Theorem 2.1 to higher dimensions, the following proposition is essential.

Proposition 3.4. *Let $(A_m) \subset U(k,l)$ be a sequence tending strongly to infinity, then there are:*

- *s natural numbers $n_1, \ldots, n_s \in \mathbb{N}$,*
- *$(2s+1)$ pairs of projective subspaces $P_1^+, \ldots, P_{2s+1}^+, P_1^-, \ldots, P_{2s+1}^-$,*
- *a pseudo-projective transformation $\Pi_+ \in \mathrm{QP}(n, \mathbb{C})$, and*
- *a set of projective equivalences $\mathcal{F} = \{\gamma_i : P_i^- \to P_i^+\}_{i=2}^{2s+1}$,*

satisfying:

1. $\mathrm{Im}(\Pi_+) = P_1^+$.
2. $\mathrm{Ker}(\Pi_+) = \mathrm{Span}(\bigcup_{j=2}^{2s+1} P_j^-)$.
3. $\dim(\mathrm{Span}(\bigcup_{j=1}^{2s+1} P_j^{\pm})) = 2s + 1 + \sum_{j=1}^{2s+1} \dim(P_j^{\pm}) = k + l - 1$.
4. *It holds $[[A_m]] \xrightarrow[m\to\infty]{} \Pi_+$, in consequence*
$$\mathcal{D}_{([[A_m]])}(x) = \Pi_+(x),$$
for each $x \in \mathbb{P}^{k+l-1} \setminus \mathrm{Ker}(\Pi_+)$.
5. *Given $j \in \{2, \ldots, 2s\}$, $y \in P_j^-$ and $x \in$* $\mathrm{Span}\left(\{y\} \cup \bigcup_{i=j+1}^{2s+1} P_i^-\right) \setminus \mathrm{Span}\left(\bigcup_{i=j+1}^{2s+1} P_i^-\right)$, *we have*
$$\mathcal{D}_{([[A_m]])}(x) = \mathrm{Span}\left(\{\gamma_j(y)\} \cup \bigcup_{i=1}^{j-1} P_i^+\right).$$
6. *For each $x \in P_{2s+1}^-$ we have*
$$\mathcal{D}_{([[A_m]])}(x) = \mathrm{Span}\left(\{\gamma_k(x)\} \cup \bigcup_{j=1}^{2s} P_j^+\right).$$

Remember that we defined $\mathcal{D}_{[[A_m]]}(x)$ as the set of all the accumulation points of all images $g_m(x_m)$ of sequences x_m converging to x for divergent g_m sequences in $\mathrm{PU}(k,l)$. The previous proposition describes the set $\mathcal{D}_{[[A_m]]}(K)$, where K is a compacts set, and gives bounds on the dimensions of this accumulation set.

Proof. In virtue of Lemma 3.1, we will restrict to the case where (γ_m) is a sequence of diagonal matrices. Take s natural numbers $n_1,\dots,n_s$, real valued sequences (α_{1m}), $(\alpha_{2m}),\dots,(\alpha_{sm})$, and block matrices $(D_{1m})\subset \mathrm{GL}(n_1,\mathbb{R}),\dots,(D_{sm})\subset \mathrm{GL}(n_s,\mathbb{R})$, and let $D_1,\dots,D_s\in \mathrm{GL}(n_s,\mathbb{R})$ be as in Definition 3.3. To make notation simpler, define

$$A_{jm}=\begin{cases} D_{jm} & \text{if } 1<j\le s,\\ \mathrm{Id} & \text{if } j=s+1,\\ D_{2s-j+2,m} & \text{if } s+1<j, \end{cases}\qquad A_j=\begin{cases} D_{j} & \text{if } 1<j\le s,\\ \mathrm{Id} & \text{if } j=s+1,\\ D_{2s-j+2} & \text{if } s+1<j, \end{cases}$$

and

$$\beta_{jm}=\begin{cases} e^{\alpha_{jm}} & \text{if } 1<j\le s,\\ 1 & \text{if } j=s+1,\\ e^{\alpha_{2s-j+2,m}} & \text{if } s+1<j. \end{cases}$$

We have

$$A_m=\begin{pmatrix} \beta_{1m}A_{1m} & & \\ & \ddots & \\ & & \beta_{2s+1,m}A_{2s+1,m}\end{pmatrix}.$$

Since we are in the case where the sequence (γ_m) consists of diagonal matrices, the spaces P_j^+ and P_j^- coincide and are given by

$$P_j^+=P_j^-=\begin{cases} \mathrm{Span}(\{[e_l]:l\in\{1,\dots,n_1\}\}) & \text{if } j=1,\\ \mathrm{Span}(\{[e_l]:l\in\{1+\sum_{i=1}^{j-1}n_i,\dots,\sum_{i=1}^{j}n_i\}\}), & \text{if } 2<j\le s\\ \mathrm{Span}(\{[e_l]:l\in\{1+k,\dots,l\}\}) & \text{if } j=s+1,\\ \mathrm{Span}(\{[e_l]:l\in\{1+l,\dots,l+n_s\}\}) & \text{if } j=s+2,\\ \mathrm{Span}(\{[e_l]:l\in\{1+l+\sum_{i=0}^{j-s-3}n_{s-i},\dots,l+\sum_{i=0}^{j-s-2}n_{s-i}\}\}) & \text{if } s+2<j\le 2s+1; \end{cases}$$

the pseudo-projective transformation is

$$\Pi_+=\begin{bmatrix} A_1 & \\ & 0\end{bmatrix},$$

and

$$\gamma_j=[[A_j]].$$

When the maps γ_n are not diagonal, the left and right actions given by the Cartan Decomposition of the elements γ_n induce different projective spaces P_j^+ and P_j^-.

We have that $\mathrm{Im}(\Pi_+)=P_1^+$, $\mathrm{Ker}(\Pi_+)=\mathrm{Span}(\bigcup_{j=2}^{2s+1}P_j^-)$ and $k+l-1=\dim(\mathrm{Span}(\bigcup_{j=1}^{2s+1}P_j^{\pm}))=2s+1+\sum_{j=1}^{2s+1}\dim(P_j^{\pm})$.

Let us show 4. Since $\gamma_m \xrightarrow[n\to\infty]{} \Pi_+$ as pseudo-projective transformations, we conclude $\gamma_m|_K \xrightarrow[n\to\infty]{} \Pi_+|_K$ uniformly. In consequence $\mathcal{D}_{(\gamma_m)}(x) = \Pi_+(x)$ for every $x \in \mathbb{P}_{\mathbb{C}}^{k+l-1} \setminus \mathrm{Ker}(\Pi_+)$.

Part 5. Given $j \in \{2, \dots, 2s\}$, $y \in P_j^-$ and $x \in \mathrm{Span}\left(\{y\} \cup \bigcup_{i=j+1}^{2s} P_i^-\right) \setminus \mathrm{Span}\left(\bigcup_{i=j+1}^{2s} P_i^-\right)$. Then there is $Y \in \mathbb{C}^{n_j} \setminus \{0\}$ and $w \in \mathbb{C}^{\tilde{n}}$ where $\tilde{n} = \sum_{i=j+1}^{2s+1} n_i$ and such that $x = [0, Y, w]$.

Let $B_\epsilon(\tilde{x})$ be the polydisc centered a $\tilde{x} = (0, Y, w)$ with radius ϵ then

$$[A_m(B_\epsilon(\tilde{x}))] = \left[\Pi_{i=1}^{j-1} \frac{\beta_{i,m}}{\beta_{j,m}} A_{im} B_\epsilon(0) \times A_{j,m} B_\epsilon(Y) \times \Pi_{i=j+1}^{2s+1} \frac{\beta_{i,m}}{\beta_{j,m}} A_{im} B_\epsilon(w)\right],$$

thus we have the following limit in the Hausdorff topology:

$$[A_m(B_\epsilon(\tilde{x}))] \xrightarrow[m\to\infty]{} B_\epsilon^\infty(\tilde{x}) = \left[\Pi_{i=1}^{j-1} \mathbb{C}^{n_i} \times A_j(B_\epsilon(Y)) \times \{0\}\right],$$

see [9] for a detailed discussion on this limit. By Lemma 3.2, we conclude

$$\mathcal{D}_{([[A_m]])}(x) = \mathrm{Span}\left(\{\gamma_j(y)\} \cup \bigcup_{i=1}^{j-1} P_i^+\right),$$

which completes this part of the proof. The last part of this lemma can be proved analogously to part 2. □

Now, after this technical step, we give some applications in the following section.

4. Complex Schottky Groups in $\mathrm{PU}(k, l)$

From the works of A. Guillot-M. Mendez (see [9]) and C. Frances (see [5]), it is known that we can construct complex Schottky groups acting in $\mathbb{P}_{\mathbb{C}}^3$ that admit representations on $\mathrm{PU}(2, 2)$. Hence a natural question is:

Under which conditions is a discrete group Γ in $\mathrm{PU}(k, l)$ a complex Schottky group?

The following are auxiliary lemmas to answer this question:

Lemma 4.1. *Let $\gamma \in \mathrm{PSL}(n, \mathbb{C})$ be a non-elliptic element. If there is a sequence $(n_m) \subset \mathbb{Z}$ of distinct elements such that there is a point p and a hyperplane $\mathcal{H}$ satisfying $\gamma^{n_m} \xrightarrow[m\to\infty]{} p$ uniformly on compact sets of $\mathbb{P}_{\mathbb{C}}^{n-1} \setminus \mathcal{H}$, then p is a fixed point of γ.*

The proof is contained in [2]. Now we show:

Lemma 4.2. *Let $([[T_m]])$ be a sequence of different elements of $\mathrm{PSL}(k + l, \mathbb{C})$ such that there is a point $p = [w_1 \wedge \cdots \wedge w_k]$ and a hyperplane $\mathcal{H}$ satisfying $[[\wedge^k T_m]] \xrightarrow[m\to\infty]{} p$ uniformly on compact sets of $\mathbb{P}(\wedge^k(\mathbb{C}^{k+l})) \setminus \mathcal{H}$. Then for all $U \in \mathrm{Gr}(k, k + l) \setminus \iota^{-1}\mathcal{H}$ we have that $T_m(U)$ converges to $W = \mathrm{Span}(w_1, \dots, w_k)$ in $\mathbb{P}_{\mathbb{C}}^{k+l}$ in the Hausdorff topology.*

Proof. To prove this lemma observe that the Plücker Embedding restricted to $\mathrm{Gr}(k, k+l)$ is an isomorphism. Then by the Commutative Diagram 1.2, we have that for every $U = \mathrm{Span}(\{u_1, \ldots, u_k\}) \in \mathrm{Gr}(k, k+l) \setminus \iota^{-1}\mathcal{H}$ the sequence $(T_m(U))$ converges to $W = \mathrm{Span}(\{w_1, \ldots, w_k\})$ as points in $\mathrm{Gr}(k-1, k+l)$. Thus $(T_m(U))$ converges to W as closed sets of $\mathbb{P}^{k+l}_{\mathbb{C}}$, in the Hausdorff topology. □

The following theorem answers the question posed at the beginning of this section. It gives a necessary condition under which a discrete group Γ in $\mathrm{PU}(k, l)$ is a complex Schottky group.

Theorem 4.3. *If a purely loxodromic free discrete subgroup of* $\mathrm{PU}(k, l)$ *is a complex Schottky group, then* $k = l$.

Proof. Let us proceed by contradiction. Suppose that $k < l$ and let $\Gamma \subset \mathrm{PU}(k, l)$ as in the hypothesis. Take a generator $\gamma \in \Gamma$ and let $\tilde{\gamma} \in U(k, l)$ be a lift of γ. Consider the Cartan Decomposition of $\tilde{\gamma}^m$, then we obtain sequences (c_m) and $(\bar{c}_m)$ in K and (A_m) in $U(k+l)$ satisfying $\tilde{\gamma}^m = c_m A_m \bar{c}_m$.

Since (c_m) and $(\bar{c}_m)$ lie in a compact set, there is a subsequence $(m_s) \subset (m)$ and elements C and $\bar{C}$ in K such that c_{m_s} converges to C and $\bar{c}_{m_s}$ converges to $\bar{C}$. Clearly, we can assume that (γ^{m_s}) tends simply to infinity and in the following we will assume that (γ^{m_s}) tends strongly to infinity. The proof of the other case is similar. We claim that there exist projective subspaces P and Q, satisfying the following properties:

1. The dimensions satisfy $\dim P = \dim Q = k - 1$.
2. The spaces P, Q are invariant under the action of γ. Moreover, P is attracting and Q is repelling.
3. If R_γ, S_γ are the disjoint open sets associated to γ given in the definition of a complex Schottky group, then either $P \subset R_\gamma$ and $Q \subset S_\gamma$, or $Q \subset R_\gamma$ and $P \subset S_\gamma$. In particular, it follows that P and Q are also disjoint and lie in distinct connected components of $\Lambda_{AP}(\Gamma)$.
4. We have $P^\perp \nsubseteq \Lambda_{AP}(\Gamma)$ and $Q^\perp \nsubseteq \Lambda_{AP}(\Gamma)$.

Set P and Q the projectivizations of the spaces $P' = C(\mathrm{Span}(\{e_1, \ldots, e_k\}))$ and $Q' = \bar{C}^{-1}(\mathrm{Span}(\{e_{l+1}, \ldots, e_{k+l}\}))$. The first part of the claim follows by construction. Let us show part (2), consider the action of $\wedge^k A_{n_m}$ on $\wedge^k \mathbb{C}^{k+l}$, then a straightforward calculation shows the matrix of $\wedge^k A_{n_m}$ with respect the standard ordered basis β of $\wedge^k \mathbb{C}^{k+l}$ is given by:

$$A_m = \begin{pmatrix} \theta_1 & & & \\ & \theta_2 & & \\ & & \ddots & \\ & & & \theta_{\binom{k}{n}} \end{pmatrix},$$

where θ_i is the product of k elements taken from the set $\{e^{\lambda_{i,m}(\gamma^{n_m})}\}$ and ordered in the lexicographical order in (i, m). In fact $\theta_1 > \theta_2 > \cdots > \theta_{\binom{k}{n}}$. Hence $[[A_m]]$ converges to $x = [e_1 \wedge \cdots \wedge e_k]$ uniformly on compact sets of $\mathbb{P}(\wedge^k(\mathbb{C}^{k+l})) \setminus \mathrm{Span}(\beta \setminus$

$\{x\}$). Therefore by Lemma 1.3, we conclude that $[[\wedge^k \tilde{\gamma}^{m_s}]]$ converges to the point $[[\wedge^k C]][e_1 \wedge \cdots \wedge e_k]$ uniformly on compact sets of $\mathbb{P}(\wedge^k(\mathbb{C}^{k+l})) \setminus [\wedge^k \bar{C}^{-1}] \operatorname{Span}(\beta \setminus \{x\})$. Finally, from Lemma 4.1 we conclude that x is a fixed point of $[[\wedge^k \tilde{\gamma}^{m_s}]]$, in consequence $P = [C] \operatorname{Span}(\{[e_1], \ldots, [e_k]\})$ is attracting and invariant under γ. In a similar way, we can prove that Q is repelling and invariant.

Part 3. On the contrary, assume that there is $x \in P \cap \mathbb{P}_{\mathbb{C}}^{k+l-1} \setminus (R_\gamma \cup S_\gamma) \neq \emptyset$, then there exists an open set U such that $x \in U \subset \mathbb{P}_{\mathbb{C}}^{k+l-1} \setminus \Lambda_{PA}(\Gamma)$. By Proposition 3.4, we conclude that

$$Q^\perp \subset \mathcal{D}_{(\gamma^{n_m})}(x) \subset \bigcap_{m \in \mathbb{N}} \gamma^m S_\gamma \subset S_\gamma.$$

Let $\gamma_1 \in \Gamma$ be a generator of Γ distinct from γ. Define $Q_1 = \gamma^{-1}\gamma_1 Q^\perp$ and observe that $Q_1 \subset R_\gamma$. As the dimensions of Q_1 and $Q^\perp$ are $l-k$, we have that $Q_1 \cap Q^\perp$ is not empty, which leads to a contradiction, because $R_\gamma \cap S_\gamma = \emptyset$.

Part 4. Assume that $P^\perp \subset \Lambda_\Gamma$. By the previous part, we can assume that $P \subset S_\gamma$. Let $\gamma_1 \in \Gamma$ be a generator of Γ distinct from γ. By Lemma 4.2 we conclude that $\gamma^{-m_s}(\gamma_1(P))$ converges to Q, therefore $\gamma_1^{-m_s}(\gamma(P^\perp))$ converges to $Q^\perp$. Hence $Q^\perp \subset \Lambda_\Gamma$. As $P \subset P^\perp$, $Q \subset Q^\perp$ and $P^\perp \cap Q^\perp \neq \emptyset$ and all of these spaces are path connected, which lead us to a contradiction.

To conclude the proof, let $p \in P^\perp \cap \Omega_\Gamma$ and $q \in Q^\perp \cap \Omega_\Gamma$. Clearly, we can assume that $p \in P^\perp \setminus P$ and $q \in P^\perp \setminus Q$. By Lemma 3.4 there exist $a, b \in P^\perp \cap Q^\perp$ such that $\operatorname{Span}(a, P) \cup \operatorname{Span}(b, Q) \subset \Lambda_\Gamma$. But $\operatorname{Span}(a, P)$, $\operatorname{Span}(b, Q)$ and $P^\perp \cap Q^\perp$ are path connected. Then we can construct a path in Λ_Γ, passing along a and b through W and connecting P with Q, which leads to a contradiction. □

With the previous results in mind let us show the main results:

Proof of Theorem 0.1. From Theorem 4.3 we know that every complex Schottky group of $\mathrm{PU}(k,l)$ acting on $\mathbb{P}_{\mathbb{C}}^{k+l-1}$ must satisfy that $k = l$. From the arguments used in the proof of Theorem 4.3 we deduce that there are elements $\gamma_1, \ldots, \gamma_n \in \Gamma$, disjoints open sets $R_1, \ldots, R_n, S_1, \ldots, S_n$ and projective spaces $P_1, \ldots, P_n$, $Q_1, \ldots, Q_n$ such that

1. the group Γ is generated by $\gamma_1, \ldots, \gamma_n$;
2. we have $\bigcup_{j=1}^n \overline{R_j \cup S_j} \neq \mathbb{P}_{\mathbb{C}}^{2k-1}$;
3. the set $\bigcup_{j=1}^n P_j \cup Q_j$ is contained in the complex anti-de Sitter space;
4. the generating set satisfies that $\gamma_j(R_j) = \mathbb{P}_{\mathbb{C}}^{2k-1} \setminus \overline{S_j}$;
5. for each $j \in \{1, \ldots, n\}$ we have $P_j \subset R_j$ and $Q_j \subset S_j$;
6. the collection of projective spaces satisfy $\dim(P_j) = \dim(Q_j) = k-1$;
7. the set $\mathbb{P}_{\mathbb{C}}^{2k-1} \setminus \overline{\Gamma\left(\bigcup_{j=1}^n (P_j \cup Q_j)\right)}$ is the largest open set on which Γ acts properly discontinuously on $\mathbb{P}_{\mathbb{C}}^{2k-1}$.

On one hand this means that Γ acts as a complex Schottky group in the complex anti-de Sitter space, and on the other hand, it is well known that a maximal open set on which Γ acts properly discontinuously is $\Gamma\left(\mathbb{P}_{\mathbb{C}}^{2k-1} \setminus \bigcup_{j=1}^n \overline{R_j \cup S_j}\right)$, see [3]. Finally, by well-known arguments of complex Schottky groups, see [16], we can ensure that $\Lambda_{PA}(\Gamma)$ is homeomorphic to the product of a Cantor set with $\mathbb{P}_{\mathbb{C}}^{k-1}$, which concludes the proof. □

We continue with the proof of Theorem 0.2.

Proof of Theorem 0.2. Using similar arguments as in the proof of Theorem 4.3, we can deduce that there are elements $\gamma_1, \ldots, \gamma_n \in \Gamma$, disjoint open sets $R_1, \ldots, R_n$, $S_1, \ldots, S_n$ in $\partial\mathbb{H}_{\mathbb{C}}^{k,k}$ and $(k-1)$-dimensional projective spaces $P_1, \ldots, P_n$, $Q_1, \ldots, Q_n$ in $\partial\mathbb{H}_{\mathbb{C}}^{k,k}$ such that:

1. the group Γ is generated by $\gamma_1, \ldots, \gamma_n$;
2. we have $\bigcup_{j=1}^{n} \overline{R_j \cup S_j} \neq \partial\mathbb{H}_{\mathbb{C}}^{k,k}$;
3. the generating set satisfies that $\gamma_j(R_j) = \partial\mathbb{H}_{\mathbb{C}}^{k,k} \setminus \overline{S_j}$;
4. for each $j \in \{1, \ldots, n\}$ we have $P_j \subset R_j$ and $Q_j \subset S_j$;
5. the projective space Q_j (resp. P_j) is an attracting (resp. repelling) fixed point for $\bigwedge^k \gamma_j$.

Thus, there are disjoints open sets $U_1, \ldots, U_n, W_1, \ldots, W_n \subset \mathbb{P}_{\mathbb{C}}^{2k-1}$ and n natural numbers $m_1, \ldots, m_n$ such that:

1. we have $\bigcup_{j=1}^{n} \overline{U_j \cup W_j} \neq \mathbb{P}_{\mathbb{C}}^{2k-1}$;
2. the generating set satisfies that $\gamma_j^{m_j}(U_j) = \mathbb{P}_{\mathbb{C}}^{2k-1} \setminus \overline{W_j}$;
3. for each $j \in \{1, \ldots, n\}$ we have $P_j \subset U_j$ and $Q_j \subset W_j$.

Taking N as the lowest common multiple of (m_i) we have that Γ^N, the group generated by $\{\gamma_1^N, \ldots, \gamma_n^N\}$, is a complex Schottky group acting on $\mathbb{P}_{\mathbb{C}}^{2k-1}$, which concludes the proof. □

4.1. An example

From the arguments in the proof of the previous Theorem we are able to answer a question by J. Seade and A. Verjovsky of whether the Kulkarni limit set always coincides with the limit set of J. Seade and A. Verjovsky.

Proposition 4.4. *There exists a complex Schottky group Γ such that the Kulkarni limit set $\Lambda_{\mathrm{Kul}}(\Gamma)$ is different to Λ_{PA}.*

Proof. Let us consider two complex Schottky groups $G = \langle \gamma_1, \gamma_2 \rangle$ and $\hat{G} = \langle \hat{\gamma}_1, \hat{\gamma}_2 \rangle$ in $U(1,1)$ such that the corresponding lifts in $\mathrm{SL}(2,\mathbb{C})$ of the maps $\gamma_1, \hat{\gamma}_1, \gamma_2$ and $\hat{\gamma}_2$ are diagonalizable and have all pairwise different eigenvalues.

Now construct a new complex Schottky group $\Gamma = \langle \sigma_1, \sigma_2 \rangle$, where σ_i is the map in $U(2,2)$ given by

$$\sigma_i = \begin{pmatrix} \gamma_i & \\ & \hat{\gamma}_i \end{pmatrix}.$$

By construction, Γ is purely loxodromic. The product of the defining curves of G and $\hat{G}$ give corresponding curves for the elements of Γ. Thus Γ is a complex Schottky group with signature $(2,2)$.

Moreover, for each γ in Γ there is a lift $\tilde{\gamma}$ in $\mathrm{SL}(4,\mathbb{C})$ such that $\tilde{\gamma}$ is diagonalizable with all the eigenvalues pairwise distinct.

By Theorem 0.1, part 2, Λ_{PA} is homeomorphic to $\mathcal{C} \times \mathbb{P}^1_{\mathbb{C}}$.

On the other hand, since Γ is generated by two elements with different eigenvalues, an application of Perron-Frobenius Theorem shows that $\Lambda(\Gamma)$ consists of the union of all eigenvectors of elements in Γ. Indeed, for every element γ of Γ we have four eigenvectors:

- two repelling eigenvectors, the most repelling eigenvector p_1 and the least repelling p_2 contained in the space P;
- two attracting eigenvectors, the most attracting q_1 and the least attracting q_2 contained in the space Q;

where P and Q are the spaces associated to γ constructed in Theorem 4.3. Let L be the line passing through p_1 and p_2. Now take a point $x \in L \setminus \{p_1, p_2\}$. Hence x does not belong to $\Lambda(\Gamma)$. By construction, L is one of the projective subspaces constructed in Proposition 3.4. By Proposition 3.4, if (x_n) is a sequence in the complement of L converging to x then the accumulation set $\mathcal{D}_{\gamma^m}(x)$ contains a space of dimension 3. But $\mathcal{D}_{\gamma^m}(x)$ is contained in $L_2(\Gamma)$. Then we have that $\Lambda_{\mathrm{Kul}}(\Gamma)$ contains subspaces of dimension 3. Hence $\Lambda_{\mathrm{Kul}}(\Gamma) \neq \Lambda_{\mathrm{PA}}$. □

In the general case, with signature (k,k), we have that Λ_{Kul} is a union of spaces of dimension $2k-1$ whereas Λ_Γ is a union of spaces of dimension k.

Acknowledgments

The authors would like to thank A. Guillot, J. Parker and M. Ucan for fruitful conversations. Also we are greatly indebted to A. Verjovsky for the stimulating conversations on the subject over the years. As well, we would like to thank the referee for the valuable input that helped us to improve this article. At last, but not the least, we thank J. Seade for his encouragement and comments to this work. The fourth author would like to thank the third author for giving her the opportunity to enunciate and prove Lemmas 3.1 and 3.2 at the seminar on complex Kleinian groups of the Institute of Mathematics of the UNAM in February and March of 2014.

References

[1] A. Cano. Schottky groups can not act on $P(2n,\mathbb{C})$ as subgroups of $PSL(2n+1,\mathbb{C})$. *Bull. Braz. Math. Soc. (N.S.)*, 39(4):573–586, 2008.

[2] A. Cano and L. Loeza. Projective Cyclic Groups in Higher Dimensions. http://arxiv.org/pdf/1112.4107.pdf, 2013.

[3] A. Cano, J. P. Navarrete, and J. Seade. *Complex Kleinian Groups*. Number 303 in Progress in Mathematics. Birkhäuser/Springer Basel AG, Basel, 2013.

[4] A. Cano and J. Seade. On the Equicontinuity Region of Discrete Subgroups of PU(1,n). *J. Geom. Anal.*, 20(2):291–305, April 2010.

[5] C. Frances. Lorentzian Kleinian groups. *Comment. Math. Helv.*, 80(4):883–910, 2005.
[6] A. W. Knapp. *Lie groups beyond an introduction.* Number 140 in Progress in Mathematics. Birkhäuser Boston, Boston, MA, second edition edition, 2002.
[7] R. S. Kulkarni. Groups with Domains of Discontinuity. *Math. Ann.*, 237(3):253–272, 1978.
[8] B. Maskit. *Kleinian groups*, volume 287 of *Grundlehren der Mathematischen Wissenschaften.* Springer-Verlag, Berlin, New York, 1988.
[9] M. Mendez. *Complex orthogonal kleinian groups of dimension three.* PhD thesis, UNAM, 2015.
[10] J. P. Navarrete. On the limit set of discrete subgroups of $PU(2,1)$. *Geom. Dedicata*, 122:1–13, 2006.
[11] Y. A. Neretin. *Lectures of Gaussian integral operators and classical groups.* EMS Series of Lectures in Mathematics. European Mathematical Society, Zürich, 2011.
[12] M. V. Nori. *The Schottky groups in higher dimensions*, volume 58. Amer. Math. Soc., Providence, RI., 1986.
[13] J. Seade and A. Verjovsky. Kleinian groups in higher dimensions. *Rev. Semin. Iberoam. Mat. Singul. Tordesillas*, 2(5):27–34, 1999.
[14] J. Seade and A. Verjovsky. Actions of discrete groups on complex projective spaces. In Contemp. Math., editor, *Laminations and foliations in dynamics, geometry and topology (Stony Brook, NY, 1998)*, volume 269. Amer. Math. Soc., Providence, RI., 2001.
[15] J. Seade and A. Verjovsky. Higher dimensional complex Kleinian groups. *Math. Ann.*, 322(2):279–300, 2002.
[16] J. Seade and A. Verjovsky. Complex Schottky groups. *Astérisque*, xx(287):251–272, 2003. Geometric methods in dynamics. II.

Vanessa Alderete
e-mail: `vanessa.alderete@im.unam.mx`

Carlos Cabrera
e-mail: `carloscabrerao@im.unam.mx`

Angel Cano
e-mail: `angelcano@im.unam.mx`

Mayra Mendez
e-mail: `mayra@matcuer.unam.mx`

Address of all authors:
UCIM UNAM, Av. Universidad s/n. Col. Lomas de Chamilpa
C.P. 62210, Cuernavaca, Morelos, México

Puiseux Parametric Equations via the Amoeba of the Discriminant

Fuensanta Aroca and Víctor Manuel Saavedra

Dedicated to José Seade on his 60th birthday

Abstract. Given an algebraic variety we get Puiseux-type parametrizations on suitable Reinhardt domains. These domains are defined using the amoeba of hypersurfaces containing the discriminant locus of a finite projection of the variety.

Mathematics Subject Classification (2000). Primary 32S05; Secondary 32B10.

Keywords. Algebraic variety, parametrization, Puiseux series.

1. Introduction

The theory of complex algebraic or analytic singularities is the study of systems of a finite number of equations in the neighborhood of a point where the rank of the Jacobian matrix is not maximal. These points are called singular points.

Isaac Newton in a letter to Henry Oldenburg [14], described an algorithm to compute term by term local parameterizations at singular points of plane curves. The existence of such parameterizations (i.e., the fact that the algorithm really works) was proved by Puiseux [12] two centuries later. This is known as the Newton-Puiseux theorem and asserts that we can find local parametric equations of the form $z_1 = t^k$, $z_2 = \varphi(t)$ where φ is a convergent power series.

Singularities of dimension greater than 1 are not necessarily parameterizable. An important class of parameterizable singularities are called quasi-ordinary. S.S. Abhyankar proved in [1] that quasi-ordinary hypersurface singularities are parameterizable by Puiseux series.

For algebraic hypersurfaces J. McDonald showed in [11] the existence of Puiseux series solutions with support in strongly convex cones. P.D. González Pérez [8] describes these cones in terms of the Newton polytope of the discriminant. In [2] F. Aroca extends this result to arbitrary codimension.

In this paper we prove that, for every connected component of the complement of the amoeba of the discriminant locus of a projection of an algebraic variety, there exist local Puiseux parametric equations of the variety. The series appearing in those parametric equations have support contained in cones which can be described in terms of the connected components of the complement of the amoeba. These cones are not necessarily strongly convex.

The results of Abhyankar [1], McDonald [11], González Pérez [8] and Aroca [2] come as corollaries of the main result.

2. Polyhedral convex cones

A set $\sigma \subseteq \mathbb{R}^N$ is said to be a **convex rational polyhedral cone** when it can be expressed in the form

$$\sigma = \{\lambda_1 u^{(1)} + \lambda_2 u^{(2)} + \cdots + \lambda_M u^{(M)} \mid \lambda_j \in \mathbb{R}_{\geq 0}\},$$

where $u^{(1)}, \ldots, u^{(M)} \in \mathbb{Z}^N$. The vectors $u^{(1)}, \ldots, u^{(M)}$ are **a system of generators** of σ and we write

$$\sigma = \langle u^{(1)}, \ldots, u^{(M)} \rangle.$$

A cone is said to be **strongly convex** if it contains no linear subspaces of positive dimension.

As usual, here $x \cdot y$ denotes the standard scalar product in $\mathbb{R}^N$. Let $\sigma \subset \mathbb{R}^N$ be a cone. Its **dual cone** $\sigma^\vee$ is the cone given by

$$\sigma^\vee := \{x \in \mathbb{R}^N \mid x \cdot u \geq 0,\ \forall u \in \sigma\}.$$

Let $\mathrm{A} \subseteq \mathbb{R}^n$ be a non-empty convex set. The **recession cone** of A is the cone

$$\mathrm{Rec}(A) := \{y \in \mathbb{R}^n \mid x + \lambda y \in \mathrm{A},\ \forall x \in A,\ \forall \lambda \geq 0\}.$$

From now on, by a cone we will mean a convex rational polyhedral cone.

Remark 2.1. Given $p \in \mathbb{R}^N$ and a cone $\sigma \subset \mathbb{R}^N$, one has,

$$p + \sigma = \{x \in \mathbb{R}^N \mid x \cdot v \geq p \cdot v,\ \forall v \in \sigma^\vee\}.$$

A cone $\sigma \subset \mathbb{R}^N$ is called a **regular cone** if it has a system of generators that is a subset of a basis of $\mathbb{Z}^N$.

Remark 2.2. Any rational polyhedral cone $\sigma \subset \mathbb{R}^N$ is a union of regular cones (see, for example, [6, section 2.6]).

For a matrix M, we denote by M^T the transpose matrix of M.

Remark 2.3. Take $u^{(1)}, \ldots, u^{(N)} \in \mathbb{R}^N$ and let M be the matrix that has as columns $u^{(i)}$ $i = 1, \ldots, N$. Suppose that the determinant of M is different from zero. If $\sigma = \langle u^{(1)}, \ldots, u^{(N)} \rangle$, then $\sigma^\vee = \langle v^{(1)}, \ldots, v^{(N)} \rangle$ where $v^{(i)}$, $i = 1, \ldots, N$ are the columns of $(\mathrm{M}^{-1})^T$ (see, for example, [4, Example 2.13.2.0.3]).

Lemma 2.4. *Let* $\sigma = \langle u^{(1)}, \dots, u^{(s)} \rangle \subset \mathbb{R}^N$ *be an s-dimensional regular cone and let* $(u^{(s+1)}, \dots, u^{(N)})$ *be a* $\mathbb{Z}$*-basis of* $\sigma^\perp \cap \mathbb{Z}^N$*. Let* M *be the matrix that has* $u^{(1)}, \dots, u^{(N)}$ *as columns and let* $v^{(1)}, \dots, v^{(N)}$ *be the columns of* $(\mathrm{M}^{-1})^T$*. Then*

$$\sigma^\vee = \langle v^{(1)}, \dots, v^{(s)}, \pm v^{(s+1)}, \dots, \pm v^{(N)} \rangle.$$

Proof. Clearly $\langle v^{(1)}, \dots, v^{(s)}, \pm v^{(s+1)}, \dots, \pm v^{(N)} \rangle \subseteq \sigma^\vee$. Let $\sigma' := \langle u^{(1)}, \dots, u^{(N)} \rangle$. Since $\langle v^{(1)}, \dots, v^{(N)} \rangle \subseteq \langle v^{(1)}, \dots, v^{(s)}, \pm v^{(s+1)}, \dots, \pm v^{(N)} \rangle$, by Remark 2.3,

$$\langle v^{(1)}, \dots, v^{(s)}, \pm v^{(s+1)}, \dots, \pm v^{(N)} \rangle^\vee \subseteq \sigma'.$$

Now let $x = \sum_{i=1}^N \lambda_i u^{(i)}$ be an element of $\langle v^{(1)}, \dots, v^{(s)}, \pm v^{(s+1)}, \dots, \pm v^{(N)} \rangle^\vee$. Since $\lambda_i u^{(i)} \cdot v^{(i)} \geq 0$ and $\lambda_i u^{(i)} \cdot -v^{(i)} \geq 0$, we have that,

$$\lambda_i = 0 \text{ for } i = s+1, \dots, N.$$

Then, $x = \sum_{i=1}^s \lambda_i u^{(i)}$. □

3. Amoebas

Consider the map

$$\begin{array}{rccc} \tau : & \mathbb{C}^N & \longrightarrow & \mathbb{R}^N_{\geq 0} \\ & (z_1, \dots, z_N) & \mapsto & (|z_1|, \dots, |z_N|). \end{array} \tag{3.1}$$

A set $\Omega \subseteq \mathbb{C}^N$ is called a **Reinhardt set** if $\tau^{-1}(\tau(\Omega)) = \Omega$.
Let log be the map defined by

$$\begin{array}{rccc} \log : & \mathbb{R}^N_{>0} & \longrightarrow & \mathbb{R}^N \\ & (x_1, \dots, x_N) & \mapsto & (\log x_1, \dots, log x_N). \end{array} \tag{3.2}$$

We will denote $\mu := \log \circ \tau$.
A Reinhardt set $\Omega \subseteq (\mathbb{C}^*)^N$ is said to be **logarithmically convex** if the set $\mu(\Omega)$ is convex.
A Laurent polynomial is a finite sum of the form $\sum_{(\alpha_1, \dots, \alpha_N) \in \mathbb{Z}^N} c_\alpha X^\alpha$ where $c_\alpha \in \mathbb{C}$. A generalization of Laurent polynomials are the Laurent series (for a further discussion of Laurent series, see, for example, [13]).

For a Laurent polynomial f we denote by $\mathcal{V}(f)$ its zero locus in $(\mathbb{C}^*)^N$. Given a Laurent polynomial f, the **amoeba** of f is the image under μ of the zero locus of f, that is,

$$\mathcal{A}_f := \mu(\mathcal{V}(f)).$$

The notion of amoeba was introduced by Gelfand, Kapranov and Zelevinsky in [7, Definition 1.4]. The amoeba is a closed set with non-empty complement and each connected component $\mathcal{F}$ of the complement of the amoeba $\mathcal{A}_f$ is a convex subset [7, Corollary 1.6]. From now on, by complement component we will mean connected component of the complement of the amoeba. For each complement component $\mathcal{F}$ of $\mathcal{A}_f$, we have that $\mu^{-1}(\mathcal{F})$ is a logarithmically convex Reinhardt domain. An example of the amoeba of a polynomial is shown in Figure 1.

FIGURE 1. Amoeba of $f(x,y) := 50x^3+83x^2y+24xy^2+y^3+392x^2+414xy+50y^2-28x+59y-100$. (Taken from wikimedia commons, file: Amoeba4 400.png; Oleg Alexandrov).

Proposition 3.1. *Let $\mathcal{F}$ be a complement component of $\mathcal{A}_f$. The fundamental group of $\mu^{-1}(\mathcal{F})$ is isomorphic to $\mathbb{Z}^N$.*

Proof. Let $x := (x_1,\dots,x_n)$ be a point of the complement component $\mathcal{F}$. By definition,

$$\begin{aligned}\mu^{-1}(x) &= \{z=(z_1,\dots,z_N)\in\mathbb{C}^N : \mathrm{Log}(z)=x\}\\ &= \{z : (\log|z_1|,\dots,\log|z_N|)=(x_1,\dots,x_N)\} = \{z : |z_i| = e^{x_i}\}.\end{aligned}$$

That is, $\mu^{-1}(x)$ is the product of N circles of radius e^{x_i}. The result follows from the fact that $\mathcal{F}$ is contractible. □

4. The Newton polytope and the order map

Let $f=\sum_{\alpha\in\mathbb{Z}^N} c_\alpha z^\alpha$ be a Laurent series. The **set of exponents** of f is the set

$$\varepsilon(f) := \{\alpha\in\mathbb{Z}^N \mid c_\alpha\neq 0\}.$$

The set $\varepsilon(f)$ is also called the **support** of f. When f is a Laurent polynomial, the convex hull of $\varepsilon(f)$ is called the **Newton polytope** of f. We will denote the Newton polytope by $\mathrm{NP}(f)$. For $p\in\mathrm{NP}(f)$, the cone given by

$$\sigma_p(\mathrm{NP}(f)) := \{\lambda(q-p) : \lambda\in\mathbb{R}_+, q\in\mathrm{NP}(f)\} = \mathbb{R}_+(\mathrm{NP}(f)-p),$$

will be called the **cone associated to** p. This cone is obtained by drawing half-lines from p through all points of $\mathcal{N}$ and then translating the result by $(-p)$ (see Figure 2).

Forsberg, Passare and Tsikh gave in [5] a natural correspondence between complement components of the amoeba $\mathcal{A}_f$ and integer points in $\mathrm{NP}(f)$ using the "order map":

$$\begin{aligned}\mathrm{ord} : \mathbb{R}^N\setminus\mathcal{A}_f &\longrightarrow \mathrm{NP}(f)\cap\mathbb{Z}^N\\ x &\mapsto \left(\frac{1}{(2\pi i)^N}\int_{\mu^{-1}(x)}\frac{z_j\partial_j f(z)}{f(z)}\frac{dz_1\cdots dz_N}{z_1\cdots z_N}\right)_{1\le j\le N}.\end{aligned}\tag{4.1}$$

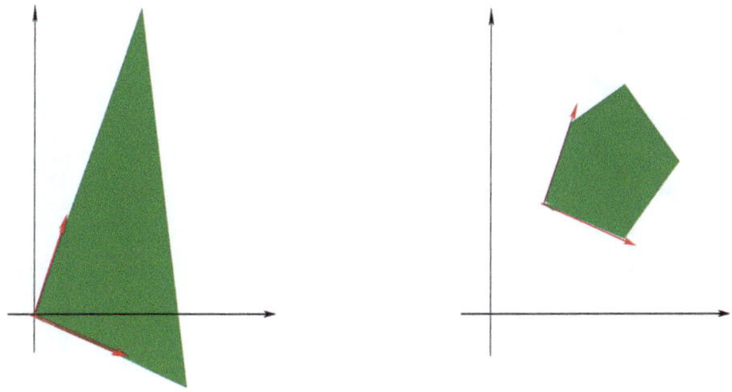

FIGURE 2. The cone associated to a vertex of the polygon on the right.

Under the order map, points in the same complement component $\mathcal{F}$ of the amoeba $\mathcal{A}_f$ have the same value. This constant value is called the **order of** $\mathcal{F}$ and it is denoted by $\operatorname{ord}(\mathcal{F})$. The order map is illustrated in Figure 3.

Proposition 4.1. *The order map induces an injective map from the set of complement components of the amoeba* $\mathcal{A}_f$ *to* $\mathrm{NP}(f) \cap \mathbb{Z}^N$. *The vertices of* $\mathrm{NP}(f)$ *are always in the image of this injection.*

Proof. See [5, Proposition 2.8]. □

Proposition 4.2. *The vertices of the Newton polytope* $\mathrm{NP}(f)$ *are in bijection with those connected components of the complement of the amoeba which contain an affine convex cone (cone with vertex) with non-empty interior.*

Proof. See [7, Corollary 1.8]. □

Forsberg, Passare and Tsikh also gave in [5] a relation between the order of a complement component of the amoeba and the recession cone of the component. They show that the recession cone of a complement component of order p is the opposite of the dual of the cone of the Newton polytope at the point p.

Proposition 4.3. *If* $\mathcal{F}$ *is a complement component of* $\mathcal{A}_f$, *then*

$$\sigma_p(\mathrm{NP}(f)) = -\operatorname{Rec}(\mathcal{F})^\vee,$$

where $p = \operatorname{ord}(\mathcal{F})$.

Proof. The proposition is just a restatement of [5, Proposition 2.6]. □

5. Toric morphisms

Let $\{u^{(1)}, \ldots, u^{(N)}\} \subset \mathbb{Z}^N$ be a N-tuple of vectors which is a basis of $\mathbb{Z}^N$. Let M be the matrix that has $u^{(1)}, \ldots, u^{(N)}$ as columns. Consider the map

$$\begin{aligned} \Phi_{\mathrm{M}} : (\mathbb{C}^*)^N &\longrightarrow (\mathbb{C}^*)^N \\ z \quad &\mapsto (z^{u^{(1)}}, z^{u^{(2)}}, \ldots, z^{u^{(N)}}). \end{aligned} \tag{5.1}$$

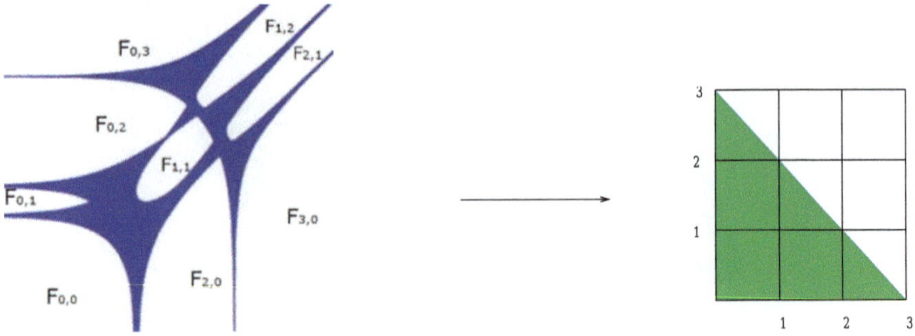

FIGURE 3. The order map between the complement components of $\mathcal{A}_f$ and NP(f) for f as in Figure 1. (Here $\mathrm{F}_{i,j}$ denotes the complement component with order (i,j)).

The map Φ_{M} is an isomorphism with inverse $\Phi_{\mathrm{M}^{-1}}$.

Lemma 5.1. *Let $\sigma \subset \mathbb{R}^N$ be a cone and $p \in \mathbb{R}^N$. If $\varrho \in (\mathbb{R}_+^*)^N$ is such that $\mu(\varrho) = p$, then*

$$\mu^{-1}(p+\sigma) = \left\{z \in (\mathbb{C}^*)^N \mid |z|^\upsilon \geq \varrho^\upsilon,\ \forall \upsilon \in \sigma^\vee\right\}.$$

Proof. We have that

$$\begin{aligned}
\mu^{-1}(p+\sigma) &\overset{\text{Remark 2.1}}{=} \left\{z \in (\mathbb{C}^*)^N \mid \mu(z)\cdot\upsilon \geq \mu(\varrho)\cdot\upsilon,\ \forall \upsilon \in \sigma^\vee\right\} \\
&= \left\{z \in (\mathbb{C}^*)^N \mid e^{\sum_{i=1}^N \upsilon_i \log|z_i|} \geq e^{\sum_{i=1}^N \upsilon_i \log \varrho_i}\right\} \\
&= \left\{z \in (\mathbb{C}^*)^N \;\middle|\; \prod_{i=1}^N e^{\upsilon_i \log|z_i|} \geq \prod_{i=1}^N e^{\upsilon_i \log \varrho_i}\right\} \\
&= \left\{z \in (\mathbb{C}^*)^N \;\middle|\; \prod_{i=1}^N |z_i|^{\upsilon_i} \geq \prod_{i=1}^N \varrho_i^{\upsilon_i},\ \forall \upsilon = (\upsilon_1,\ldots,\upsilon_N) \in \sigma^\vee\right\}. \quad \square
\end{aligned}$$

Proposition 5.2. *Let* M *be as in* (5.1) *and let $\sigma \subset \mathbb{R}^N$ be a cone. Given $p \in \mathbb{R}^N$, one has*

$$\mu(\Phi_{\mathrm{M}}(\mu^{-1}(p+\sigma))) = q + \mathrm{M}^T\sigma,$$

where $\{q\} = \mu(\Phi_{\mathrm{M}}(\mu^{-1}(p)))$.

Proof. We have that

$$\begin{aligned}
&\Phi_{\mathrm{M}}(\mu^{-1}(p+\sigma)) \\
&\overset{\text{Lemma 5.1}}{=} \left\{z \in (\mathbb{C}^*)^N \mid |\Phi_{\mathrm{M}^{-1}}(z)|^\upsilon \geq \Phi_{\mathrm{M}^{-1}}(\rho)^\upsilon,\ \forall \upsilon \in \sigma^\vee; \Phi_{\mathrm{M}}(\rho) = \varrho\right\} \\
&= \left\{z \in (\mathbb{C}^*)^N \mid |z|^{\mathrm{M}^{-1}\upsilon} \geq \rho^{\mathrm{M}^{-1}\upsilon},\ \forall \upsilon \in \sigma^\vee\right\} \\
&= \left\{z \in (\mathbb{C}^*)^N \mid |z|^w \geq \rho^w,\ \forall w \in \mathrm{M}^{-1}\sigma^\vee\right\}
\end{aligned}$$

and

$$\begin{aligned}\mu(\Phi_{\mathrm{M}}(\mu^{-1}(p+\sigma))) &= \left\{\mu(z)\in\mathbb{R}^N \mid w\cdot\mu(z)\geq\log(\rho^w),\ \forall w\in\mathrm{M}^{-1}\sigma^\vee\right\}\\ &= \left\{y\in\mathbb{R}^N \mid w\cdot y\geq\log(\rho^w),\ \forall w\in\mathrm{M}^{-1}\sigma^\vee\right\}\\ &\overset{\text{Remark 2.1}}{=} q+(\mathrm{M}^{-1}\sigma^\vee)^\vee = q+\mathrm{M}^T\sigma.\end{aligned}$$

□

Corollary 5.3. *Let $\Omega\subset(\mathbb{C}^*)^N$ be a Reinhardt domain and let $\sigma\subset\mathbb{R}^N$ be a cone. If $\sigma\subset\mathrm{Rec}(\mu(\Omega))$, then $\mathrm{M}^T\sigma\subset\mathrm{Rec}(\mu(\Phi_{\mathrm{M}}(\Omega)))$.*

Proposition 5.4. *Let $\sigma\subset\mathbb{R}^N$ be a cone. Let φ be a Laurent series and suppose that $\varepsilon(\varphi)\subset p+\sigma$ where $p\in\mathbb{R}^N$. Then $\varepsilon(\varphi\circ\Phi_{\mathrm{M}})\subset\mathrm{M}p+\mathrm{M}\sigma$.*

Proof. It is enough to make the substitution. □

6. Series development on Reinhardt domains

It is well known that the Taylor development of a holomorphic function on a disc centered at the origin is a series with support in the non-negative orthant. In this section we will get a similar result for holomorphic functions on $\xi_d^{-1}(\Omega)$ where Ω is a Reinhardt domain.

Proposition 6.1. *If $f(x)$ is a holomorphic function on a logarithmically convex Reinhardt domain, then there exists a (unique) Laurent series converging to $f(x)$ in this domain.*

Proof. See, for example, [13, Theorem 1.5.26]. □

Lemma 6.2. *Let $\Omega\subset(\mathbb{C}^*)^N$ be a Reinhardt domain. Let $(e^{(1)},\dots,e^{(N)})$ be the canonical basis of $\mathbb{R}^N$. If $\langle -e^{(1)},\dots,-e^{(s)}\rangle\subset\mathrm{Rec}(\mu(\Omega))$, then for every $w\in\Omega$, the s-dimensional polyannulus*

$$\mathbb{D}^*_{\tau(w),s} := \left\{z\in(\mathbb{C}^*)^N \mid |z_i|\leq|w_i|; 1\leq i\leq s \text{ and } z_i=w_i; s+1\leq i\leq N\right\}$$

is contained in Ω.

Proof. Consider the cone $\sigma:=\langle -e^{(1)},\dots,-e^{(s)}\rangle$ and take $w\in\Omega$. By Lemma 2.4, the dual cone of σ is $\sigma^\vee=\langle -e^{(1)},\dots,-e^{(s)},e^{s+1},-e^{s+1},\dots,e^N,-e^N\rangle$. Since the cone $\sigma\subset\mathrm{Rec}(\mu(\Omega))$, we have that $\mu(w)+\sigma\subset\mu(\Omega)$. Since Ω is a Reinhardt domain, then

$$\begin{aligned}\Omega\supseteq\mu^{-1}(\mu(w)+\sigma) &\overset{\text{Lemma 5.1}}{=} \left\{z\in(\mathbb{C}^*)^N \mid |z|^v\geq|w|^v,\ \forall v\in\sigma^\vee\right\}\\ &= \{z\in(\mathbb{C}^*)^N \mid |z_i|\leq|w_i|, \text{ for } i=1,\dots,s;\\ &\quad \text{and } |z_i|=|w_i|, \text{ for } i=s+1,\dots,N\}\supset\mathbb{D}^*_{\tau(w),s}.\end{aligned}$$

□

Given a natural number d, set

$$\begin{array}{rccc}\xi_d: & \mathbb{C}^N & \longrightarrow & \mathbb{C}^N\\ & (z_1,\dots,z_N) & \mapsto & (z_1^d,\dots,z_N^d)\end{array} \tag{6.1}$$

Lemma 6.3. *Let Ω be a Reinhardt domain and suppose that*

$$(-\mathbb{R}_{\geq 0})^s \times \{0\}^{N-s} \subset \mathrm{Rec}(\mu(\Omega)).$$

Let f be a bounded holomorphic function on $\xi_d^{-1}(\Omega)$. Then the set of exponents of the Laurent series expansion of f is contained in $(\mathbb{R}_{\geq 0})^s \times \mathbb{R}^{N-s}$.

Proof. Let $\varphi = \sum_{I \in \mathbb{Z}^N} a_I z^I$ be the Laurent series expansion of f on $\xi_d^{-1}(\Omega)$. Take $w \in \Omega$. By Lemma 6.2 we have that $\mathbb{D}^*_{\tau(w),s} \subseteq \Omega$. Therefore, φ is convergent and bounded on $\mathbb{D}^*_{\tau(\xi_d^{-1}(\tau(w))),s}$. Let π be the map defined by,

$$\begin{array}{rccl} \pi: & (\mathbb{C}^*)^N & \longrightarrow & (\mathbb{C}^*)^{N-s} \\ & (z_1,\ldots,z_N) & \mapsto & (z_{s+1}, z_{s+2}, \ldots, z_N). \end{array}$$

The series in s variables, $\varphi_w := \sum_{\alpha \in \mathbb{Z}^s} \psi_\alpha(\pi(w)) z^\alpha$ where

$$\psi_\alpha(\pi(w)) = \sum_{(\alpha, i_{s+1}, \ldots, i_N) \in \varepsilon(\varphi)} a_I w_{s+1}^{i_{s+1}} \cdots w_N^{i_N},$$

is convergent and bounded on the polyannulus

$$\mathbb{D}^* := \{z \in (\mathbb{C}^*)^s \mid |z_i| \leq \sqrt[d]{|w_i|}; 1 \leq i \leq s\}.$$

By the Riemann removable singularity theorem (see, for example, [13, Theorem 4.2.1]), there exists a (unique) holomorphic map that extends φ_w on $\mathbb{D}$ and φ_w is its Taylor development on that disc. Then φ_w cannot have negative exponents. □

Proposition 6.4. *Let Ω be a Reinhardt domain. Let $\sigma \subset \mathbb{R}^N$ be a cone. Suppose that $\sigma \subseteq \mathrm{Rec}(\mu(\Omega))$. Let f be a bounded holomorphic map on $\xi_d^{-1}(\Omega)$. The set of exponents of the Laurent series expansion φ of f on $\xi_d^{-1}(\Omega)$ is contained in $-\sigma^\vee$.*

Proof. By Remark 2.2, we can assume that σ is a regular cone. Let $\{v^{(1)}, \ldots, v^{(s)}\}$ be the generator set of σ. Let $(v^{(s+1)}, \ldots, v^{(N)})$ be a basis of $\sigma^\perp \cap \mathbb{Z}^N$. Let A be the matrix that has as columns $v^{(i)}$ for $i = 1, \ldots, N$. Set $\mathrm{M} := -(A^{-1})^T$. Since $\sigma \subset \mathrm{Rec}(\mu(\Omega))$, by Corollary 5.3, we have that

$$-A^{-1}\sigma = \langle -e^{(1)}, \ldots, -e^{(s)} \rangle \subseteq \mathrm{Rec}(\mu(\Phi_{\mathrm{M}}(\Omega))).$$

The series $h := \varphi \circ \Phi_{\mathrm{M}}^{-1}$ is convergent in $\Phi_{\mathrm{M}}(\xi_d^{-1}(\Omega))$ then, by Lemma 6.3,

$$\varepsilon(h) \subset \langle e^{(1)}, \ldots, e^{(s)}, \pm e^{(s+1)}, \ldots, \pm e^{(N)} \rangle.$$

Therefore, by Proposition 5.4 and Lemma 2.4,

$$\varepsilon(\varphi) \subseteq \mathrm{M}\langle e^{(1)}, \ldots, e^{(s)}, \pm e^{(s+1)}, \ldots, \pm e^{(N)} \rangle = -\sigma^\vee.$$

□

7. Parameterizations compatible with a projection

In what follows $\mathbb{X} \subseteq \mathbb{C}^{N+M}$ will denote an irreducible algebraic variety of dimension N such that the canonical projection

$$\begin{array}{rccc} \Pi : & \mathbb{X} & \longrightarrow & \mathbb{C}^N \\ & (z_1, \ldots, z_{N+M}) & \mapsto & (z_1, \ldots, z_N), \end{array} \tag{7.1}$$

is finite (that is, proper with finite fibers). In this case, there exists an algebraic set $\mathcal{A} \subset \mathbb{C}^N$ such that, the restriction

$$\Pi : \mathbb{X} \setminus \Pi^{-1}(\mathcal{A}) \longrightarrow \Pi(\mathbb{X}) \backslash \mathcal{A}$$

is locally biholomorphic (see, for example, [3, Proposition 3.7] or [9, §9]). The intersection of all sets $\mathcal{A}$ with this property is called the **discriminant locus** of Π. We denote the discriminant locus of Π by Δ.

Theorem 7.1. *Let $\mathbb{X}$, Π and Δ be as above and let $\mathcal{V}(\delta)$ be an algebraic hypersurface of $\mathbb{C}^N$ containing Δ. Given a complement component $\mathcal{F}$ of $\mathcal{A}_\delta$, set $\Omega := \mu^{-1}(\mathcal{F})$. For every connected component $\mathcal{C}$ of $\Pi^{-1}(\Omega) \cap \mathbb{X}$, there exists a natural number d and a holomorphic morphism $\Psi : \xi_d^{-1}(\Omega) \to \mathbb{C}^M$ such that*

$$\mathcal{C} = \{(\xi_d(z), \Psi(z)) \mid z \in \xi_d^{-1}(\Omega)\}.$$

Proof. Note that $\Pi : \mathbb{X} \cap \Pi^{-1}(\Omega) \longrightarrow \Omega$ is locally biholomorphic. Let $\mathcal{C}$ be a connected component of $\mathbb{X} \cap \Pi^{-1}(\Omega)$ and let d be the cardinal of the generic fiber of $\Pi|_{\mathcal{C}}$. Since both $\Pi|_{\mathcal{C}}$ and $\xi_d|_{\xi_d^{-1}(\Omega)}$ are locally biholomorphic, the pairs $(\mathcal{C}, \Pi)$ and $(\xi_d^{-1}(\Omega), \xi_d)$ are a d-sheeted and a d^N-sheeted covering of Ω respectively. Choose a point $z_0 \in \Omega$, a point $z_1 \in \xi_d^{-1}(z_0)$ and a point $z_2 \in \Pi^{-1}(z_0) \cap \mathcal{C}$. Take the induced monomorphisms on the fundamental groups:

$$\begin{array}{ccc} \pi_1(\xi_d^{-1}(\Omega), z_1) & & \pi_1(\mathcal{C}, z_2) \\ & \searrow^{\xi_{d*}} & \downarrow \Pi_* \\ & & \pi_1(\Omega, z_0) \end{array}$$

Note that:

i) An element $\gamma \in \pi_1(\Omega, z_0)$ is in the subgroup $\xi_{d*}\pi_1(\xi_d^{-1}(\Omega), z_1)$ if and only if $\gamma = \alpha^d$ for some $\alpha \in \pi_1(\Omega, z_0)$.
ii) The index of $\Pi_*(\pi_1(\mathcal{C}, z_2))$ in $\pi_1(\Omega, z_0)$ is equal to d (see, for example, [10, V§7]).

By Proposition 3.1, $\pi_1(\Omega, z_0)$ is abelian, then the cosets of $\Pi_*(\pi_1(\mathcal{C}, z_2))$ in $\pi_1(\Omega, z_0)$ form a group of order d. By i) and ii), we have that for any α in $\pi_1(\Omega, z_0)$, the element α^d belongs to $\Pi_*(\pi_1(\mathcal{C}, z_2))$. Then,

$$\xi_{d*}(\pi_1(\xi_d^{-1}(\Omega), z_1)) \subseteq \Pi_*(\pi_1(\mathcal{C}, z_2)).$$

Applying the lifting lemma (see [10, Theorem 5.1]) we obtain a unique holomorphic morphism φ, such that $\varphi(z_1) = z_2$ and the following diagram commutes:

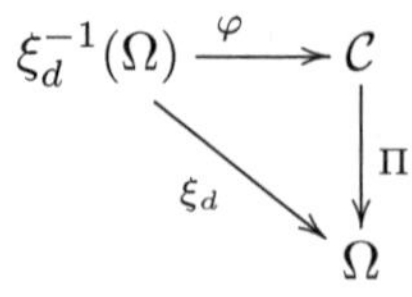

The result follows from the fact that φ is a holomorphic morphism. □

Remark 7.2. For every $P \in \Pi^{-1}(z_0) \cap \mathcal{C}$ there exists a unique φ as in the proof of Theorem 7.1 such that $\varphi(z_1) = P$. It follows that there exist d different morphisms φ's.

8. The series development of the parameterizations

Let $\mathcal{C}$ be a connected component of $\Pi^{-1}(\Omega) \cap \mathbb{X}$ where Π, Ω and $\mathbb{X}$ are as in Theorem 7.1. By the same theorem, there exists $\varphi : \xi_d^{-1}(\Omega) \to \mathbb{C}^{N+M}$ of the form

$$\begin{array}{rccl} \varphi : & \xi_d^{-1}(\Omega) & \longrightarrow & \mathbb{X} \\ & (z_1, \dots, z_N) & \mapsto & (z_1^d, \dots, z_N^d, \varphi_1, \dots, \varphi_M), \end{array} \tag{8.1}$$

where $\varphi_i : \xi_d^{-1}(\Omega) \longrightarrow \mathbb{C}$ is a holomorphic function for $i = 1, \dots, M$. Since $\xi_d^{-1}(\Omega)$ is a Reinhardt domain, by Proposition 6.1 we have:

Proposition 8.1. *For every connected component of* $\Pi^{-1}(\Omega) \cap \mathbb{X}$ *there exist a natural number* d *and Laurent series* ϕ_i *converging to* φ_i *in* $\xi_d^{-1}(\Omega)$ *for* $i = 1, \dots, M$*, such that* $(z_1^d, \dots, z_N^d, \phi_1(z), \dots, \phi_M(z)) \in \mathbb{X}$*, for all* $z \in \xi_d^{-1}(\Omega)$.

In fact, if k is the degree of the projection Π, by Remark 7.2, there are k M-tuples $(\phi_1, \dots, \phi_M)$ of convergent Laurent series such that

$$(z_1^d, \dots, z_N^d, \phi_1(z), \dots, \phi_M(z)) \in \mathbb{X}, \ \forall\, z \in \xi_d^{-1}(\Omega).$$

Now we describe the support set of the above Laurent series ϕ_i.

Proposition 8.2. *Let* $\mathcal{F}$ *be a complement component of* $\mathcal{A}_\delta$ *where* δ *is a polynomial as in Theorem* 7.1. *Let* ϕ_i *for* $i = 1, \dots, M$ *be the Laurent series that converges to* φ_i *as in Proposition* 8.1. *Then* $\varepsilon(\phi_i) \subseteq \sigma_p(\mathrm{NP}(\delta))$ *for all* $i = 1, \dots, M$*, where* $p = \mathrm{ord}(\mathcal{F})$.

Proof. By Proposition 4.3, we have that $-\sigma_p(\mathrm{NP}(\delta))^\vee = \mathrm{Rec}(\mathcal{F})$. Since every φ_i is a bounded holomorphic function on $\xi_d^{-1}(\mu^{-1}(\mathcal{F}))$, the result follows from Proposition 6.4. □

Theorem 8.3. *Let* $\mathbb{X}$ *be an algebraic set in* $\mathbb{C}^{N+M}$ *with* $\underline{0} \in \mathbb{X}$ *and* $\dim(\mathbb{X}) = N$. *Let* $\mathcal{V}(\delta)$ *be an algebraic hypersurface containing the discriminant locus of the* *projection* $\pi : \mathbb{X} \longrightarrow \mathbb{C}^N$. *Then, given a complement component* $\mathcal{F}$ *of* $\mathcal{A}_\delta$, *there exist local parametric equations of* $\mathbb{X}$ *of the form*

$$z_i = t_i^d \quad i = 1, \ldots, N, \qquad z_{N+j} = \phi_j(t_1, \ldots, t_N) \quad j = 1, \ldots, M,$$

where d *is a natural number, the* ϕ_j *are convergent Laurent series in* $\xi_d^{-1}(\mu^{-1}(\mathcal{F}))$ *and their support is contained in the cone* $-\mathrm{Rec}(\mathcal{F})^\vee$.

Proof. It is just a restatement of the Theorem 7.1 and Proposition 8.2. □

Corollary 8.4 (Aroca, [2]). *Let* $\mathbb{X}$ *be an algebraic variety of* $\mathbb{C}^{N+M}$, $\underline{0} \in \mathbb{X}$, $\dim(\mathbb{X}) = N$. *Let* U *be a neighborhood of* $\underline{0}$, *and let* π *be the restriction to* $\mathbb{X} \cap U$ *of the projection* $(z_1, \ldots, z_{N+M}) \mapsto (z_1, \ldots, z_N)$. *Assume* π *is a finite morphism. Let* δ *be a polynomial vanishing on the discriminant locus of* π. *For each cone* σ *of* $NP(\delta)$ *associated to a vertex, there exist* $k \in \mathbb{N}$ *and* M *convergent Laurent series* $s_1, \ldots, s_M$, *such that*

$$\varepsilon(s_i) \subseteq \sigma, \quad i = 1, \ldots, M, \quad \text{and} \quad f(z_1^k, \ldots, z_N^k, s_1(z), \ldots, s_M(z)) = 0$$

for any f *vanishing on* $\mathbb{X}$, *and any* z *in the domain of convergence of the* s_i.

Proof. By Proposition 4.1, for every vertex V of $NP(\delta)$ there exists a complement component of $\mathcal{A}_\delta$ with order V. Then, by Theorem 8.3 there exist convergent Laurent series with support in the cone associated to the complement component of order V. □

Corollary 8.5 (McDonald, [11]). *Let* $F(x_1, \ldots, x_N, y) = 0$ *be an algebraic equation with complex coefficients. There exists a fractional power series expansion (Puiseux series)* $\phi(x_1, \ldots, x_N)$ *such that*

$$F(x_1, \ldots, x_N, \phi) = 0$$

and the support of ϕ *is contained in some strongly convex polyhedral cone.*

Proof. The result follows by applying a similar argument as in the proof of Corollary 8.4. □

Remark 8.6. P.D. González Pérez showed in [8], with an additional hypothesis, that the supporting cone in Corollary 8.5 can be chosen to be a cone of the Newton polytope of the discriminant of the polynomial defining the hypersurface with respect to y. Thus, in this sense, by the proof of Corollary 8.5 we also get this result.

Corollary 8.7 (Abhyankar-Jung [1]). *Let* $\underline{0}$ *be a quasi-ordinary singularity of a complex algebraic set* $\mathbb{X} \subseteq \mathbb{C}^{N+M}$, $\dim(\mathbb{X}) = N$. *Then, there exists a natural number* d *and* M *convergent power series* $\phi_1, \ldots, \phi_M$ *such that*

$$z_i = t_i^d, \quad i = 1, \ldots, N, \quad z_{N+j} = \phi_j(t_1, \ldots, t_N), \quad j = 1, \ldots, M,$$

are parametric equations of $\mathbb{X}$ *about* $\underline{0}$.

Proof. By definition of quasi-ordinary singularity, the discriminant locus of the projection π is contained in the coordinate hyperplanes, then it is contained in the algebraic hypersurface defined by $\beta(z) := z_1 \cdots z_N$. Note that $NP(\beta)$ has just one cone contained in the non-negative orthant. The result follows from Theorem 8.3. □

Example. Consider the hypersurface defined by $f := z^2 - x - y + 1$ in $\mathbb{C}^3$. The discriminant of the projection to the (x, y)-plane is $\Delta := x + y - 1$. By the generalized binomial theorem, we have a series expansion of $\Delta^{1/2}$

$$\varphi_1 := \sum_{k=0}^{\infty} \binom{1/2}{k} y^k (-1 + x)^{1/2-k} = \sum_{k=0}^{\infty} \binom{1/2}{k} \sum_{j=0}^{\infty} \binom{1/2-k}{j} (-1)^j y^k x^{1/2-k-j}.$$

We know by Theorem 8.3 that φ_1 converges in $\mu^{-1}(\mathrm{F})$ for some complement component F of the amoeba. Since φ_1 converges in the region $|x| > 1$, $|y| < |x| - 1$ and this region is mapped under μ to the complement component $\mathbf{A}$ of the amoeba (see figure 4), we have that φ_1 converges in $\mu^{-1}(\mathbf{A})$. Since $\sigma_{(1,0)}$ is the unique cone of NP(Δ) such that a translation of $-\sigma^{\vee}_{(1,0)}$ is contained in the complement component $\mathbf{A}$, by Proposition 4.2 this complement component is associated to the vector $(1, 0)$. According to the Theorem 8.3 we must have that

$$\varepsilon(\varphi_1) \subset \langle(-1, 0), (-1, 1)\rangle$$

which is true, because

$$\varepsilon(\varphi_1) = \{(1/2 - k - j, k) \mid j, k \in \mathbb{N} \cup \{0\}\}$$

and

$$(1/2 - k - j, k) = j - 1/2(-1, 0) + k(-1, 1).$$

Reasoning analogously as before, we get another series expansion of $\Delta^{1/2}$,

$$\varphi_2 := \sum_{k=0}^{\infty} \binom{1/2}{k} x^k (-1 + y)^{1/2-k} = \sum_{k=0}^{\infty} \binom{1/2}{k} \sum_{j=0}^{\infty} \binom{1/2-k}{j} (-1)^j x^k y^{1/2-k-j},$$

which converges in the region $|y| > 1$, $|x/y - 1| < 1$. Therefore, φ_2 converges in $\mu^{-1}(\mathbf{B})$. As before, by Proposition 4.2 we can see that this complement component is associated to the vector $(0, 1)$. Therefore, by Theorem 8.3 we must have that

$$\varepsilon(\varphi_2) \subset \langle(0, -1), (1, -1)\rangle.$$

This is true because

$$\varepsilon(\varphi_2) = \{(k, 1/2 - k - j) \mid j, k \in \mathbb{N} \cup \{0\}\}$$

and

$$(k, 1/2 - k - j) = j - 1/2(0, -1) + k(1, -1).$$

Analogously, for

$$\varphi_3 := \sum_{k=0}^{\infty} \binom{1/2}{k} (-1)^{1/2-k} (x + y)^k$$

we have that φ_3 converges in $\mu^{-1}(\mathbf{C})$. The complement component $\mathbf{C}$ is associated to the vector $(0,0)$ and the support of φ_3 is contained in the non-negative orthant.

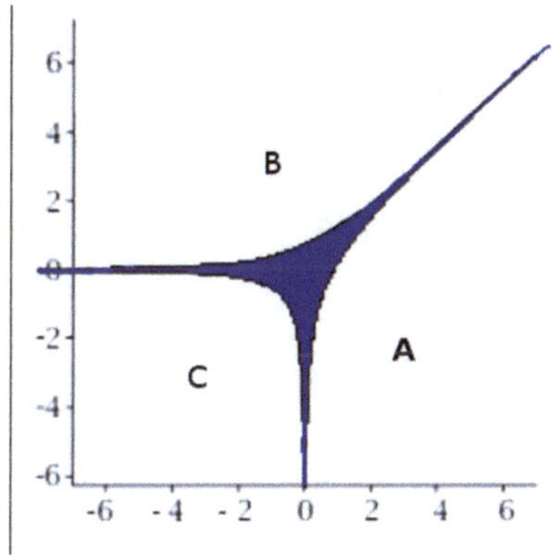

FIGURE 4. The amoeba of $x+y-1$ (taken from wikimedia commons; Oleg Alexandrov).

Acknowledgment

This work has been supported by PAPIIT IN104713 y IN108216, ECOS NORD M14M03, LAISLA and CONACYT

References

[1] Abhyankar, S. (1955). *On the ramification of algebraic functions.* American Journal of Mathematics, 77(3), 575-592.

[2] Aroca, F. (2004). *Puiseux parametric equations of analytic sets.* Proceedings of the American Mathematical Society, 132(10), 3035-3045.

[3] Chirka, E. M. (2012). Complex analytic sets (Vol. 46). Springer Science & Business Media.

[4] Dattorro, Jon. *Convex optimization & Euclidean distance geometry.* Lulu. com, 2010.

[5] Forsberg, Mikael, Mikael Passare, and August Tsikh. *Laurent determinants and arrangements of hyperplane amoebas.* Advances in mathematics 151.1 (2000): 45-70.

[6] Fulton, W. *Introduction to toric varieties* (No. 131). Princeton University Press, 1993.

[7] Gelfand, Israel M., Mikhail Kapranov, and Andrei Zelevinsky. *Discriminants, resultants, and multidimensional determinants.* Springer Science & Business Media, 2008.

[8] González Pérez, P.D. (2000). *Singularités quasi-ordinaires toriques et polyèdre de Newton du discriminant.* Canadian Journal of Mathematics-Journal Canadien de Mathématiques, 52(2), 346-368.

[9] Lojasiewicz, Stanislaw. *Introduction to complex analytic geometry.* Birkhäuser, 2013.

[10] Massey, W. S. (1991). *A basic course in algebraic topology.* Vol. 127. Springer Science & Business Media, 1991.

[11] McDonald, J. (1995). *Fiber polytopes and fractional power series.* Journal of Pure and Applied Algebra, 104(2), 213-233.

[12] Puiseux, V. *Recherches sur les fonctions algébriques.* Journal de Mathématiques pures et appliquées (1850): 365-480.

[13] Scheidemann, Volker. *Introduction to complex analysis in several variables.* Basel: Birkhäuser Verlag, 2005.

[14] Whiteside, D. T., & Mahoney, M. S. (1970). *The Mathematical Papers Of Isaac Newton,* Vol. III: 1670-1673. Physics Today, 23, 63.

Fuensanta Aroca and Víctor Manuel Saavedra
Instituto de Matemáticas
Universidad Nacional Autónoma de México
Unidad Cuernavaca
A.P.273-3 C.P. 62251
Cuernavaca, Morelos
México
e-mail: `fuen@matcuer.unam.mx`
`victorm@matcuer.unam.mx`

Some Open Questions on Arithmetic Zariski Pairs

Enrique Artal Bartolo and José Ignacio Cogolludo-Agustín

Dedicado con cariño a Pepe, singular matemático y amigo

Abstract. In this paper, complement-equivalent arithmetic Zariski pairs will be exhibited answering in the negative a question by Eyral-Oka [14] on these curves and their groups. A complement-equivalent arithmetic Zariski pair is a pair of complex projective plane curves having Galois-conjugate equations in some number field whose complements are homeomorphic, but whose embeddings in $\mathbb{P}^2$ are not.

Most of the known invariants used to detect Zariski pairs depend on the étale fundamental group. In the case of Galois-conjugate curves, their étale fundamental groups coincide. Braid monodromy factorization appears to be sensitive to the difference between étale fundamental groups and homeomorphism class of embeddings.

Mathematics Subject Classification (2000). Primary 14N20, 32S22, 14F35; Secondary 14H50, 14F45, 14G32.

Keywords. Zariski pairs, number fields, fundamental group.

Introduction

In this work some open questions regarding Galois-conjugated curves and arithmetic Zariski pairs will be answered and some new questions will be posed. The techniques used here combine braid monodromy calculations, group theory, representation theory, and the special real structure of Galois-conjugated curves.

A Zariski pair [2] is a pair of plane algebraic curves $\mathcal{C}_1, \mathcal{C}_2 \in \mathbb{P}^2 \equiv \mathbb{CP}^2$ whose embeddings in their regular neighborhoods are homeomorphic $(T(\mathcal{C}_1), \mathcal{C}_1) \cong (T(\mathcal{C}_2), \mathcal{C}_2)$ but their embeddings in $\mathbb{P}^2$ are not $(\mathbb{P}^2, \mathcal{C}_1) \ncong (\mathbb{P}^2, \mathcal{C}_2)$. The first condition is given by a discrete set of invariants which we refer to as *purely combinatorial*

Partially supported by the Spanish Government MTM2013-45710-C2-1-P and Grupo Consolidado Geometría E15 from the Gobierno de Aragón/Fondo Social Europeo.

in the following sense. The *combinatorics* of a curve $\mathcal{C}$ with irreducible components $\mathcal{C}^1,\ldots,\mathcal{C}^r$ is defined by the following data:

($\mathcal{C}1$) the degrees $d_1,\ldots,d_r$ of $\mathcal{C}^1,\ldots,\mathcal{C}^r$;
($\mathcal{C}2$) the topological types $\mathcal{T}_1,\ldots,\mathcal{T}_s$ of the singular points $P_1,\ldots,P_s\in\operatorname{Sing}(\mathcal{C})$;
($\mathcal{C}3$) each $\mathcal{T}_i$ is determined by the topological types $\mathcal{T}_i^1,\ldots,\mathcal{T}_i^{n_i}$ of its local irreducible branches $\delta_i^1,\ldots,\delta_i^{n_i}$ and by the local intersection numbers $(\delta_i^j,\delta_i^k)_{P_i}$ of each pair of irreducible branches. The final data for the definition of the combinatorics is the assignment of its global irreducible component for each local branch δ_i^j.

The first Zariski pair was found by O. Zariski in [28, 30] and it can be described as a pair of irreducible sextics with six ordinary cusps. This example has two main features. On the one hand, the embeddings of the curves in $\mathbb{P}^2$ are not homeomorphic because their complements are not. On the other hand, one of the curves of the pair satisfies a nice global algebraic property (which is not part of its combinatorics): its six singular points lie on a conic. The first fact can be proved directly by showing that the fundamental groups of their complements are not isomorphic. Also, using [29] it is possible to prove this by means of a weaker, but more tractable, invariant which was later called the Alexander polynomial of the curve by Libgober [20] which is sensitive to global aspects such as the position of the singularities. The second feature is the fact that one of the sextics is a curve of *torus type*, i.e., a curve whose equation is of the form $f_2^3+f_3^2=0$, where f_j is a homogeneous polynomial of degree j.

Since then, many examples of Zariski pairs (and tuples) have been found by many authors, including J. Carmona, A. Degtyarev, M. Marco, M. Oka, G. Rybnikov, I. Shimada, H. Tokunaga, and the authors, (see [7] for precise references).

By the work of Degtyarev [9] and Namba [22], Zariski pairs can appear only in degree at least 6, and this is why the literature of Zariski pairs of sextics is quite extensive. Given a pair of curves, it is usually easy to check that they have the same combinatorics. What is usually harder to prove is whether or not they are homeomorphic. Note that two curves which admit an equisingular deformation are topologically equivalent and this is why the first step to check whether a given combinatorics may admit Zariski pairs is to find the connected components of the space of realizations of the combinatorics. Namely, given a pair of curves with the same combinatorics, a necessary condition for them to be a Zariski pair is that they are not connected by an equisingular deformation, in the language of Degtyarev, they are not *rigidly isotopic*.

Most of the effective topological invariants used in the literature to prove that a pair of curves is a Zariski pair can be reinterpreted in algebraic terms, in other words, they only depend on the algebraic (or étale) fundamental group, defined as the inverse limit of the system of subgroups of finite index of a fundamental group. This is why, in some sense when it comes to Zariski pairs, the most difficult candidates to deal with are those of an arithmetic nature, i.e., curves $\mathcal{C}_1,\mathcal{C}_2$ whose equations have coefficients in some number field $\mathbb{Q}(\xi)$ and they are Galois conjugate. Note that Galois-conjugate curves have the same étale fundamental group.

There are many examples of pairs of Galois-conjugate sextic curves which are not rigidly isotopic. The first example of an arithmetic Zariski pair was found in [4] in degree 12 and it was built up from a pair of Galois-conjugate sextics (see also [1, 19, 23] for similar examples on compact surfaces).

In another direction, the equivalence class of embeddings, i.e., the homeomorphism class of pairs $(\mathbb{P}^2, \mathcal{C})$, can be refined by allowing only homeomorphisms that are holomorphic at neighborhoods of the singular points of the curve (called *regular* by Degtyarev [11]). Also, one can allow only homeomorphisms that can be extended to the exceptional divisors on a resolution of singularities. Curves that have the same combinatorics and belong in different classes are called *almost*-Zariski pairs in the first case and NC-Zariski pairs in the second case.

Interesting results concern these other Zariski pairs, for instance Degtyarev proved in [11] that sextics with simple singular points and not rigidly isotopic are *almost* Zariski pairs, and among them there are plenty of arithmetic pairs.

Shimada developed in [24, 25] an invariant denoted N_C which is a topological invariant of the embedding, but not of the complements. He found the first examples of arithmetic Zariski pairs for sextics. None of these examples is of torus type.

In [14], Eyral and Oka study a pair of Galois-conjugated curves of torus type. They were able to find presentations of the fundamental groups of their complements and was conjectured that these groups are not isomorphic, in particular this would produce an arithmetic Zariski pair. The invariant used by Shimada to find arithmetic Zariski pairs of sextics does not distinguish Eyral-Oka curves. Also, Degtyarev [10] proposed alternative methods to attack the problem, but it is still open as originally posed by Eyral-Oka.

This paper answers some questions on the Eyral-Oka example. The first part of the conjecture is solved in the negative by proving that the fundamental groups of both curve complements are in fact isomorphic. The question about them being an arithmetic Zariski pair remains open but, using the techniques in [7], several arithmetic Zariski pairs can be exhibited by adding lines to the original curves. It is right hence to conjecture that they form an arithmetic Zariski pair themselves. Moreover, some of these Zariski pairs are *complement-equivalent Zariski pairs*, (cf. [7]) that is, their complements are homeomorphic (actually analytically and algebraically isomorphic in this case) but no homeomorphism of the complements extends to the curves.

Also, a very relevant fact about these curves that makes computations of braid monodromies, and hence fundamental groups, very effective from a theoretical point of view is that they are not only real curves, but *strongly real curves*, that is, their singular points are also real plus the real picture and the combinatorics are enough to describe the embedding. Some of the special techniques used for strongly real curves were outlined in [6]. In this work we will describe them in more detail.

The paper is organized as follows. In Section 1 the projective Eyral-Oka curves will be constructed. Their main properties are described and one of the main results of this paper is proved: after adding a line to the projective Eyral-Oka curve we obtain the affine Eyral-Oka curve and we show that their complements are homeomorphic in Theorem 1.10. Section 2 is devoted to giving a description of the braid monodromy factorization of the affine Eyral-Oka curves as well as a theoretical description of the fundamental groups of their complements in Theorem 2.9 which allow us to show that the fundamental groups of the projective Eyral-Oka curve complements are isomorphic in Corollary 2.10. Finally, in Section 3 we define a new invariant of the embedding of fibered curves and use it to produce examples of complement-equivalent arithmetic Zariski pairs in Theorem 3.7.

1. Construction of Eyral-Oka curves

In [14] M. Oka and C. Eyral proposed a candidate for an arithmetic Zariski pair of sextics. This candidate is the first one formed by curves of torus type, i.e., which can be written as $f_2^3 + f_3^2 = 0$, for f_j a homogeneous polynomial in $\mathbb{C}[x, y, z]$ of degree j.

Eyral-Oka curves are irreducible, they have degree 6, and their singularities are given by: two points of type $\mathbb{E}_6$, one $\mathbb{A}_5$, and one $\mathbb{A}_2$. The equisingular stratum of such curves is described in [14], however for the sake of completeness, an explicit construction of this space will be provided here. In particular, this realization space has two connected components of dimension 0 up to projective transformation.

To begin with proving the basic properties of these curves, let us fix a sextic curve $\mathcal{C} : f(x, y, z) = 0$ with the above set of singularities.

Lemma 1.1. *The curve $\mathcal{C}$ is rational and irreducible.*

Proof. Recall that $\mathbb{E}_6$ singularities have a local equation of the form $x^3 + y^4$ for a choice of generators of the local ring $\mathcal{O}_{\mathbb{C}^2,0}$. In particular it is an irreducible singularity whose δ-invariant is 3 and thus it can only be present in an irreducible curve of degree at least 4. Since the total degree of $\mathcal{C}$ is 6, this implies that both $\mathbb{E}_6$ singularities have to be on the same irreducible component. Again, by a genus argument, the irreducible component containing both $\mathbb{E}_6$ singularities needs to have degree at least 5, but then the existence of the irreducible singularity $\mathbb{A}_2$ implies that $\mathcal{C}$ cannot be a quintic and a line (note also that no quintic with two $\mathbb{E}_6$ singularities exists, because of Bézout's Theorem). Hence, if it exists, it has to be irreducible. Also note that the total δ-invariant of the singular locus $2\mathbb{E}_6 + \mathbb{A}_5 + \mathbb{A}_2$ is 10, which implies that the sextic has to be rational. □

We are going to prove now that $\mathcal{C}$ is of torus type. Using an extension of the de Franchis method [15] to rational pencils (see [8, 27]), it follows that $\mathcal{C}$ is of torus type if and only if the cyclotomic polynomial of order 6, $\varphi_6(t) = t^2 - t + 1$, divides the Alexander polynomial $\Delta_{\mathcal{C}}(t)$ of the curve $\mathcal{C}$. Moreover, a torus decomposition is unique (up to scalar multiplication) if, in addition, the multiplicity of $\varphi_6(t)$ in

$\Delta_{\mathcal{C}}(t)$ is exactly 1. The Alexander polynomial of a curve $V(F)=\{F(x,y,z)=0\}$ was introduced by Libgober in [20]. It can be interpreted as the characteristic polynomial of the monodromy action on the first homology group of the cyclic covering of $\mathbb{P}^2 \setminus V(F)$ defined as the affine surface $t^d - F(x,y,z) = 0$ in $\mathbb{C}^3$. Following the notation in [2] (see also [20, 13, 21]) and ideas coming back from Zariski [29], the Alexander polynomial can be computed as follows.

Proposition 1.2 ([2, Proposition 2.10]). *Let $V(F)$ be a reduced curve of degree d. All the roots of the Alexander polynomial $\Delta_F(t)$ of $V(F)$ are d-th roots of unity. Let $\zeta_d^k := \exp(\frac{2i\pi k}{d})$. Then the multiplicity of ζ_d^k as a root of $\Delta_F(t)$ equals the number $d_k + d_{d-k}$, where d_k is the dimension of the cokernel of the natural map*

$$\rho_k : H^0(\mathbb{P}^2; \mathcal{O}_{\mathbb{P}^2}(k-3)) \longrightarrow \bigoplus_{P\in \mathrm{Sing}(V(F))} \frac{\mathcal{O}_{\mathbb{P}^2,P}}{\mathcal{J}_{P,d,k}},$$

where $\mathcal{J}_{P,d,k} \subset \mathcal{O}_{\mathbb{P}^2,P}$ is an ideal which depends on the germ of $V(F)$ at $P \in \mathrm{Sing}(V(F))$ and $\frac{k}{d}$.

Remark 1.3. Note that d_k can also be described as

$$d_k = \sum_{P\in \mathrm{Sing}\, V(F)} \dim \frac{\mathcal{O}_{\mathbb{P}^2,P}}{\mathcal{J}_{P,d,k}} - \binom{k-1}{2} + \dim\ker \rho_k. \tag{1.1}$$

In fact, $\ker \rho_k = H^0(\mathbb{P}^2; \mathcal{J}_{d,k}(k-3))$ the global sections of an ideal sheaf supported on $\mathrm{Sing}\, V(F)$ whose stalk at P is $\mathcal{J}_{P,d,k}$. Curves in this ideal sheaf will be said to *pass through* the ideal $\mathcal{J}_{P,d,k}$ for all $P \in \mathrm{Sing}(V(F))$ or simply *pass through* $\mathcal{J}_{d,k}$.

We can be more precise in the description of the ideal $\mathcal{J}_{P,d,k}$ by means of an embedded resolution $\sigma : \hat{X} \to \mathbb{P}^2$ of the point P as a singular point of $V(F)$. Assume for simplicity that $P = [0:0:1]$ and let $E_1^P, \ldots, E_n^P$ be the exceptional divisors over P. Let N_i be the multiplicity of $\sigma^*(F(x,y,1))$ along E_i^P and let $\nu_i - 1$ be the multiplicity of $\sigma^*(dx \wedge dy)$ along E_i^P. Then,

$$\mathcal{J}_{P,d,k} := \left\{ h \in \mathcal{O}_{\mathbb{P}^2,P} \,\middle|\, \text{the multiplicity of } \sigma^* h \text{ along } E_i \text{ is } > \left\lfloor \frac{kN_i}{d} \right\rfloor - \nu_i \right\}.$$

It is an easy exercise to compute these ideals for the singular points of $\mathcal{C}$.

Lemma 1.4. *Let $\mathfrak{m}_P$ be the maximal ideal of $\mathcal{O}_{\mathbb{P}^2,P}$ and let ℓ_P be the local equation of the tangent line of $\mathcal{C}$ at P. Then, the ideal $\mathcal{J}_{P,6,5} \subset \mathcal{O}_{\mathbb{P}^2,P}$ equals*

(1) $\mathfrak{m}_P$ *if P is an $\mathbb{A}_2$-point,*
(2) $\langle \ell_P \rangle + \mathfrak{m}_P^2$ *if P is either an $\mathbb{A}_5$-point or an $\mathbb{E}_6$-point,*

whereas $\mathcal{J}_{P,6,1} = \mathcal{O}_{\mathbb{P}^2,P}$ at any singular point P.

Proposition 1.5. *The multiplicity m of $\varphi_6(t)$ as a factor of $\Delta_{\mathcal{C}}(t)$ equals 1. In particular $\mathcal{C}$ admits exactly one torus decomposition.*

Proof. By Proposition 1.2, $m = d_1 + d_5 = d_5$ since the target of morphism ρ_1 is trivial. On the other hand, using equation (1.1) and Lemma 1.4 one obtains $d_5 = 1 + \dim \ker \rho_5$. Finally, note that $\ker \rho_5 = 0$, since otherwise a conic curve would *pass through* the ideal $\mathcal{J}_{6,5}$, contradicting Bézout's Theorem. □

Therefore the result below follows.

Proposition 1.6. *The curve $\mathcal{C}$ is of torus type and it has a unique toric decomposition.* □

In a torus curve $V(F)$, where $F = f_2^3 + f_3^2 = 0$, the intersection points of the conic $f_2 = 0$ and the cubic $f_3 = 0$ are singular points of $V(F)$. It is an easy exercise to check that singularities of type $\mathbb{A}_2$, $\mathbb{A}_5$, and $\mathbb{E}_6$ can be obtained locally as $u^3 + v^2$ where $u = 0$ is the germ of a conic and $v = 0$ is the germ of a cubic only as follows:

(T1) For $\mathbb{A}_2$, the curves $f_2 = 0$ and $f_3 = 0$ are smooth and transversal at the point.

(T2) For $\mathbb{A}_5$, the curve $f_3 = 0$ is smooth at the point and its intersection number with $f_2 = 0$ is 2, for instance $(v + u^2)^3 + v^2 = 0$.

(T3) For $\mathbb{E}_6$, the curve $f_2 = 0$ is smooth at the point, the curve $f_3 = 0$ is singular and their intersection number is 2, for instance $u^3 + (u^2 + v^2)^2 = 0$.

If $\operatorname{Sing} V(F) = V(f_2) \cap V(f_3)$, then $V(F)$ is called a *tame* torus curve, otherwise $V(F)$ is *non-tame.*

Lemma 1.7. *The curve $\mathcal{C}$ is a non-tame torus curve $\mathcal{C} = V(f_2^3 - f_3^2)$.*

Moreover $V(f_2)$ is a smooth conic and $V(f_3)$, $f_3 = \ell \cdot q$ is a reducible cubic where $V(q)$ is a smooth conic tangent to $V(f_2)$ only at one point and the line $V(\ell)$ passes through the remaining two points of intersection of the conics.

In particular, the only non-tame singularity is the $\mathbb{A}_2$-point.

Proof. It follows from the explanation above (and Bézout's Theorem) that the only possible combination of singularities at the intersection points of $V(f_2)$ and $V(f_3)$ is $\mathbb{A}_5 + 2\mathbb{E}_6$; they are the singularities of any generic element of the pencil $F_{\alpha,\beta} = \alpha f_2^3 + \beta f_3^2$. Note in particular, that the genus of a generic element of the pencil is 1, and its resolution provides an elliptic fibration.

Since $V(f_3)$ has two double points (at the points of type $\mathbb{E}_6$), it must be reducible and the line $V(\ell)$ joining these two points is one of the components. Let $q := \frac{f_3}{\ell}$. Recall that $\mathcal{C}$ is tangent to $V(f_3)$ at the point of type $\mathbb{A}_5$; in fact, it must be tangent to $V(q)$. Using again Bézout's Theorem $V(q)$ must be smooth.

Let us resolve the pencil. It is easily seen that it is enough to perform the minimal embedded resolution of the base points of the pencil. We obtain a map $\tilde{\Phi} : \tilde{X} \to \mathbb{P}^1$ where $\chi(\tilde{X}) = 14$ and the generic fiber is elliptic. The curve $V(f_3)$ produces a singular fiber, see Figure 1, with four irreducible components: the strict transforms of the line and the conic, and the first exceptional components A_1, B_1 of blow-ups of the $\mathbb{E}_6$-points. We can blow-down the -1-rational curves (the strict transforms of the line and the conic) in order to obtain a relatively minimal map

$\Phi : X \to \mathbb{P}^1$. The above fiber becomes a Kodaira singular fiber of type I_2, while $\mathcal{C}$ becomes a singular fiber of type II.

For the fiber coming from $V(f_2)$, its type changes depending on whether $V(f_2)$ is smooth or reducible: it is of type $\tilde{E}_6$ (smooth) or $\tilde{E}_7$ (reducible), as it can be seen from Figure 1. An Euler characteristic argument on this elliptic fibration shows that $V(f_2)$ has to be smooth. □

After a projective change of coordinates, we can assume that $P = [0:1:0]$ (the $\mathbb{A}_5$-point), $Q_1 = [1:0:0]$, $Q_2 = [0:0:1]$ (the $\mathbb{E}_6$-points), $\ell = y$ and $q = xz - y^2$, where $V(\ell) \cap V(f_2) = \{Q_1, Q_2\}$ and $V(f_2) \cap V(f_3) = \{P, Q_1, Q_2\}$. Note moreover that only the projective automorphism $[x:y:z] \mapsto [z:y:x]$ and the identity globally fix the above points and curves. The equation of f_2 must be:

$$(x-y)(z-y) - uy(x-2y+z),$$

for some $u \in \mathbb{C}^*$.

Proposition 1.8. *Any Eyral-Oka curve $\mathcal{C}$ is projectively equivalent to*

$$\mathcal{C}_a : y^2(xz-y^2)^2 - 48(26a+45)f_2(x,y,z)^3 = 0. \tag{1.2}$$

where $f_2(x,y,z) = (x-y)(z-y) + 4(a+2)y(x-2y+z)$ and $a^2 = 3$.

Moreover, the curves $\mathcal{C}_+ := \mathcal{C}_{\sqrt{3}}$ and $\mathcal{C}_- := \mathcal{C}_{-\sqrt{3}}$ are not projectively equivalent (in particular, they are not rigidly isotopic).

Proof. For a generic value of u, the meromorphic function $\frac{f_2^3}{\ell^2 q^2}$ has two critical values outside $0, \infty$. Computing a discriminant we find the values of u for which only one double critical value arises, obtaining the required equation f_2.

The computation of the critical value gives the equations in the statement. The result follows from the fact that $\mathcal{C}_\pm$ are invariant by $[x:y:z] \mapsto [z:y:x]$. □

Remark 1.9. If L denotes the tangent line to $V(f_2)$ (and $V(q)$) at P, then note that $(\mathcal{C} \cdot L)_P = 4$. This can be computed using the equations but it is also a direct consequence of the construction of $\mathcal{C}$. Since $\mathcal{C}$ at P has an $\mathbb{A}_5$ singularity and L is smooth, then using Noether's formula of intersection $(\mathcal{C} \cdot L)_P$ can be either 4 or 6. The latter case would imply that $(V(f_3) \cdot L)_P = (V(q) \cdot L)_P = 3$, contradicting Bézout's Theorem.

This construction was already given in [14]. Let us end this section with another particular feature of these curves.

Let us perform a change of coordinates such that $L = V(z)$ is the tangent line to $\mathcal{C}$ at P and the normalized affine equation of $\mathcal{C}$ in x, y is symmetric by the transformation $x \mapsto -x$. One obtains:

$$0 = h_a(x,y) = y^2\left(y - \frac{(x^2-1)}{4}\right)^2 + \frac{1}{2}\left(\frac{2a}{3}y - \frac{2a+3}{24}(x^2-1)\right)^3. \tag{1.3}$$

A direct computation shows that $h_a(x, -y + \frac{x^2-1}{4}) = h_{-a}(x,y)$, i.e., these affine curves are equal. After this change of variables the singularities have the following coordinates:

$$P = [24 : -\sqrt{26a+45} : 0], Q_1 = [1:0:1], Q_2 = [-1:0:1], R = [0 : a+1 : -8], \tag{1.4}$$

where P, Q_1, and Q_2 are defined as above and R is the $\mathbb{A}_2$ singularity.

Let us interpret it in a computation-free way. Recall from Remark 1.9 that $(\mathcal{C} \cdot L)_P = 4$. As in the proof of Lemma 1.7, we blow up the indeterminacy of the pencil map $\mathbb{P}^2 \dashrightarrow \mathbb{P}^1$ defined as $[x : y : z] \mapsto [f_2^3 : f_3^2]$, whose fibers are denoted by $V(F_{\alpha,\beta})$, for $F_{\alpha,\beta} = \alpha f_2^3 + \beta f_3^2$. A picture of these fibration is depicted in Figure 1. Most of the exceptional components of this blow-up are part of fibers. The last components A_4, B_4 over the $\mathbb{E}_6$-points are sections while the last component E_3 over the $\mathbb{A}_5$-point is a 2-section, that is, the elliptic fibration restricted to this divisor is a double cover of $\mathbb{P}^1$. It is ramified at the intersections with E_2 (in the fiber of $V(F_{1,0})$) and the strict transform of $V(q)$ (in the fiber of $V(F_{0,1})$); they have both multiplicity 2.

In Figure 1, we show also the strict transform of L, the tangent line at the $\mathbb{A}_5$-point. One check that this strict transform becomes a 2-section; one ramification point is the intersection with E_2 (in the fiber of $V(F_{1,0})$) and the other one is the intersection with the strict transform of $V(\ell)$ (in the fiber of $V(F_{0,1})$). In particular, there is no more ramification and hence L intersects all other fibers $V(F_{\alpha,\beta})$ at two distinct points.

It is clear in Figure 1 that the combinatorics of E_3 and L coincide. In other words, interchanging the roles of E_3 and L and blowing down accordingly, then one obtains a birational transformation of $\mathbb{P}^2$ recovering a sextic curve $\mathcal{C}_-$ with the same combinatorics as $\mathcal{C}$ and a line L_- (the transformation of E_3) which is tangent at the $\mathbb{A}_5$-point. Note that this birational transformation exchanges the line $V(\ell)$ (resp. the conic $V(q)$) and the corresponding conic $V(q_-)$ (resp. line $V(\ell_-)$). In particular, this implies that the transformation cannot be projective. Thus, this transformation exchanges the curves $\mathcal{C}_+ (:= \mathcal{C})$, $\mathcal{C}_-$ and the lines $L_+ (:= L)$, L_- resulting in the following.

Theorem 1.10. *The complements* $\mathbb{P}^2 \setminus (L_\pm \cup \mathcal{C}_\pm)$ *are analytically isomorphic.* □

2. Fundamental group of Eyral-Oka curves

The main tool to compute the fundamental group of the complement of a plane curve is the Zariski-van Kampen method. In fact, in this method the computation of the fundamental group of $\mathbb{C}^2 \setminus \mathcal{C}^{\text{aff}}$ for a suitable affine part of the projective curve $\mathcal{C}$ is obtained first. Namely, a line L is chosen (the line at infinity) so that $\mathbb{C}^2 \equiv \mathbb{P}^2 \setminus L$ is defined and thus $\mathbb{C}^2 \setminus \mathcal{C}^{\text{aff}} \equiv \mathbb{P}^2 \setminus (L \cup \mathcal{C})$. Once $\pi_1(\mathbb{C}^2 \setminus \mathcal{C}^{\text{aff}})$ is obtained, the fundamental group of $\mathbb{P}^2 \setminus \mathcal{C}$ can be recovered after factoring out by (the conjugacy class of) a meridian of L [16]. This is particularly simple if $L \pitchfork \mathcal{C}$,

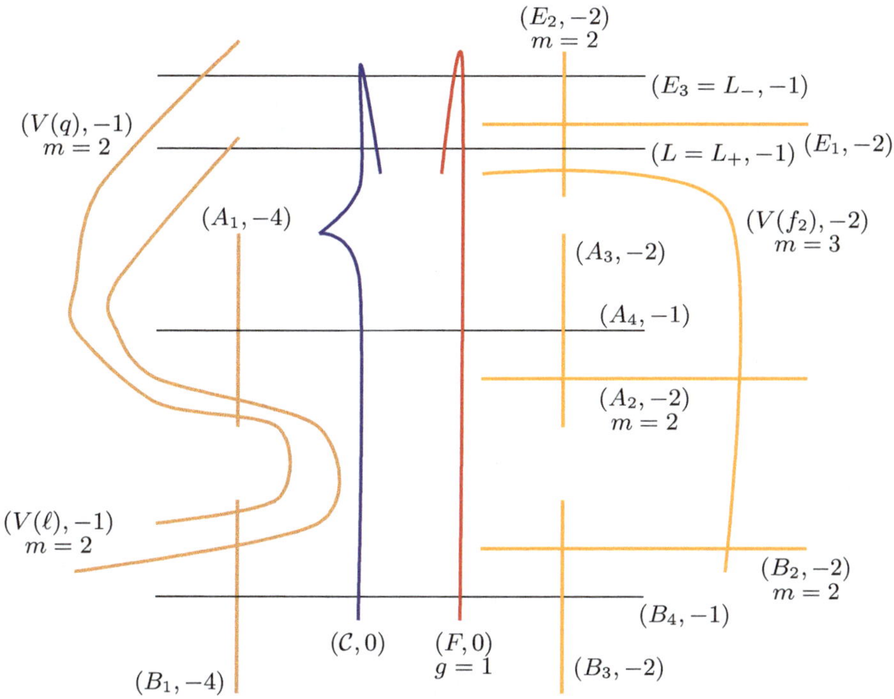

FIGURE 1. Elliptic fibration

but the argument also follows for arbitrary lines. Applying Theorem 1.10, one only needs to compute the fundamental group for one of the affine curves, since they are isomorphic. Finally, factoring out by (the conjugacy class of) a meridian of L or E_3 will make the difference between the groups of the respective curves $\mathcal{C}_\pm$.

The Zariski-van Kampen method uses a projection $\mathbb{C}^2 \to \mathbb{C}$, say the vertical one. In Figure 4, we have drawn a real picture of the affine curves $\mathcal{C}_+^{\text{aff}}$ in $\mathbb{P}^2 \setminus L_+$. For each vertical line, we have also drawn the real part of the complex-conjugate part as dotted lines.

First, we study the situation at infinity.

2.1. The topology at infinity

In order to understand the topology at infinity, let us simplify the construction of the elliptic fibration, carried out at the end of the previous section by minimizing the amount of blowing ups and blowing downs as follows.

Let us consider a sequence of blow-ups as in Figure 2 which yields a birational morphism $\sigma^+ := \sigma_3^+ \circ \sigma_2^+ \circ \sigma_1^+ : X_3 \to \mathbb{P}^2$.

(B1) The first picture represents a neighborhood of L in $\mathbb{P}^2$.

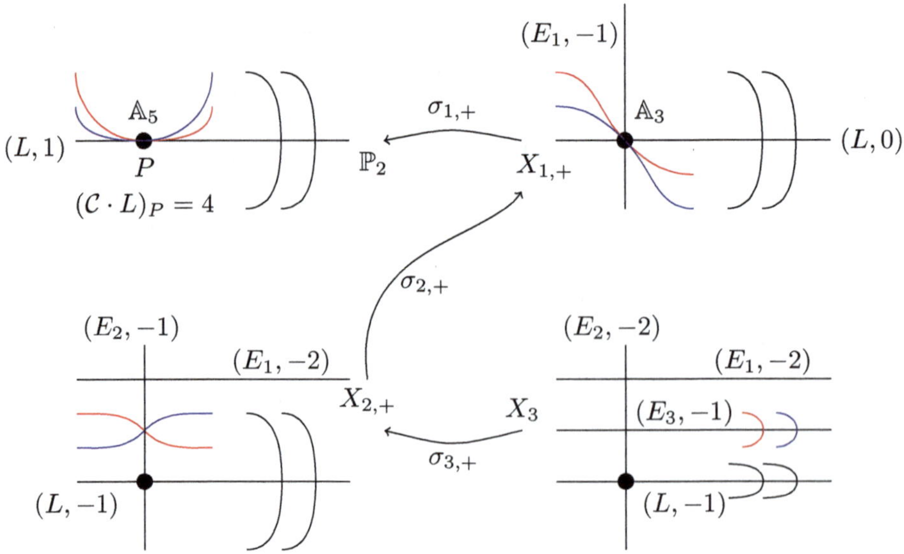

FIGURE 2. Sequence of blowups

(B2) The second picture is a neighborhood of the total transform of L by the blowing-up of P. Let us denote by E_1 the exceptional divisor (this notation will also be used for its strict transforms). Note that $E_1 \cap L$ is a point of type $\mathbb{A}_3$ in $\mathcal{C}$ which is transversal to both divisors.

(B3) The third part is a neighborhood of the total transform of the divisor E_1+L by the blow-up of $E_1 \cap L$. In this case E_2 denotes the new exceptional component, which intersects $\mathcal{C}$ at a nodal point not lying on $L \cup E_1$.

(B4) The fourth picture is obtained by blowing up that nodal point. For convenience, $E_3 = L_-$ will denote the new exceptional component. Note that the divisors L_- and $L = L_+$ are combinatorially indistinguishable.

As a consequence there is an analogous sequence of blow-ups $\sigma_- := \sigma_{3,-} \circ \sigma_{2,-} \circ \sigma_{1,-} : X_3 \to \mathbb{P}^2$ where $(\sigma_{i,-})$ are the blow-ups whose exceptional components are (the strict transforms of) $E_3 = L_-$ $(i=3)$, E_2 $(i=2)$ and E_1 $(i=1)$. We obtain a birational map

$$\Phi := \sigma_-^{-1} \circ \sigma_+ : \mathbb{P}^2 \dashrightarrow \mathbb{P}^2. \tag{2.1}$$

Consider the birational transformation $\sigma_{2,+} \circ \sigma_{1,+}$ from the sequence of the first two blow-ups shown in Figure 2 (first at the $\mathbb{A}_5$-point P, second at the $\mathbb{A}_3$-point $E_1 \cap L$). We compose it with the blowing-down of $L \subset X_{2,+}$, which corresponds to a blowing-up $\tau_+ : X_{2,+} \to \Sigma_2$, where Σ_2 is the ruled surface with fiber and base $\mathbb{P}^1$ and one section E_1 with self-intersection -2. This process yields birational transformation $\sigma := \sigma_{2,+} \circ \sigma_{1,+} \circ \tau^{-1} : \Sigma_2 \dashrightarrow \mathbb{P}^2$ which is depicted in Figure 3, with special attention to the neighborhood at the E_2 divisor (which is a fiber in Σ_2). An analogous blow-up $\tau_- : X_{2,-} \to \Sigma_2$ of the other node can be done.

Lemma 2.1. *The birational transformation σ converts the projection from P into the projection $\pi : \Sigma_2 \to \mathbb{P}^1$ such that the strict transform of $\mathcal{C}$ is disjoint with the section E_1.*

The exceptional components of the blowing-ups of the nodal points of the strict transform of $\mathcal{C}$ on E_2 (see Figure 3*) are the strict transforms of the lines $L_\pm$.*

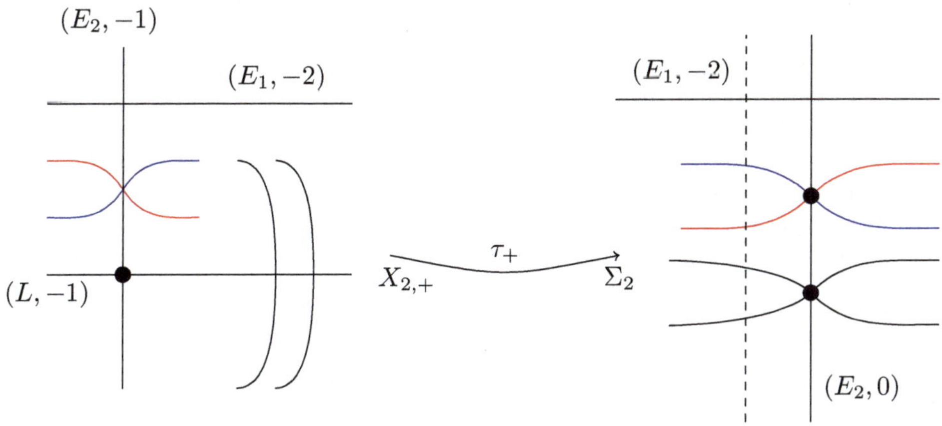

FIGURE 3. Ruled surface Σ_2 near infinity.

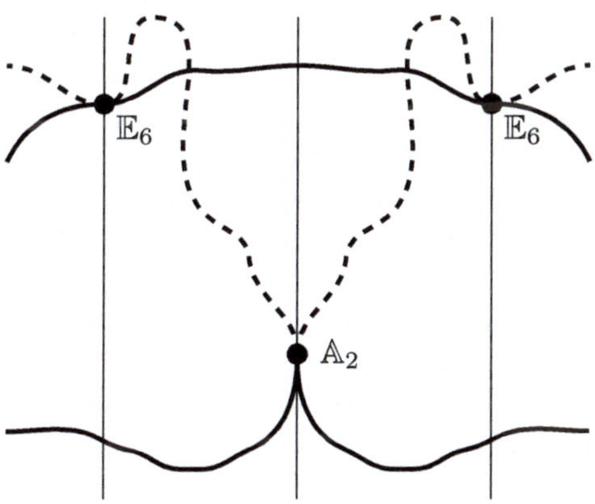

FIGURE 4. Real affine picture of Eyral-Oka's curve

Let us explain how we have constructed Figure 4. Recall (1.3) and (1.4) for the definition of the equation of the curve and description of the singular points respectively.

(F1) We compute the discriminant of $h_a(x,y)$ of (1.3) with respect to y. This allows to check that the $\mathbb{A}_2$ point is on the line $x=0$ and the $\mathbb{E}_6$ point Q_1 (resp. Q_2) is on the line $x=1$ (resp. $x=-1$).

(F2) We factorize the polynomials $h_a(x_0,y)$ for $x_0=0,\pm 1$ and we obtain which intersection points are up and down.

(F3) For $x_0=0,\pm 1$, let y_0 be the y-coordinate of the singular point. For the $\mathbb{A}_2$ point, $y_0=-\frac{a+1}{8}$ we check that the cusp is tangent to the vertical line, and the Puiseux parametrization is of the form $y=y_0+y_1x^{\frac{2}{3}}+\dots$. We obtain that $y_1<0$ and it implies that the real part of the complex solutions is bigger than the real solution, near $x=0$.

(F4) We proceed in a similar way for the point of type $\mathbb{E}_6$; in this case, a Puiseux parametrization is of the form $y=y_0+y_1x+y_2x^{\frac{4}{3}}+\dots$, and $y_2<0$.

(F5) With this data, we draw Figure 4. Note that between the vertical fibers $x=0,1$, we have an odd number of crossings. We show later that the actual number is irrelevant for the computations.

(F6) If we look the situation at ∞ in Σ_2 (Figure 3), we check that the imaginary branches are still up. Hence, from $x=1$ to ∞ there is an even number of crossings with the real parts and, as before, the actual number is irrelevant.

Remark 2.2. The two real branches that go to infinity are the real part of the branches of $\mathcal{C}_-$ at P_-, while the two conjugate complex branches belong to the branches of $\mathcal{C}_+$ at P_+.

2.2. Strongly real curves and braid monodromy factorization

The curve in Figure 4 is said to be *strongly real* since it is real, all its affine singularities are real, and thus Figure 4 contains all the information to compute the braid monodromy of $\mathcal{C}^{\text{aff}}$ and, as a consequence, the fundamental group of its complement.

Let $f(x,y)\in\mathbb{C}[x,y]$ be a monic polynomial in y. The braid monodromy of f with respect to its vertical projection is a group homomorphism $\nabla:\mathbb{F}_r\to\mathbb{B}_d$, where $d:=\deg_y f$ and r is the number of distinct roots $\{x_1,\dots,x_r\}$ of the discriminant of f with respect to y (i.e., the number of non-transversal vertical lines to $f=0$). In our case, $r=3$ and $d=4$. In order to calculate ∇, one starts by considering $x=x_0$ a transversal vertical line and $\{y_1,\dots,y_d\}$ the roots of $f(x_0,y)=0$. By the continuity of roots, any closed loop γ in $\mathbb{C}$ based at x_0 and avoiding the discriminant defines a braid based at $\{y_1,\dots,y_d\}$ and denoted by $\nabla(\gamma)$. Since the vertical projection produces a locally trivial fibration outside the discriminant, the construction of the braid only depends on the homotopy class of the loop γ. This produces the well-defined morphism ∇.

Moreover, the morphism ∇ can be used to define an even finer invariant of the curve called the braid monodromy factorization, via the choice of a special geometrically-based basis of the free group $\mathbb{F}_r$. Note that the group $\mathbb{F}_r$ can be identified with $\pi_1(\mathbb{C}\setminus\{x_1,\dots,x_r\};x_0)$ and a basis can be chosen by meridians γ_i around x_i such that $(\gamma_r\cdot\ldots\cdot\gamma_1)^{-1}$ is a meridian around the point at infinity (this is known as a *pseudo-geometric basis*). A braid monodromy factorization of f is then given by the r-tuple of braids $(\nabla(\gamma_1),\ldots,\nabla(\gamma_r))$.

The morphism ∇ is enough to determine the fundamental group of the complement to the curve, however a braid monodromy factorization is in fact a topological invariant of the embedding of the fibered curve resulting from the union of the original curve with the preimage of the discriminant (see Theorem 3.2).

In order to compute a braid monodromy factorization, two important choices are required. First a pseudo-geometric basis in $\pi_1(\mathbb{C}\setminus\{x_1,\ldots,x_r\};x_0)\equiv\mathbb{F}_r$ and second, an identification between the braid group based at $\{y_1,\ldots,y_d\}$ and the standard Artin braid group $\mathbb{B}_d$. This is done with the following choices, see [4].

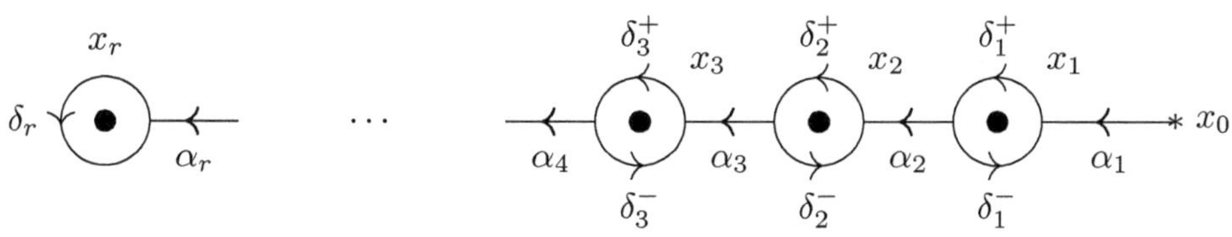

FIGURE 5. Pseudo-geometric basis

(C1) For a strongly real curve, a pseudo-geometric basis is chosen as in Figure 5, where the points are arranged on the base line so that $x_r<\cdots<x_1<x_0$. Let

$$\underset{1\leq i<r}{\delta_i:=\delta_i^+\cdot\delta_i^-},\qquad \underset{1<i\leq r}{\beta_i:=\prod_{j=1}^{i-1}\alpha_j\cdot\delta_j^+}.$$

The basis is:

$$\gamma_1:=\alpha_1\cdot\delta_1^+\cdot\delta_1^-\cdot\alpha_1^{-1},\quad \gamma_i:=(\beta_i\cdot\alpha_i)\cdot\delta_i^+\cdot\delta_i^-\cdot(\beta_i\cdot\alpha_i)^{-1},\ 1<i\leq r. \tag{2.2}$$

Applied to our case, paths $\gamma_1,\gamma_2,\gamma_3$ are required around the points $-1,0,1$ respectively.

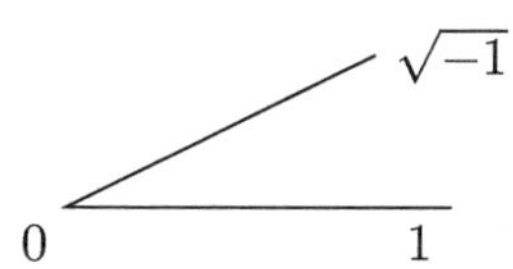

FIGURE 6. Complex line

(C2) The identification of the braid group on $\{y_1,\ldots,y_d\}$ is made using a lexicographic order of the roots on their real parts ($\Re y$) and imaginary parts ($\Im y$) such that $\Re y_1\geq\cdots\geq\Re y_d$ and $\Im y<\Im y'$ whenever $\Re y=\Re y'$. A very useful fact about this canonical construction is that it allows one to identify the braids over any path in Figure 5 (whether open or closed) with braids in $\mathbb{B}_d$. These conventions can be understood from Figure 6. Namely, projecting the braids onto the real line $\mathbb{R}$, and for complex conjugate numbers we slightly deform the projection such that the positive imaginary part number goes to the right and the negative one to the left. In a crossing, the upward strand is the one with a smaller imaginary part.

In our case, note that the braid group is $\mathbb{B}_4$ generated by the Artin system σ_i, $i = 1, \ldots, 3$, the positive half-twist interchanging the i-th and $(i+1)$-th strands.

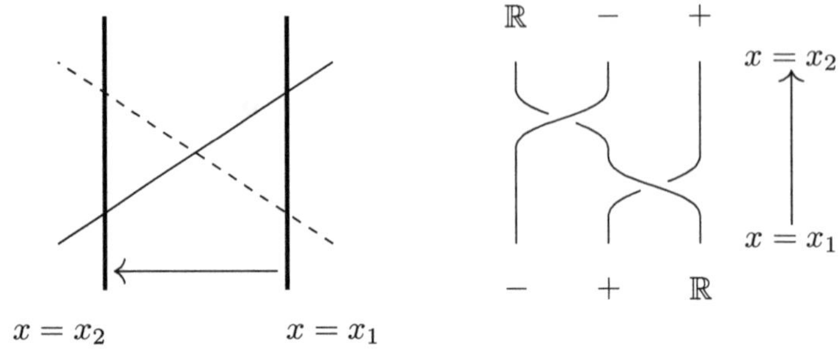

FIGURE 7. Crossing of a real branch with a couple of complex conjugate branches

(C3) Given a strongly real curve one can draw its real picture. This real picture might be missing complex conjugate branches. For those, one can draw their real parts as shown in Figure 4 with dashed curves. This picture should pass the *vertical line test*, that is, each vertical line should intersect the picture in d points counted appropriately, that is, solid line intersections count as one whereas dashed line intersections count as two.

At this point, the braids can be easily recovered as long as the dashed lines have no intersections as follows:

- At intersections of solid lines one has a singular point. The local braid over δ^+ and δ^- can be obtained via the Puiseux pairs of the singularity.
- At an intersection of a solid and a dashed line as in Figure 7, the local braid on three strands $\sigma_1^{-1} \cdot \sigma_2$ is obtained as a lifting of the open path α crossing the intersection, where the generators σ_i are chosen locally and according the identification given in (C2). In the reversed situation (that is, when the solid and dashed lines are exchanged), the inverse braid is obtained. This justifies the assertions in (F5) and (F6).

In our case the following braids in $\mathbb{B}_4$ are obtained:

$$\alpha_1 \mapsto 1, \quad \alpha_2 \mapsto \sigma_2^{-1} \cdot \sigma_1, \quad \alpha_3 \mapsto \sigma_1^{-1} \cdot \sigma_2.$$

In order to finish the computation of the braid monodromy factorization of $\mathcal{C}^{\text{aff}}$, we need to compute the braids associated to $\delta_i^{\pm}$, $i = 1, 2$, and δ_3. Next lemma provides the key tools.

Lemma 2.3. *Let $f(x, y) = y^3 - x$; following the above conventions, the braid in $\mathbb{B}_3$ obtained from the path $\alpha : t \mapsto x = \exp(2\sqrt{-1}\pi t)$, $t \in [0, 1]$, equals $\sigma_2 \cdot \sigma_1$. For $g(x, y) = y^3 + x$, the braid associated with α equals $\sigma_1 \cdot \sigma_2$.*

Proof. Note that for $x = 1$, the values of the roots of the y-polynomial $f(1, y)$ are $1, \zeta, \bar{\zeta}$, for $\zeta := \exp\left(\frac{2\sqrt{-1}\pi}{3}\right)$, and thus the associated braid is nothing but the rotation of angle $\frac{2}{3}\pi$.

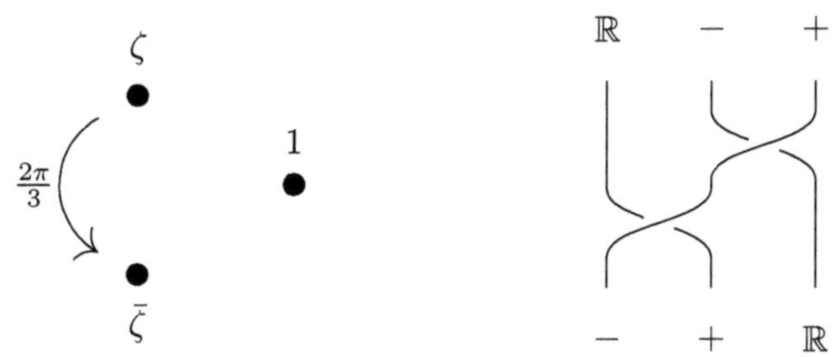

FIGURE 8. Braid for $y^3 = x$.

The result follows from the identification described in (C2) and Figure 8. □

Applying this in our situation one obtains (see Figure 4):

$$\delta_1^{\pm} \mapsto (\sigma_1 \cdot \sigma_2)^2 \quad \Longrightarrow \delta_1, \delta_3 \mapsto (\sigma_1 \cdot \sigma_2)^4 \qquad \delta_2^{\pm} \mapsto \sigma_2 \cdot \sigma_3 \quad \Longrightarrow \delta_2 \mapsto (\sigma_2 \cdot \sigma_3)^2. \tag{2.3}$$

Combining all the braids obtained above, one can give the monodromy factorization.

Proposition 2.4. *The braid monodromy factorization of $\mathcal{C}^{\text{aff}}$ is (τ_1, τ_2, τ_3) where:*

$$\begin{aligned} \tau_1 :=& (\sigma_1 \cdot \sigma_2)^4, \\ \tau_2 :=& (\sigma_1 \cdot \sigma_2 \cdot \sigma_1^2) \cdot (\sigma_2 \cdot \sigma_3)^2 \cdot (\sigma_1 \cdot \sigma_2 \cdot \sigma_1^2)^{-1}, \\ \tau_3 :=& (\sigma_2 \cdot \sigma_1^2 \cdot \sigma_2 \cdot \sigma_3) \cdot (\sigma_2 \cdot \sigma_1)^4 \cdot (\sigma_2 \cdot \sigma_1^2 \cdot \sigma_2 \cdot \sigma_3)^{-1}. \end{aligned} \tag{2.4}$$

Remark 2.5. Note that the closure of $\mathcal{C}^{\text{aff}}$ in the ruled surface Σ_2 is disjoint from the negative section E_1. As stated in [18, Lemma 2.1], the product of all braids (associated to paths whose product in the complement of the discriminant in $\mathbb{P}^1$ is trivial) equals $(\Delta^2)^2$. Hence $\Delta^4 \cdot (\tau_3 \cdot \tau_2 \cdot \tau_1)^{-1}$ is the braid associated to two disjoint nodes, see Figure 3. The following equality is a straightforward exercise:

$$(\sigma_1^2 \cdot \sigma_3^2) \cdot (\tau_3 \cdot \tau_2 \cdot \tau_1) = \Delta^4.$$

2.3. A presentation of the fundamental group

Our next step will be to compute $G := \pi_1(\mathbb{C}^2 \setminus \mathcal{C}^{\text{aff}})$. The main tool towards this, as mentioned before, entails considering a braid monodromy factorization and its action on a free group. Before stating Zariski-van Kampen's Theorem precisely, let us recall this natural right action of $\mathbb{B}_d$ on $\mathbb{F}_d$ with basis $g_1, \ldots, g_d$ which will be denoted by g^σ for a braid $\sigma \in \mathbb{B}_d$ and an element $g \in \mathbb{F}_d$. It is enough to describe it for g_i a system of generators in $\mathbb{F}_d$ and σ_j an Artin system of $\mathbb{B}_d$:

$$g_i^{\sigma_j} := \begin{cases} g_{i+1} & \text{if } i = j, \\ g_i \cdot g_{i-1} \cdot g_i^{-1} & \text{if } i = j+1, \\ g_i & \text{otherwise.} \end{cases} \tag{2.5}$$

The following is the celebrated Zariski-van Kampen Theorem, which allows for a presentation of the fundamental group of an affine curve complement from a given braid monodromy factorization.

Theorem 2.6 (Zariski-van Kampen Theorem). *If $(\tau_1, \ldots, \tau_r) \in \mathbb{B}_d^r$ is a braid monodromy factorization of an affine curve $\mathcal{C}$, then:*

$$\pi_1(\mathbb{C}^2 \setminus \mathcal{C}^{\text{aff}}) = \langle g_1, \ldots, g_d \mid g_i = g_i^{\tau_j}, \quad 1 \leq j \leq r, \quad 1 \leq i < d \rangle .$$

If $\tau_i = \alpha_i^{-1} \cdot \beta_i \cdot \alpha_i$, and β_i is a (usually positive) braid involving strands $k_i + 1, \ldots, k_i + m_i$, then:

$$\pi_1(\mathbb{C}^2 \setminus \mathcal{C}^{\text{aff}}) = \left\langle g_1, \ldots, g_d \mid g_i^{\alpha_j} = (g_i^{\beta_j})^{\alpha_j}, \quad 1 \leq j \leq r, \quad k_j < i < k_j + m_j \right\rangle . \tag{2.6}$$

Remark 2.7. In our case, $r = 3$, $d = 4$, $k_1 = k_3 = 0$, $m_1 = m_3 = 2$, $k_2 = 1$, $m_2 = 1$, and

$$\begin{aligned} \alpha_1 &= 1 & \beta_1, &= (\sigma_1 \cdot \sigma_2)^4, \\ \alpha_2 &= (\sigma_1 \cdot \sigma_2 \cdot \sigma_1^2)^{-1}, & \beta_2 &= (\sigma_2 \cdot \sigma_3)^2, \\ \alpha_3 &= (\sigma_2 \cdot \sigma_1^2 \cdot \sigma_2 \cdot \sigma_3)^{-1}, & \beta_3 &= (\sigma_2 \cdot \sigma_1)^4. \end{aligned} \tag{2.7}$$

The previous sections where a braid monodromy factorization allow us to give a presentation of the fundamental group of an affine curve complement.

Corollary 2.8. *Let $\mathcal{C}^{\text{aff}}$ be the affine Eyral-Oka curve as described at the beginning of Section 2, consider a braid monodromy factorization as described in Proposition 2.4 and (2.7). Then the group $G = \pi_1(\mathbb{C}^2 \setminus \mathcal{C}^{\text{aff}})$ admits a presentation as*

$$\begin{aligned} \Big\langle g_1, \ldots, g_4 : g_1^{\beta_1} = g_1, g_2^{\beta_1} = g_2, g_2^{\alpha_2} = (g_2^{\beta_2})^{\alpha_2}, \\ g_3^{\alpha_2} = (g_3^{\beta_2})^{\alpha_2}, g_1^{\alpha_3} = (g_1^{\beta_3})^{\alpha_3}, g_2^{\alpha_3} = (g_2^{\beta_3})^{\alpha_3} \Big\rangle . \end{aligned} \tag{2.8}$$

Presentation (2.8) contains 6 relations for a total length of 40. For the sake of clarity, instead of showing an explicit presentation, we will describe the group G in a more theoretical way that would allow to understand its structure. The following description characterizes the group G completely.

Theorem 2.9. *The fundamental group G can be described as follows:*

(1) *Its derived subgroup $G' \subset G$ can be decomposed as a semidirect product $K \rtimes V$, where:*
  (a) *The subgroup V is the Klein group $\langle a, b \mid a^2 = b^2 = a \cdot b \cdot a^{-1} \cdot b^{-1} = 1\rangle$.*
  (b) *The subgroup K is the direct product of a rank-2 free group and a cyclic group of order 2 with presentation*

$$\langle x, y, w \mid w^2 = x \cdot w \cdot x^{-1} \cdot w = y \cdot w \cdot y^{-1} \cdot w = 1\rangle.$$

  (c) *The action of V on K is given by:*

$$x^a = x, \quad y^a = y \cdot w, \quad w^a = w, \quad x^b = x \cdot w, \quad y^b = y, \quad w^b = w.$$

  In particular, w is central in G'.

(2) *There exists a meridian g of $\mathcal{C}^{\text{aff}}$ such that $G = G' \rtimes \mathbb{Z}$, where $\mathbb{Z}$ is identified as $\langle g \mid -\rangle$ and the action is defined by:*

$$g{\cdot}x{\cdot}g^{-1} = y^{-1}, \quad g{\cdot}y{\cdot}g^{-1} = y{\cdot}x{\cdot}b, \quad g{\cdot}w{\cdot}g^{-1} = w, \quad g{\cdot}a{\cdot}g^{-1} = b, \quad g{\cdot}b{\cdot}g^{-1} = a{\cdot}b.$$

(3) *There is a central element z such that $z \cdot g^6 = [y, x]$. The center of G is generated by z, w.*

(4) *There is an automorphism of G sending z to $z \cdot w$.*

Proof. A presentation of this sort can be obtained using `Sagemath` [26] (which contains `GAP4` [17] as main engine for group theory). We indicate the steps of the proof:

(G1) The original presentation (2.8) (with four generators and six relations) can be simplified to have only two generators and four relations, both generators being meridians of $\mathcal{C}^{\text{aff}}$. Any of such meridians can play the role of g in (2).

(G2) From the simplified presentation above, one can find a central element $z \in G$ whose image by the standard abelianization morphism is -6. Recall the abelianization of G is $\mathbb{Z}$. Moreover, the abelianization can be fixed by setting the image of any meridian to be 1, this is what we call the standard abelianization.

Since z is central, note that $G' \cong (G/\langle z\rangle)'$, see e.g. [12]. Since the latter derived group is of index 6 in $G/\langle z\rangle$, we can apply Reidemeister-Schereier method to find a finite presentation of G' with 5 generators $x, y, a, b, w \in G'$.

(G3) From the previous steps it is a tedious computation to verify the structure of G' indicated in the statement as well to check the conjugation action of g. In particular, that w is central in G' and $z = g^{-6} \cdot [y, x]$.

(G4) Note that the center of G is the group generated by z and w and that $z, z{\cdot}w$ are the only central elements which are sent to -6 by the standard abelianization. Moreover, it is straightforward to prove that

$$g \mapsto g, \quad x \mapsto y^{-1} \cdot a, \quad y \mapsto y \cdot x \cdot w, \quad a \mapsto b, \quad b \mapsto a \cdot b, \quad w \mapsto w$$

defines an automorphism of G such that $z \mapsto z \cdot w$. ∎

Going back to the discussion about the topology at infinity, one can detect the meridians of the tangent line $L = L_+$ and the meridian corresponding to the exceptional divisor $E_3 = L_-$. These are required to recuperate the original fundamental groups of the complement to the projective Eyral-Oka sextics.

Corollary 2.10. *The central element z is a meridian of L_+ while $z \cdot w$ is a meridian of L_-. In particular, the groups $\pi_1(\mathbb{P}^2 \setminus \mathcal{C}_\pm)$ are isomorphic.*

Proof. Starting from the Zariski-van Kampen presentation (2.8), and using the blow-up blow-down process described in Figure 2, it is straightforward that a meridian of L_+ (resp. L_-) is given by $e^2 \cdot (g_2 \cdot g_1)$ (resp. $e^2 \cdot (g_4 \cdot g_3)$), where $e = (g_4 \cdot \ldots \cdot g_1)^{-1}$. The result follows from tracing these meridians along the steps described in Theorem 2.9. □

Corollary 2.10 answers negatively a question in [14]. In the following section this curve will be used to construct arithmetic Zariski pairs that are complement equivalent.

3. Zariski pairs and braid monodromy factorizations

In Corollary 2.10, we have proved that the fundamental groups of the Eyral-Oka curve complements are isomorphic, and hence this invariant cannot be used to decide whether these two curves, which are not rigidly equivalent, form an arithmetic Zariski pair.

Degtyarev [11] proved that any two non-rigidly equivalent equisingular sextic curves with simple singularities cannot have *regularly homeomorphic* embeddings, where a regular homeomorphism is a homeomophism that is holomorphic at the singular points.

In particular, by Degtyarev's result, Eyral-Oka curves are close to being an arithmetic Zariski pair. Shimada was able to refine Degtyarev's arguments in [24, 25] and developed an $N_\mathcal{C}$-invariant that was able to exhibit that some of these candidates to Zariski pairs were in fact so. Unfortunately, the $N_\mathcal{C}$-invariant coincides for the Galois-conjugate projective Eyral-Oka curves.

As we showed in §1, the curves $\mathcal{C}_\pm \cup L_\pm$ have homeomorphic complements (even more, analytically isomorphic) via the birational morphism shown in Figure 2. The Cremona transformation that connects both complements is not a homeomorphism of the pairs $(\mathbb{P}^2, \mathcal{C}_\pm \cup L_\pm)$, so these curves are candidates to be complement equivalent Zariski pairs.

We are not able to decide on that problem, but we are more successful when adding more lines to the original curves $\mathcal{C}_\pm \cup L_\pm$. Let us consider the two lines $L^2_\pm$ joining the points $P_\pm$ (of type $\mathbb{A}_5$) and the $\mathbb{A}_2$-points in $\mathcal{C}_\pm$. Analogously, we denote by $L^{6,j}_\pm$, $j = 1, 2$, the four lines joining the points $P_\pm$ and $Q_{j,\pm}$ (the $\mathbb{E}_6$-points in $\mathcal{C}_\pm$). Note that these extra lines correspond to the preimage of the discriminants in $\mathbb{C} \subset \mathbb{P}^1$ of the projections of $\mathcal{C}_\pm$ from $P_\pm$ (see Figure 4).

Proposition 3.1. *There is an analytic isomorphism*

$$\mathbb{P}^2 \setminus (\mathcal{C}_+ \cup L_+ \cup L_+^2 \cup L_+^{6,1} \cup L_+^{6,2}) \to \mathbb{P}^2 \setminus (\mathcal{C}_- \cup L_- \cup L_-^2 \cup L_-^{6,1} \cup L_-^{6,2}).$$

Proof. Let us consider the birational transformation $\Phi : \mathbb{P}^2 \dashrightarrow \mathbb{P}^2$ of (2.1), graphically described in Figure 2. This transformation defines an analytic isomorphism $\Phi_|$ from $\mathbb{P}^2 \setminus (\mathcal{C}_+ \cup L_+)$ onto $\mathbb{P}^2 \setminus (\mathcal{C}_- \cup L_-)$.

Let us consider the line L_+^2; its strict transform is a fiber of the ruled surface $X_{1,+}$ passing through the preimage by $\sigma_{1,+}$ of the $\mathbb{A}_2$-point of $\mathcal{C}_+$; this strict transform intersects E_1 outside $E_1 \cap L$. In particular, it is not affected by the blowups and blowdowns $\sigma_{2,+}, \sigma_{3,+}, \sigma_{3,-}^{-1}, \sigma_{2,-}^{-1}$. Hence, the blowdown $\sigma_{1,-}^{-1}$ sends it to the line L_-^2, since it passes through P_- and the $\mathbb{A}_2$-point of $\mathcal{C}_-$. Hence

$$\Phi_|(L_+^2 \setminus (L_+ \cup \mathcal{C}_+)) = (L_-^2 \setminus (L_- \cup \mathcal{C}_-).$$

A similar argument yields

$$\Phi_| \left(\left(L_+^{6,1} \cup L_+^{6,2} \right) \setminus (L_+ \cup \mathcal{C}_+) \right) = \left(L_-^{6,1} \cup L_-^{6,2} \right) \setminus (L_- \cup \mathcal{C}_-).$$

The restriction of $\Phi_|$ yields the desired isomorphism. □

Our goal is to prove that $\mathcal{C}_\pm \cup L_\pm \cup L_\pm^2 \cup L_\pm^{6,1} \cup L_\pm^{6,2}$ is a complement-equivalent arithmetic Zariski pair.

3.1. Fibered curves and braid monodromy factorization

These curves are called *fibered* (see [4]) since their complements induce a locally trivial fibration on a finitely punctured $\mathbb{P}^1$ (the complement of the discriminant). A fibered curve has a *horizontal part* (the curve that intersects the generic fibers in a finite number of points) and a *vertical part* (the preimage of the discriminant). As mentioned above, the braid monodromy of its horizontal part is a topological invariant of a fibered curve.

Let us recall this result. Consider $\mathcal{C} \subset \mathbb{P}^2$ a projective curve, $L \subset \mathbb{P}^2$ be a line and $P \in L$. Let us assume that, if $P \in \mathcal{C}$, then L is the tangent cone of $\mathcal{C}$. Consider $L_1, \ldots, L_r$ the lines in the pencil through P (besides L) which are non-transversal to $\mathcal{C}$. The curve $\mathcal{C}^\varphi := \mathcal{C} \cup L \cup \bigcup_{j=1}^r L_j$ is the *fibered curve* associated with $(\mathcal{C}, L, P)$.

Consider now the braid monodromy factorization $(\tau_1, \ldots, \tau_r) \in \mathbb{B}_d^r$ of the affine curve $\mathcal{C}^{\text{aff}} := \mathcal{C} \setminus L$, with respect to the projection based at P, where d is the difference between $\deg \mathcal{C}$ and the multiplicity of $\mathcal{C}$ at P. We are ready to state the result.

Theorem 3.2 ([3, Theorem 1]). *Let us suppose the existence of a homeomorphism* $\Phi : (\mathbb{P}^2, \mathcal{C}_1^\varphi) \to (\mathbb{P}^2, \mathcal{C}_2^\varphi)$ *such that:*

1. *The homeomorphism is orientation preserving on* $\mathbb{P}^2$ *and on the curves.*
2. $\Phi(P_1) = P_2$, $\Phi(L_1) = L_2$.

Then, the braid monodromies of the two triples $(\mathcal{C}_1, L_1, P_1)$ *and* $(\mathcal{C}_2, L_2, P_2)$ *are equivalent.*

To understand the statement, let us recall the notion of equivalence of braid monodromies. Let $(\tau_1, \ldots, \tau_r) \in \mathbb{B}_d^r$ be a braid monodromy factorization; for its construction we have identified the Artin braid group $\mathbb{B}_d$ with the braid group based at some specific d points of $\mathbb{C}$; two such identifications differ by conjugation, i.e.,

$$(\tau_1, \ldots, \tau_r) \sim (\tau_1^\tau, \ldots, \tau_r^\tau), \quad \forall \tau \in \mathbb{B}_d.$$

There is a second choice, the choice of a pseudo-geometric basis in $\mathbb{F}_r$. Two such bases differ by what is called a *Hurwitz move*. The Hurwitz action of $\mathbb{B}_r$ on G^r (where G is an arbitrary group) is defined as follows. Let us denote by $s_1, \ldots, s_{r-1}$ the Artin generators of $\mathbb{B}_r$ (we replace σ by s to avoid confusion when G is a braid group). Then:

$$(g_1, \ldots, g_r)^{s_i} \mapsto (g_1, \ldots, g_{i-1}, g_{i+1}, g_{i+1} \cdot g_i \cdot g_{i+1}^{-1}, \ldots, g_{i+2}, \ldots, g_r).$$

Definition 3.3. Two braid monodromies in $\mathbb{B}_d^r$ are *equivalent* if they belong to the same orbit by the action of $\mathbb{B}_r \times \mathbb{B}_d$ described above.

Note that it is hopeless to apply directly Theorem 3.2 to our case: the braid monodromies are equal! In [5], we refined Theorem 3.2 to work with *ordered line arrangements*: the classical braid groups were replaced everywhere by pure braid groups. We are going to state now an intermediate refinement of Theorem 3.2.

Let us think about our case. If we color in a different way the two first strands and the two last strands, we take into account, that the first ones are the branches of the node in Σ_2 which provides L_+, while the last ones provide L_-. Let us set that (τ_1, τ_2, τ_3) is the braid monodromy factorization for $\mathcal{C}_+$ with this coloring. To compare both curves, the braid monodromy factorization of $\mathcal{C}_-$ would have the strands associated to L_- in the first place; this is accomplished, considering:

$$(\tilde{\tau}_1, \tilde{\tau}_2, \tilde{\tau}_3) := (\tau_1^\tau, \tau_2^\tau, \tau_3^\tau), \qquad \tau = (\sigma_2 \cdot \sigma_3 \cdot \sigma_1)^2$$

since the braid τ exchanges the two pairs of strands.

Definition 3.4. Let A be a partition of the set $\{1, \ldots, n\}$. The *braid group* $\mathbb{B}(A)$ relative to A is the subgroup of $\mathbb{B}_n$ consisting of the braids that respect the given partition.

Remark 3.5. For instance, note that both the total and the discrete partition provide recognizable groups: $\mathbb{B}_n = \mathbb{B}(\{\{1, \ldots, n\}\})$ whereas $\mathbb{B}(\{\{1\}, \ldots, \{n\}\})$ provides the pure braid group.

The proof of the following result follows along the same lines as that of [3, Theorem 1].

Theorem 3.6. *Let A_d, A_r be partitions of $\{1, \ldots, d\}$ and $\{1, \ldots, r\}$, respectively, such that A_d induces partitions on $L_i \cap C_i$. Assume that there exists a homeomorphism $\Phi : (\mathbb{P}^2, \mathcal{C}_1^\varphi) \to (\mathbb{P}^2, \mathcal{C}_2^\varphi)$ satisfying the hypotheses in Theorem 3.2, and also satisfying:*

(1) *The blocks of lines through P_1, P_2 associated to the partition are respected.*

(2) *The partitions on* $L_i \cap C_i$ *are respected.*

Then, the triples $(\mathcal{C}_1, L_1, P_1)$ *and* $(\mathcal{C}_2, L_2, P_2)$ *have braid monodromy factorizations* $(\tau_1^j, \ldots, \tau_r^j) \in \mathbb{B}_d^r$, $j = 1, 2$ *(respecting the above partitions) which are equivalent by the action of* $\mathbb{B}(A_d) \times \mathbb{B}(A_r)$.

Theorem 3.7. *There is no homeomorphism*

$$(\mathbb{P}^2, \mathcal{C}_+ \cup L_+ \cup L_+^2 \cup L_+^{6,1} \cup L_+^{6,2}) \to (\mathbb{P}^2, \mathcal{C}_- \cup L_- \cup L_-^2 \cup L_-^{6,1} \cup L_-^{6,2}).$$

Proof. Let us assume that such a homeomorphism Φ exists. From the topological properties of $\mathbb{P}^2$, it must respect the orientation of $\mathbb{P}^2$. The intersection form in $\mathbb{P}^2$ implies that Φ either respects or reverses all the orientations on the irreducible components of the curves. Since the equations are real, in the latter case one can compose Φ with the complex conjugation. This composition respects the orientations of the irreducible components. Therefore, one may assume that Φ respects the orientation of the curves.

By the topology of the curve at the singularities, it is easy to see that $\Phi(\mathcal{C}_+) = \mathcal{C}_-$, $\Phi(L_+) = L_-$ and $\Phi(L_+^2) = L_-^2$. In particular, $\Phi(P_+) = P_-$. Also note that $\Phi(L_+^{6,1} \cup L_+^{6,2}) = L_-^{6,1} \cup L_-^{6,2}$.

Moreover, the homeomorphism must respect the two branches of the $\mathbb{A}_5$ point and, hence, the two other points in $\mathcal{C}_\pm \cap L_\pm$ (globally).

Let us consider the partition $A_4^d = \{\{1, 2\}, \{3, 4\}\}$ for the strands of the braids. In the base, we consider the partition $A_3^r = \{\{1, 3\}, \{2\}\}$. Then, from Theorem 3.6, the braid monodromies $T := (\tau_1, \tau_2, \tau_3)$ and $\tilde{T} := (\tilde{\tau}_1, \tilde{\tau}_2, \tilde{\tau}_3)$ are equivalent under the action of $\mathbb{B}(A_4^d) \times \mathbb{B}(A_3^r)$.

We are going to show that this does not happen and, in particular, the expected homeomorphism does not exist.

There is no algorithm ensuring that two braid monodromy factorizations are equivalent. In order to look for necessary conditions, we consider a finite representation $\varphi : \mathbb{B}_4 \to F$, where F is a finite group. We need to check if $\varphi(T)$ and $\varphi(\tilde{T})$ are equivalent under the action of $\varphi(\mathbb{B}(A_4^d)) \times \mathbb{B}(A_3^r)$. Since the orbits are finite, this approach should lead to an answer.

Let us denote $F_A := \varphi(\mathbb{B}(A_4^d))$; let $\hat{F} := F^3/F_A$, i.e., the quotient of the cartesian product F^3 under the diagonal conjugation action of F_A. The group $\mathbb{B}(A_3^r)$ acts by Hurwitz moves on it. We want to check if the classes $[T], [\tilde{T}] \in \hat{F}$ are in the same orbit under this action. Note that in general, this can be computationally expensive.

There is a natural way to obtain representations of the braid group. Consider the reduced Burau representation $\phi : \mathbb{B}_4 \to \mathrm{GL}(3, \mathbb{Z}[t^{\pm 1}])$. Let R be either $\mathbb{Z}/m$, for some $m \in \mathbb{N}$, or $\mathbf{F}_q$, q some prime power. Let s be a unit in R; then we define

$$\varphi : \mathbb{B}_4 \to F := \varphi(\mathbb{B}_4) \subset GL(3, R)$$

by considering the natural map $\mathbb{Z} \to R$ and specializing t to s. Let us do it for $R = \mathbb{Z}/4$ and $s \equiv -1 \bmod 4$. We have:

$$
\begin{aligned}
\mathbb{B}_4 &\xrightarrow{\varphi} F \subset \mathrm{GL}(3, \mathbb{Z}/4)\\
\sigma_1 &\longmapsto \begin{pmatrix} 1 & 0 & 0\\ 1 & 1 & 0\\ 1 & 0 & 1 \end{pmatrix}\\
\sigma_2 &\longmapsto \begin{pmatrix} 2 & 3 & 0\\ 1 & 0 & 0\\ 0 & 0 & 1 \end{pmatrix}\\
\sigma_3 &\longmapsto \begin{pmatrix} 1 & 0 & 0\\ 0 & 2 & 3\\ 0 & 1 & 0 \end{pmatrix},
\end{aligned}
$$

where F is a finite group of order 768. The class of $\tilde{T}$ is

$$
[\tilde{T}] = \left[\begin{pmatrix} 2 & 2 & 3\\ 0 & 3 & 1\\ 3 & 0 & 3 \end{pmatrix}, \begin{pmatrix} 3 & 3 & 0\\ 0 & 1 & 3\\ 2 & 1 & 0 \end{pmatrix}, \begin{pmatrix} 3 & 3 & 2\\ 3 & 0 & 2\\ 1 & 0 & 1 \end{pmatrix}\right]
$$

and the orbit of $[T]$ is

$$
\left[\begin{pmatrix} 0 & 3 & 2\\ 0 & 2 & 3\\ 1 & 2 & 2 \end{pmatrix}, \begin{pmatrix} 3 & 0 & 3\\ 1 & 3 & 0\\ 3 & 2 & 2 \end{pmatrix}, \begin{pmatrix} 1 & 0 & 1\\ 0 & 2 & 1\\ 0 & 1 & 1 \end{pmatrix}\right], \left[\begin{pmatrix} 3 & 0 & 3\\ 1 & 3 & 0\\ 3 & 2 & 2 \end{pmatrix}, \begin{pmatrix} 0 & 1 & 2\\ 1 & 3 & 2\\ 2 & 1 & 1 \end{pmatrix}, \begin{pmatrix} 1 & 0 & 1\\ 0 & 2 & 1\\ 0 & 1 & 1 \end{pmatrix}\right], \left[\begin{pmatrix} 0 & 3 & 2\\ 0 & 2 & 3\\ 1 & 2 & 2 \end{pmatrix}, \begin{pmatrix} 1 & 0 & 1\\ 0 & 2 & 1\\ 0 & 1 & 1 \end{pmatrix}, \begin{pmatrix} 2 & 3 & 0\\ 1 & 3 & 2\\ 0 & 3 & 3 \end{pmatrix}\right],
$$
$$
\left[\begin{pmatrix} 0 & 1 & 2\\ 1 & 3 & 2\\ 2 & 1 & 1 \end{pmatrix}, \begin{pmatrix} 0 & 3 & 2\\ 0 & 2 & 3\\ 1 & 2 & 2 \end{pmatrix}, \begin{pmatrix} 1 & 0 & 1\\ 0 & 2 & 1\\ 0 & 1 & 1 \end{pmatrix}\right], \left[\begin{pmatrix} 1 & 0 & 1\\ 0 & 2 & 1\\ 0 & 1 & 1 \end{pmatrix}, \begin{pmatrix} 1 & 2 & 1\\ 1 & 3 & 0\\ 1 & 0 & 0 \end{pmatrix}, \begin{pmatrix} 2 & 3 & 0\\ 1 & 3 & 2\\ 0 & 3 & 3 \end{pmatrix}\right], \left[\begin{pmatrix} 0 & 1 & 2\\ 1 & 3 & 2\\ 2 & 1 & 1 \end{pmatrix}, \begin{pmatrix} 1 & 0 & 1\\ 0 & 2 & 1\\ 0 & 1 & 1 \end{pmatrix}, \begin{pmatrix} 1 & 2 & 1\\ 1 & 3 & 0\\ 1 & 0 & 0 \end{pmatrix}\right].
$$

It is easily checked that $\tilde{T}$ is not conjugate to any element of the orbit of T.

The group $\mathbb{B}(A_3^r)$ is generated by $s_1^{-2}, s_2^{-2}, s_1 \cdot s_2 \cdot s_1^{-1}$; they induce the following permutations in the orbit of T:

$$
[(1,2,4)(3,5,6), (1,3,5)(2,4,6), (1,3,4)(2,5,6)],
$$

showing that it is actually an orbit. Since we have shown that the braid monodromies are not conjugate, we deduce that no homeomorphism exists. The computations have been done with `Sagemath` [26] and `GAP4` [17]. □

3.2. Final comments

Eyral-Oka curves give rise to other arithmetic Zariski pairs, namely using the projections from the singular points of type $\mathbb{E}_6$ and $\mathbb{A}_2$. When projecting from a point of type $\mathbb{E}_6$ it does not matter which one because of the symmetry of the curves which exchanges both points.

One can compute the braid monodromy factorizations using again the fact that they are strongly real curves. In these cases, it is more involved to prove that the braid monodromy factorizations are not equivalent. In a future work we will use the computed representations to distinguish the braid monodromies using diagonal representations.

Acknowledgment

Many thanks to our anonymous referee for their comments that have certainly improved the exposition of this paper.

References

[1] H. Abelson, *Topologically distinct conjugate varieties with finite fundamental group*, Topology **13** (1974), 161–176.

[2] E. Artal, *Sur les couples de Zariski*, J. Algebraic Geom. **3** (1994), no. 2, 223–247.

[3] E. Artal, J. Carmona, and J.I. Cogolludo-Agustín, *Braid monodromy and topology of plane curves*, Duke Math. J. **118** (2003), no. 2, 261–278.

[4] ———, *Effective invariants of braid monodromy*, Trans. Amer. Math. Soc. **359** (2007), no. 1, 165–183.

[5] E. Artal, J. Carmona, J.I. Cogolludo-Agustín, and M.Á. Marco, *Topology and combinatorics of real line arrangements*, Compos. Math. **141** (2005), no. 6, 1578–1588.

[6] E. Artal, J. Carmona, J.I. Cogolludo-Agustín, and H. Tokunaga, *Sextics with singular points in special position*, J. Knot Theory Ramifications **10** (2001), no. 4, 547–578.

[7] E. Artal, J.I. Cogolludo-Agustín, and H. Tokunaga, *A survey on Zariski pairs*, Algebraic geometry in East Asia—Hanoi 2005, Adv. Stud. Pure Math., vol. 50, Math. Soc. Japan, Tokyo, 2008, pp. 1–100.

[8] J.I. Cogolludo-Agustín and A. Libgober, *Mordell-Weil groups of elliptic threefolds and the Alexander module of plane curves*, J. Reine Angew. Math. **697** (2014), 15–55.

[9] A.I. Degtyarëv, *Isotopic classification of complex plane projective curves of degree* 5, Leningrad Math. J. **1** (1990), no. 4, 881–904.

[10] ———, *Fundamental groups of symmetric sextics*, J. Math. Kyoto Univ. **48** (2008), no. 4, 765–792.

[11] ———, *On deformations of singular plane sextics*, J. Algebraic Geom. **17** (2008), no. 1, 101–135.

[12] ———, *Plane sextics with a type E_8 singular point*, Tohoku Math. J. (2) **62** (2010), no. 3, 329–355.

[13] H. Esnault, *Fibre de Milnor d'un cône sur une courbe plane singulière*, Invent. Math. **68** (1982), no. 3, 477–496.

[14] C. Eyral and M. Oka, *Fundamental groups of the complements of certain plane non-tame torus sextics*, Topology Appl. **153** (2006), no. 11, 1705–1721.

[15] M. de Franchis, *Sulle superficie algebriche le quali contengono un fascio irrazionale di curve*, Palermo Rend. **20** (1905), 49–54.

[16] T. Fujita, *On the topology of noncomplete algebraic surfaces*, J. Fac. Sci. Univ. Tokyo Sect. IA Math. **29** (1982), no. 3, 503–566.

[17] The GAP Group, *GAP – Groups, Algorithms, and Programming, Version 4.7.7*, 2015.

[18] V.M. Kharlamov and Vik.S. Kulikov, *On braid monodromy factorizations*, Izv. Ross. Akad. Nauk Ser. Mat. **67** (2003), no. 3, 79–118.

[19] ———, *Automorphisms of Galois coverings of generic m-canonical projections*, Izv. Ross. Akad. Nauk Ser. Mat. **73** (2009), no. 1, 121–156.

[20] A. Libgober, *Alexander polynomial of plane algebraic curves and cyclic multiple planes*, Duke Math. J. **49** (1982), no. 4, 833–851.

[21] F. Loeser and M. Vaquié, *Le polynôme d'Alexander d'une courbe plane projective*, Topology **29** (1990), no. 2, 163–173.

[22] M. Namba, *Geometry of projective algebraic curves*, Marcel Dekker Inc., New York, 1984.

[23] J.-P. Serre, *Exemples de variétés projectives conjuguées non homéomorphes*, C. R. Acad. Sci. Paris Sér. I Math. **258** (1964), 4194–4196.

[24] I. Shimada, *On arithmetic Zariski pairs in degree 6*, Adv. Geom. **8** (2008), no. 2, 205–225.

[25] ———, *Non-homeomorphic conjugate complex varieties*, Singularities—Niigata–Toyama 2007, Adv. Stud. Pure Math., vol. 56, Math. Soc. Japan, Tokyo, 2009, pp. 285–301.

[26] W.A. Stein et al., *Sage Mathematics Software (Version 6.6)*, The Sage Development Team, 2015, `http://www.sagemath.org`.

[27] H. Tokunaga, $(2,3)$ *torus sextics and the Albanese images of* 6*-fold cyclic multiple planes*, Kodai Math. J. **22** (1999), no. 2, 222–242.

[28] O. Zariski, *On the problem of existence of algebraic functions of two variables possessing a given branch curve*, Amer. J. Math. **51** (1929), 305–328.

[29] ———, *On the irregularity of cyclic multiple planes*, Ann. of Math. (2) **32** (1931), no. 3, 485–511.

[30] ———, *The topological discriminant group of a Riemann surface of genus p*, Amer. J. Math. **59** (1937), 335–358.

Enrique Artal Bartolo and José Ignacio Cogolludo-Agustín
Departamento de Matemáticas, IUMA
Universidad de Zaragoza
C. Pedro Cerbuna 12
50009 Zaragoza
Spain
e-mail: `artal@unizar.es`
`jicogo@unizar.es`

Logarithmic Vector Fields and the Severi Strata in the Discriminant

Paul Cadman, David Mond and Duco van Straten

Abstract. The discriminant, D, in the base of a miniversal deformation of an irreducible plane curve singularity, is partitioned according to the genus of the (singular) fibre, or, equivalently, by the sum of the delta invariants of the singular points of the fibre. The members of the partition are known as the *Severi strata*. The smallest is the δ-constant stratum, $D(\delta)$, where the genus of the fibre is 0. It is well known, by work of Givental' and Varchenko, to be Lagrangian with respect to the symplectic form Ω obtained by pulling back the intersection form on the cohomology of the fibre via the period mapping. We show that the remaining Severi strata are also co-isotropic with respect to Ω, and moreover that the coefficients of the expression of Ω with respect to a basis of $\Omega^2(\log D)$ are equations for $D(\delta)$. Similarly, the coefficients of $\Omega^{\wedge k}$ with respect to a basis for $\Omega^{2k}(\log D)$ are equations for $D(\delta - k + 1)$. These equations allow us to show that for E_6 and E_8, $D(\delta)$ is Cohen-Macaulay (this was already shown by Givental' for A_{2k}), and that, as far as we can calculate, for A_{2k} all of the Severi strata are Cohen-Macaulay.

Mathematics Subject Classification (2000). 32S30 14B07 14H50 (53D17).

Keywords. Explicit equations, Delta-constant, intersection form.

1. Introduction: the discriminant and its Severi strata

Two of the most basic invariants of a plane curve singularity $(C, 0)$ are its *Milnor number* μ and its *delta invariant* δ. If $f : (\mathbb{C}^2, 0) \to (\mathbb{C}, 0)$ is a holomorphic map defining $(C, 0) = f^{-1}(0)$, then $\mu(C)$ is the dimension of the Jacobian algebra $\mathcal{O}_{\mathbb{C}^2,0}/J_f$ and equals the dimension of the vanishing cohomology. If $n : \widetilde{C} \longrightarrow C$ denotes the normalisation of $(C, 0)$, then $\delta(C)$ is the dimension $n_* \mathcal{O}_{\widetilde{C}} / \mathcal{O}_C$ and equals the number of double points appearing in a generic perturbation of the map n. These invariants are related by the relation

$$\mu = 2\delta + r - 1$$

where r denotes the number of branches of $(C,0)$. See [10, pp. 206–211]. The number μ also appears as the number of parameters of an $\mathscr{R}_e$ miniversal deformation $F:(\mathbb{C}^2\times\mathbb{C}^\mu,0)\to(\mathbb{C},0)$ of the function $f:(\mathbb{C}^2,0)\to(\mathbb{C},0)$ defining $(C,0)$. The restriction $\pi: X:=F^{-1}(0)\to S=(\mathbb{C}^\mu,0)$ is a versal deformation of the plane curve singularity $(C,0)$. The fibre C_u over $u\in S$ is the curve defined by zero level of the deformed function $f_u:=F(\ ,u)$ and *discriminant* $D\subset S$ is the set of parameter values u for which the fibre C_u is singular. This set is stratified by the types of singularities that appear in the fibres. In this paper we will focus on the so-called *Severi strata*, where the sum of the delta-invariants add up to a value $\geq k$:

$$D(k)=\{u\in S:\delta(C_u)\geq k\}$$

where $\delta(C_u)=\sum_{x\in C_u}\delta(C_u,x)$. Clearly $D(0)=S$ and $D(1)=D$, and as $D(i)$ is contained in $D(i-1)$ we obtain a chain

$$D(\delta)\subset D(\delta-1)\subset\ldots\subset D(1)\subset D(0)=S.$$

The smallest non-empty Severi stratum, $D(\delta)$, is the classical *δ-constant stratum*. The term "stratum" here is a bit of a misnomer, since the Severi strata are not in general smooth.

It is a classical fact, going back at least to Cayley [5], that any curve singularity with $\delta=k$ can be deformed into a curve with precisely k A_1 points, a fact which explains the name *virtual number of double points* for the δ-invariant. For a very nice proof see the paper by C. Scott [20]. Thus the set $D(kA_1)$ of parameter values u for which C_u has precisely k A_1 singularities is *dense* in $D(k)$. Moreover, $D(k)$ is smooth at such points, for there, by openness of versality, $D(k)$ is a normal crossing of k local smooth components of the discriminant D. A curve singularity with δ-invariant $k>1$ is also adjacent to a curve with one A_2 singularity and $k-1$ A_1 singularities. Hence $D(k)_{\text{reg}}=D(kA_1)$. We refer to [23] for more background on this.

In the famous Anhang F to his *Vorlesungen über Algebraische Geometrie* [21], Severi considered the variety of plane curves of degree d with a given number of nodes which he used to argue for the irreducibility of the space of all curves of a given genus. A complete argument along these lines with given much later by J. Harris, [12], and by Harris and Diaz in [6], which started the interest in the local case. This seems to justify the name *Severi-strata* for the $D(k)$'s, which was introduced in [22] . Recently, these strata have been the subject of several papers and their geometry appears to hide some great mysteries. In [7] the multiplicity of $D(\delta)$ was shown to be equal to the Euler number of the compactified Jacobian of $(C,0)$. This was further explored in [22], where multiplicities of the other $D(k)$ were related to the puntual Hilbert-schemes $\mathrm{Hilb}^n(C,0)$. Most surprisingly, these invariants turn out to be related to the HOMFLY-polynomial of the knot in the 3-sphere defined by $(C,0)$, [17].

If the curve $(C,0)$ is irreducible, its Milnor fibre C_u has just one boundary component, and it follows that the intersection form I_u on $H^1(C_u;\mathbb{C})$ is non-degenerate. In [9], Givental' and Varchenko used the period map associated to a non-degenerate 1-form η on the total space of the Milnor fibration of F, together with the Gauss-Manin connection, to pull back the intersection form from the cohomology bundle $\mathscr{H}^*$ over S to get a symplectic form Ω on $S \setminus D$, and proved

Theorem 1.1. (a) *Ω extends to a symplectic form on S, and*
(b) *the δ-constant stratum $D(\delta)$ in the discriminant is Lagrangian with respect to Ω.*

Below we complement their results and show the following theorems.

Theorem 1.2. *All of the Severi strata are coisotropic with respect to Ω.*

The form Ω can also be used to obtain equations defining the Severi-strata. Let $\wedge^k\Omega$ be the k-fold wedge product of Ω. Although it is a regular form, it can be considered as an element of $\Omega_S^{2k}(\log D)$. Let I_k be the ideal generated by its coefficients with respect to a basis of $\Omega_S^{2k}(\log D)$.

Theorem 1.3. *For $k = 1, \ldots, \delta$, the Severi stratum $D(k)$ is defined by the ideal $I_{\delta-k+1}$:*

$$D(k) = V(I_{\delta-k+1}).$$

Equivalently, if $\chi_1, \ldots, \chi_\mu$ form a basis for the free module of logarithmic vector fields $\Theta_S(-\log D)$, then $D(k)$ is defined by the ideal generated by the Pfaffians of size $2\delta - 2k + 2$ of the skew matrix $(\Omega(\chi_i, \chi_j))_{1 \leq i,j, \leq \mu}$.

We do not know whether in general the ideals I_k are radical. Our calculations suggest that they are, but we have not been able to show this. In [10, II. Proposition 2.57] it was shown that the strata are analytic.

Givental' proved in [8] that for curve singularities of type A_{2k+1}, $D(\delta)$ is Cohen-Macaulay and it can be conjectured that this is always the case,[24]. In the first author's PhD thesis, [4], Theorem 1.3 was used to show that $D(\delta)$ is Cohen Macaulay also for E_6 and E_8. Calculations using Theorem 1.3 suggest that the remaining Severi strata are Cohen-Macaulay in the case of A_{2k}, but show that for E_6 the stratum $D(2)$ is not Cohen-Macaulay.

In the process of proving these theorems we noticed that Ω determines a natural rank 2 maximal Cohen-Macaulay module over the discriminant D, which seems to be of independent interest.

2. Preliminaries

Let $f : (\mathbb{C}^{n+1}, 0) \to (\mathbb{C}, 0)$ define an isolated singularity $(C, 0)$ and let

$$g_1, g_2, \ldots, g_\mu = 1 \in \mathcal{O}_{\mathbb{C}^{n+1},0}$$

be functions giving a basis for the Jacobian algebra $\mathcal{O}/J_f$. We consider a good representative of a miniversal deformation of f of the form

$$F : B \times S \to \mathbb{C}, \quad F(x, u) = f(x) + \sum_{i=1}^{\mu} u_i g_i(x) \,,$$

where B is a Milnor ball for C and S is a sufficiently small ball in $\mathbb{C}^\mu$,[16]. The set $X := F^{-1}(0)$ comes with a map $\pi : X \longrightarrow S$, with C_u as fibre over $u \in S$.

2.1. The critical space Σ

The relative critical set Σ of F is defined to be

$$\Sigma = \left\{(x,u) \in B \times S : \frac{\partial F}{\partial x_i}(x,u) = 0, \;\; i = 0, \ldots, n\right\}.$$

It is smooth and the projection $\pi : \Sigma \to S$ is a μ-fold branched cover: its fibre over $u \in S$ consists of the critical points of $F(-,u)$. As the partial derivatives form a regular sequence,

$$\mathcal{O}_\Sigma = \mathcal{O}_{B\times S} \,/ (\partial F/\partial x_0, \ldots, \partial F/\partial x_n)$$

is a free $\mathcal{O}_S$-module of rank μ. Miniversality of F is equivalent to the statement that the Kodaira-Spencer map

$$dF : \Theta_S \to \mathcal{O}_\Sigma, \quad \vartheta \mapsto \vartheta(F) = dF(\vartheta)$$

is an isomorphism. The set $X \cap \Sigma$ is the union over $u \in S$ of the set of singular points of C_u, and its image under π is the discriminant, D. For brevity we denote $X \cap \Sigma$ by $\tilde{D}$. It is indeed the normalisation of D.

2.2. D as a free divisor

Let $\bar{F} : (B \times S, (0,0)) \to (\mathbb{C} \times S, (0,0))$ be the unfolding of f corresponding to the deformation F. Then $\Sigma \subset B \times \mathbb{C}^\mu$ is the (absolute) critical locus of $\bar{F}$. We write $\Delta = \bar{F}(\Sigma) \subset \mathbb{C} \times S$ for the set of critical values of $\bar{F}$. It is well known that Σ is the normalisation of Δ: it is smooth, and the map $\bar{F}| : \Sigma \to \Delta$ is generically one-to-one. Then $D = \Delta \cap \{0\} \times S$. As usual, $\Theta_{\mathbb{C}\times S}(-\log \Delta)$ denotes the $\mathcal{O}_{\mathbb{C}\times S}$-module of vector fields on $\mathbb{C} \times S$ which are tangent to Δ, and $\Theta_S(-\log D)$ denotes the $\mathcal{O}_S$-module of vector fields on S which are tangent to D.

Proposition 2.1. (i) *$\Theta_{\mathbb{C}\times S}(-\log \Delta)$ is the $\mathcal{O}_{\mathbb{C}\times S}$-module of vector fields on $\mathbb{C}\times S$ which are $\bar{F}$-liftable to $B \times S$.*

(ii) *$\Theta_S(-\log D)$ is the $\mathcal{O}_S$-module of vector fields on S which are π-liftable to $V(F)$.*

Proof. ([16]) (i) Let $\vartheta \in \Theta_{\mathbb{C}\times S}(-\log \Delta)$. Since $F| : \Sigma \to \Delta$ is the normalisation of Δ, there is a vector field $\tilde{\vartheta}_0$ on Σ lifting ϑ. For any extension $\tilde{\vartheta}_1$ of $\tilde{\vartheta}_0$ to $B \times S$, $\omega F(\vartheta) - tF(\tilde{\vartheta}_1)$ vanishes on Σ, and since the Jacobian ideal $(\partial F/\partial x_0, \ldots, \partial F/\partial x_n)$ is radical, there exists a second vector field ξ on $B\times S$ such that $\omega F(\tilde{\vartheta}_1) - tF(\tilde{\vartheta}_1) = tF(\xi)$. Then $\tilde{\vartheta}_1 + \xi$ is an $\bar{F}$-lift of ϑ.

Conversely, suppose $\tilde{\vartheta}$ is a $\bar{F}$-lift of ϑ. Then $\tilde{\vartheta}$ must be tangent to Σ, for the integral flows $\tilde{\Phi}_t$ and Φ_t of $\tilde{\vartheta}$ and ϑ satisfy $\Phi_1 \circ \bar{F} = \bar{F} \circ \Phi_t$, showing that $\tilde{\Phi}_t$ defines an isomorphism $\bar{F}^{-1}(u) \to \bar{F}^{-1}(\Phi_t(u))$, and must therefore map singular points of $\bar{F}^{-1}(u)$ to singular points of $\bar{F}^{-1}(\Phi_t(u))$. It follows that ϑ is tangent to Δ.

(ii) Let $i_0 : S \to \mathbb{C} \times S$ be the inclusion $u \mapsto (0,u)$. Then $D = i_0^{-1}(\Delta)$. Now i_0 is logarithmically transverse to Δ, i.e., transverse to the distribution $\Theta_{\mathbb{C}\times S}(-\log \Delta)$. If F is the standard deformation $f(x) + \sum_i u_i g_i$, with $g_\mu = 1$, then this transversality is obvious: the vector field $\partial/\partial t + \partial/\partial u_\mu$ on $\mathbb{C} \times S$ has $\bar{F}$-lift $\partial/\partial u_\mu$, and therefore lies in $\Theta_{\mathbb{C}\times S}(-\log \Delta)$. Any other miniversal deformation is parametrised $\mathscr{R}$-equivalent to the standard deformation, so the transversality holds there too.

Identifying $\mathbb{C}^\mu$ with $\{0\} \times \mathbb{C}^\mu$, from the logarithmic transversality of i_0 to Δ it follows that $\Theta_S(-\log D)$ is equal to $\theta_{\mathbb{C}\times S}(-\log \Delta) \bigcap \theta_{\mathbb{C}\times S}(-\log(\{0\} \times S))$ restricted to $\mathbb{C}^\mu$, and that every vector field in $\Theta_S(-\log D)$ extends to a vector field in $\Theta_{\mathbb{C}\times S}(-\log \Delta)$. Clearly, any lift to $\mathbb{C}^{n+1} \times S$ of a vector field in $\theta_{\mathbb{C}\times S}(-\log \Delta) \bigcap \theta_{\mathbb{C}\times S}(-\log(\{0\} \times S))$ must be tangent to $V(F)$, and any vector field whose $\bar{F}$-lift is tangent to $V(F)$ must lie in $\theta_{\mathbb{C}\times S}(-\log \Delta) \bigcap \theta_{\mathbb{C}\times S}(-\log(\{0\} \times S))$. □

Therefore we have a diagram

$$\begin{array}{ccccccccc} 0 & \longleftarrow & \mathcal{O}_{\tilde{D}} & \longleftarrow & \Theta_S & \hookleftarrow & \Theta_S(-\log D) & & \\ & & \downarrow{=} & & \downarrow{dF} & & \downarrow{\frac{dF}{F}} & & \\ 0 & \longleftarrow & \mathcal{O}_{\tilde{D}} & \longleftarrow & \mathcal{O}_\Sigma & \overset{F}{\longleftarrow} & \mathcal{O}_\Sigma & \longleftarrow & 0 \end{array} \tag{2.1}$$

where the vertical maps are isomorphisms. This diagram can be used to find a basis for $\Theta_S(-\log D)$. The germs $FdF(\partial/\partial u_i)$ generate $(F)\,\mathcal{O}_\Sigma$, therefore if

$$dF(\chi_i) = FdF\left(\frac{\partial}{\partial u_i}\right), \tag{2.2}$$

then the χ_i generate $\Theta_S(-\log D)$. This shows that $\Theta_S(-\log D)$ is a locally free module, so that D is a free divisor.

2.3. Stratification of D

The discriminant D is stratified in various ways. Of special relevance to us are the local $\mathscr{R}$ and $\mathscr{K}$ strata.

Suppose as before that $F : B \times S \to \mathbb{C}$ is a good representative of a versal deformation of f, where B is open in $\mathbb{C}^{n+1}$ and S is open in $\mathbb{C}^\mu$. Write $f_u = F(\_, u)$, and suppose that $p_1, \ldots, p_k$ are the critical points of f_u lying on $f_u^{-1}(0)$. For each point p_i, the germ

$$F : (B \times S, (p_i, u)) \to (\mathbb{C}, 0)$$

is an $\mathscr{R}_e$-versal deformation of the germ of f_u at p_i, by openness of versality. Hence there is a germ of submersion h_i from (S, u) to the base of an $\mathscr{R}_e$-miniversal deformation

$$G_i : (B \times \mathbb{C}^{\mu_i}, (p_i, 0)) \to (\mathbb{C}, 0)$$

of this germ, such that the germ of deformation $F : (B \times S, (p_i, u)) \to (\mathbb{C}, 0)$ is isomorphic to $h_i^*(G_i)$. We set

$$\mathscr{R}_i(u) = h_i^{-1}(0).$$

This is independent of the choice of miniversal deformation G_i and submersion h_i, since any two choices can be related by a commutative diagram of spaces and maps. Again by openness of versality, the $\mathscr{R}_i(u)$, $i = 1, \ldots, k$ are in general position with respect to one another, and we set

$$\mathscr{R}(u) = \bigcap_{i=1}^{k} \mathscr{R}_i(u).$$

This is the *$\mathscr{R}$ stratum through u*. It is smooth of dimension $\mu - \sum_i \mu(f_u, p_i)$.

If in the above definition we replace $F : B \times S \to \mathbb{C}$ by the projection $V(F) \to S$, and replace each G_i by a $\mathscr{K}_e$-miniversal deformation H_i of the hypersurface singularity (C_u, p_i), then we obtain the $\mathscr{K}$-strata $\mathscr{K}_i(u)$ and their intersection $\mathscr{K}(u)$, the *$\mathscr{K}$-stratum through u*, which is once again smooth, by openness of versality, and has dimension $\mu - \sum_i \tau(C_u, p_i)$. Since $\mathscr{R} \subset \mathscr{K}$, $\mathscr{R}(u) \subset \mathscr{K}(u)$.

If, for example, the fibre C_u has k A_1 singularities and no other singular points, then $\mathscr{R}(u) = \mathscr{K}(u)$ and its germ at u coincides with the germ at u of the set of points u' such that $C_{u'}$ has k A_1 points and no other singularities.

Definition 2.2. The logarithmic tangent space $T_u^{\log D} S$ is the vector subspace of $T_u S$ spanned at u by the logarithmic vector fields.

Proposition 2.3. *One has the equality of vector spaces*

$$T_u^{\log D} S = T_u \mathscr{K}(u).$$

Proof. We have the exact sequence

$$0 \to \Theta_S(-\log D) \to \Theta_S \to \pi_*(\mathscr{O}_{\tilde{D}}) \to 0$$

which gives

$$\frac{\Theta_S}{\Theta_S(-\log D)} \simeq \pi_*(\mathscr{O}_{\tilde{D}})$$

and so

$$\frac{T_u \mathbb{C}^\mu}{T_u^{\log D} S} \simeq \frac{\Theta_S}{\Theta_S(-\log D) + \mathfrak{m}_{S,u}\Theta_{S,u}} \simeq \bigoplus_i T^1_{\mathscr{K}_e}(f_u, x_i)$$

This means that to show

$$T_u^{\log D} S = T_u \mathscr{K}(u)$$

we need show only one inclusion. If $\vartheta \in \Theta_S(-\log D)_u$, then it has a lift $\tilde{\vartheta}$ tangent to $V(F)$. The integral flows of ϑ and $\tilde{\vartheta}$, φ_t on (S, u) and $\tilde{\varphi}_t$ on $V(F)$, satisfy $\pi \circ \tilde{\varphi}_t = \varphi_t \circ \pi$. It follows that $\tilde{\varphi}_t$ maps C_u to $C_{\varphi_t(u)}$, and therefore for each singular point p_i in C_u, the curve germ $\{\varphi_t(u) : t \in (\mathbb{C}, 0)\}$ lies in the set $\mathscr{K}_i(u)$ defined above. Hence $\{\varphi_t(u) : t \in (\mathbb{C}, 0)\} \subset \bigcap_i \mathscr{K}_i(u) = \mathscr{K}(u)$, and $\vartheta(0) \in T_u \mathscr{K}(u)$. □

2.4. Isomorphism $\mathscr{O}_\Sigma \to \Omega_F$

A choice of a nowhere-vanishing relative $(n+1)$-form $\omega \in \Omega^{n+1}_{B \times S/S}$ determines an isomorphism

$$\mathscr{O}_\Sigma \simeq \Omega_F^{n+1}, \quad g \mapsto g\omega$$

where

$$\Omega_F^{n+1} := \Omega^{n+1}_{B \times S/S} / dF \wedge \Omega^n_{B \times S/S}.$$

Such an isomorphism leads to many additional structures. First of all, there is a canonical perfect pairing, the *residue pairing*,

$$\mathrm{Res} : \Omega_F^{n+1} \times \Omega_F^{n+1} \to \mathcal{O}_S,$$

from which one obtains a perfect pairing on $\mathcal{O}_\Sigma$.

$$\langle\, \_ , \_ \,\rangle : \mathcal{O}_\Sigma \times \mathcal{O}_\Sigma \to \mathcal{O}_S\,.$$

Furthermore, because Ω^1_S and $\Omega^1_S(\log D)$ are $\mathcal{O}_S$-dual to Θ_S and $\Theta_S(-\log D)$, such a choice of ω also determines isomorphisms

$$\alpha : \Omega^1_S \to \mathcal{O}_\Sigma \quad \text{and} \quad \beta : \Omega^1_S(\log D) \to \mathcal{O}_\Sigma$$

via the formulas

$$\langle dF(\vartheta), \alpha(\xi)\rangle = \xi(\vartheta), \quad \text{and} \quad \langle \frac{dF}{F}(\vartheta), \beta(\xi)\rangle = \xi(\vartheta).$$

As a result $\Theta_S, \Theta_S(-\log D), \Omega^1_S$ and $\Omega^1_S(\log D)$ are all identified with $\mathcal{O}_\Sigma$ and hence with one another. For example we have the isomorphism

$$k^{-1} \circ \beta : \Omega^1_S(\log D) \to \Theta_S,$$

where $k : \Theta_S \to \mathcal{O}_\Sigma$ is the Kodaira-Spencer map dF.

Note that for any $a, b, c \in \mathcal{O}_\Sigma$, the pairing satisfies

$$\langle a, bc\rangle = \langle ab, c\rangle,$$

and so multiplication by F on $\mathcal{O}_\Sigma$ is self-adjoint:

$$\langle g, Fh\rangle = \langle Fg, h\rangle.$$

As a result, if $\breve{g}_i$, $i = 1, \ldots, \mu$ denotes the $\mathcal{O}_S$ basis of $\mathcal{O}_\Sigma$ dual to the basis $g_i = \partial F/\partial u_i$, $i = 1, \ldots, \mu$, then replacing $FdF(\partial/\partial u_i)$ in (2.2) by $\breve{g}_i$, we get generators $\chi_1, \ldots, \chi_\mu$ for $\Theta_S(-\log D)$ whose matrix of coefficients with respect to the $\partial/\partial u_j$ is the symmetric matrix with i, j entry $\chi_{ij} = \langle \breve{g}_i, F\breve{g}_j\rangle$.

In our calculations in section 7 we always use such a basis. We note that if $\omega_1, \ldots, \omega_\mu$ is the basis for $\Omega^1(\log D)$ dual to $\chi_1, \ldots, \chi_\mu$ then

$$k^{-1}\beta(\omega_i) = \frac{\partial}{\partial u_i}, \quad \text{and} \quad k^{-1}\alpha(du_i) = \chi_i. \tag{2.3}$$

3. The Gauß-Manin connection

The study of the Gauß-Manin connection for hypersurface singularities was initiated by BRIESKORN in [3] and has since then developed into a very detailed theory. We can only outline the parts of the theory that are relevant to our application. For a more detailed accounts we refer to [11], [16], [2], [15], [13] and the original papers quoted there.

3.1. The cohomology bundle and its extensions

The spaces $H^n(X_u) = H^n(X_u; \mathbb{C})$ fit together into the cohomology bundle

$$H^* = \bigcup_{u \in S \setminus D} H^n(X_u)$$

over $S \setminus D$. It is a flat vector bundle and the associated sheaf of holomorphic sections

$$\mathscr{H}^* = H^* \otimes_{\mathbb{C}} \mathcal{O}_{S \setminus D}$$

is equipped with a natural flat connection, the Gauss Manin connection,

$$\nabla : \mathscr{H}^* \to \mathscr{H}^* \otimes_{\mathcal{O}_S} \Omega^1_{S \setminus D}. \tag{3.1}$$

As usual, we write

$$\nabla_\vartheta : \mathscr{H}^* \longrightarrow \mathscr{H}^*$$

for the action of a vector field $\vartheta \in \Theta_{S \setminus D}$. The sheaf $\mathscr{H}^*$ over $S \setminus D$ has various extensions to S. Most relevant to us is the parameterised version of Brieskorn's module H':

$$\mathscr{H}' := \pi_*(\Omega^n_{X/S})/d\pi_*(\Omega^{n-1}_{X/S}). \tag{3.2}$$

A section of $\mathscr{H}'$ over $U \subset S$ is represented by a (relative) holomorphic n-form η on $\pi^{-1}(U) \subset X$. If $U \subset S \setminus D$ and $u \in U$, the restriction of η to the smooth fibre X_u is a closed form n-form and thus determines a cohomology class

$$[\eta|_{X_u}] \in H^n(X_u).$$

In this way one obtains an isomorphism $\mathscr{H}'(U) \to \mathscr{H}^*(U)$ and thus $\mathscr{H}'$ can be considered as an extension of $\mathscr{H}^*$, that is, there is a map of $\mathcal{O}_S$-modules

$$\mathscr{H}' \longrightarrow j_* \mathscr{H}^*,$$

which is an isomorphism over $S \setminus D$, where $j : S \setminus D \hookrightarrow S$ is the inclusion. The sheaf $\mathscr{H}'$ is a locally free sheaf of rank μ, but for a general $\vartheta \in \Theta_S$ the Gauß-Manin connection maps $\mathscr{H}'$ into a bigger extension $\mathscr{H}'' \supset \mathscr{H}'$. This second Brieskorn module $\mathscr{H}''$ can be defined as

$$\mathscr{H}'' := \pi_* \omega_{X/S} / d\pi_*(d\Omega^{n-1}_{X/S}),$$

where $\omega_{X/S}$ denotes the relative dualising module, [16, page 158]. Elements from $\omega_{X/S}$ are most conviently described as residues of $n+1$-forms, that is, as Gelfand-Leray forms. There is an exact sequence

$$0 \longrightarrow \mathscr{H}' \longrightarrow \mathscr{H}'' \longrightarrow \Omega^{n+1}_{X/S} \longrightarrow 0. \tag{3.3}$$

When we restrict to logarithmic vector fields, the connection maps $\mathscr{H}'$ and $\mathscr{H}''$ to themselves, so we have logarithmic connections

$$\nabla : \mathscr{H}' \longrightarrow \mathscr{H}' \otimes_{\mathcal{O}_S} \Omega^1_S(\log D),$$
$$\nabla : \mathscr{H}'' \longrightarrow \mathscr{H}'' \otimes_{\mathcal{O}_S} \Omega^1_S(\log D)$$

extending the Gauss-Manin connection (3.1). (As there is no possibility of confusion, we denote all these maps by the same symbol ∇.)

The action of $\chi \in \Theta_S(-\log D)$ on a local section $[\eta]$ represented by a relative n-form η is given by the Lie derivative with respect to a lift $\widetilde{\chi}$ of χ:

$$\nabla_\chi \eta = [Lie_{\widetilde{\chi}}(\eta)]$$

([16, p. 148]).

3.2. $\mathscr{H}'$ and the cohomology of singular fibres

We have seen that for $u \in S \setminus D$, the restriction of a global relative n-form η to a smooth fibre X_u determines a cohomology class

$$[\eta|_{X_u}] \in H^n(X_u).$$

If $u \in D$, then the fiber X_u is singular, but the form η still can be integrated over n-cycles in X_u and gives rise to a well defined cohomology class in $H^n(X_u)$. We sketch the argument. Suppose γ_1 and γ_2 are n-cycles in C_u and Γ is a $n+1$-chain in X_u with $\partial\Gamma = \gamma_1 - \gamma_2$. After subdivision, we can write $\Gamma = \Gamma' + \Gamma''$ where Γ' is a $n+1$-chain in the smooth part of C_u and $\Gamma'' = \Gamma \cap \bigcup_i B_\varepsilon(p_i)$, where the p_i are the singular points of C_u. Then

$$\int_{\gamma_1} \eta - \int_{\gamma_2} \eta = \int_{\partial\Gamma'} \eta + \int_{\partial\Gamma''} \eta.$$

The first integral on the right-hand side vanishes by Stokes's Theorem. The contribution $\int_{\partial\Gamma''} \eta$ tends to 0 as $\varepsilon \to 0$, as the integrand is regular and one is integrating over ever smaller sets.

In general, if Z is any analytic space with singularities we can look at the de Rham complex $(\Omega^\bullet_Z, d)$ of Kähler forms, and integration over p-cycles is well-defined and determines a *de Rham evaluation map*

$$DR : H^p(\Gamma(Z, \Omega^\bullet_Z)) \to H^p(Z, \mathbb{C}).$$

If Z is a Stein space, then this map is even *surjective*. The reason is the following: because Z is Stein, the group at the left hand side is equal to the p-th hypercohomology group $\mathbb{H}^p$ of the de Rham complex $(\Omega^\bullet_Z, d)$. The map of complexes $\mathbb{C}_Z \to (\Omega^\bullet_Z, d)$ (induced by the inclusion map $\mathbb{C}_Z \to \mathcal{O}_Z$) induces a map

$$\alpha : H^p(Z;\mathbb{C}) = \mathbb{H}^p(\mathbb{C}_Z) \to \mathbb{H}^p((\Omega^\bullet_Z, d)) = H^p(\Gamma(Z, \Omega^\bullet_Z)),$$

and it is shown in [16], p. 141, that DR is a *section* of the map α, i.e., $DR \circ \alpha = \mathrm{Id}$. In particular, DR is surjective.

Proposition (8.5) of [16] provides a relative version of this argument, that we specialise to our situation of $\pi : X \longrightarrow S$. For this we look at the (truncated) relative de Rham complex

$$0 \longrightarrow \mathcal{O}_X \longrightarrow \Omega^1_{X/S} \longrightarrow \ldots \longrightarrow \Omega^{n-1}_{X/S} \longrightarrow \Omega^n_{X/S}.$$

The cohomology sheaves are $\pi^{-1}\mathcal{O}_S$ in degree 0 and

$$\mathcal{H}^n := \Omega^n_{X/S}/d\Omega^{n-1}_{X/S},$$

a sheaf supported on $\widetilde{D}$, in degree n. The direct image $(\pi_*\Omega^\bullet_{X/S}, d)$ also has two non-vanishing cohomologies, namely $\pi_*\pi^{-1}\mathcal{O}_S$ in degree 0 and $\mathscr{H}'$ in degree n. The two hypercohomology spectral sequences now produces a short exact sequence

$$0 \longrightarrow R^n\pi_*(\mathbb{C}_X) \otimes \mathcal{O}_S \xrightarrow{\alpha} \mathscr{H}' \xrightarrow{\beta} \pi_*\mathcal{H}^n \longrightarrow 0$$

([16, Proposition 8.5]). Restriction to a (geometrical) fibre over u gives an exact sequence

$$0 \longrightarrow H^n(X_u) \longrightarrow \mathscr{H}'|_u \longrightarrow \pi_*\mathcal{H}^n_u \longrightarrow 0.$$

In the middle we have a vector space of dimension μ, at the right-hand side a direct sum of vector spaces of dimension μ_i, the Milnor numbers of the singularities appearing in the fibre over u. So indeed

$$\dim H^n(X_u) = \mu - \sum \mu_i.$$

The composition

$$R^n\pi_*(\mathbb{C}_X) \longrightarrow R^n\pi_*(\mathbb{C}_X) \otimes \mathcal{O}_S \xrightarrow{\alpha} \mathscr{H}'$$

is for any $u \in S$ a *section* to the deRham-evaluation map

$$DR_u : \mathscr{H}'_u \longrightarrow H^n(X_u, \mathbb{C}).$$

Corollary 3.1. *For all $u \in S$, the deRham evaluation map*

$$\mathscr{H}'_u \to H^n(X_u); \eta \to [\eta|_{X_u}]$$

is surjective.

3.3. The period map

The theory of the period map was developed independently by VARCHENKO and K. SAITO around the same time and has numerous applications. The basic idea is quite simple. Let us first fix a relative n-form η and a point $u \in S \setminus D$ and a horizontal basis $\gamma_1(s), \gamma_2(s), \ldots, \gamma_\mu(s) \in H_n(X_s)$ for points s in a neighbourhood U of u. The *period map*

$$P_\eta : U \longrightarrow \mathbb{C}^\mu, \quad s \mapsto \left(\int_{\gamma_1(s)} \eta, \int_{\gamma_2(s)} \eta, \ldots, \int_{\gamma_\mu(s)} \eta\right)$$

sends a point s to the tuple of periods of the form η. By further parallel transport one extends P_η to a (multi-valued) map

$$P_\eta : S \setminus D \longrightarrow \mathbb{C}^\mu$$

between spaces of the same dimension μ. The form η is called *non-degenerate* if it is non-degenerate at all points $u \in S \setminus D$, which means that P_η is a local isomorphism near u. Of course, this can be tested by looking at the derivative of this map, which can be identified with the map

$$\nabla P_{\eta,u} : T_uS \to H^1(X_u), \quad \vartheta \mapsto [\nabla_\vartheta \eta | X_u] \in H^n(X_u)$$

which is the geometrical fibre at u of the sheaf map

$$\Theta_{S\setminus D} \longrightarrow \mathscr{H}^*, \quad \vartheta \mapsto \nabla_\vartheta \eta.$$

This map extends to a sheaf map

$$\Theta_S \longrightarrow \mathscr{H}'', \quad \vartheta \mapsto \nabla_\vartheta \eta$$

which is an *isomorphism* in case η is non-degenerate.

Proposition 3.2. *A non-degenerate relative n-form η gives rise to a commutative diagram*

$$\begin{array}{ccccccccc} 0 & \longrightarrow & \mathscr{H}' & \longrightarrow & \mathscr{H}'' & \longrightarrow & \Omega^{n+1}_{X/S} & \longrightarrow & 0 \\ & & \uparrow & & \uparrow & & \uparrow & & \\ 0 & \longrightarrow & \Theta_S(-\log D) & \longrightarrow & \Theta_S & \longrightarrow & \mathscr{O}_{\widetilde{D}} & \longrightarrow & 0 \end{array}$$

with exact rows and where the vertical maps are the isomorphisms described in the last paragraph and where the map at the right-hand side is induced by multiplication by $\omega = d\eta$.

This diagram can be found in [19, p. 1248].

From this we get immediately the following

Theorem 3.3. *If η is non-degenerate, then for each point $u \in S$ one obtains an isomorphism*

$$\nabla P_{\eta,u} : T_u^{\log D} S \longrightarrow \mathscr{H}'_u.$$

The composition with the de Rham evaluation map gives a surjection

$$DR \circ \nabla P_{\eta,u} : T_u^{\log D} S \longrightarrow H^n(X_u).$$

In fact the restriction of $\mathrm{DR} \circ \nabla P_{\eta,u}$ to $T_u\mathscr{R}(u)$ is an isomorphism. This statement was shown by Varchenko to hold in special cases and conjectured to hold in general, [25]. A proof basically along these lines was sketched to us in a letter by Hertling, [14].

4. The case of curves

We specialise to the case $n = 1$, so $C := X_0$ is a plane curve singularity. If C has r branches then by the formula of Milnor

$$\mu = 2\delta - r + 1,$$

and for $u \in S \setminus D$ the fibre $C_u := X_u$ is a smooth Riemann surface of genus $\delta - r + 1$ with r boundary circles. In the case where C is irreducible, then $\mu = 2\delta$ and for $u \notin D$, C_u is a smooth Riemann surface of genus δ. For $u \in D$ the curve C_u is a singular, say with singularities (C_u, p_i), $i = 1, 2, \ldots, N$, then its normalisation $\widetilde{C}_u$ has genus

$$\delta(C) - \delta(C_u)$$

where $\delta(C_u) = \sum_{i=1}^N \delta(C_u, p_i)$.

4.1. Intersection form

Now assume that C is irreducible. For fixed $u \in S$ let $C_u^* = C_u/\partial C_u$ be the closed Riemann surface obtained by shrinking ∂C_u to a point, and let $\widetilde{C}_u$ and $\widetilde{C}_u^*$ be the normalisations of C_u and C_u^*.

The diagram

$$\begin{array}{ccc} \widetilde{C}_u & \hookrightarrow & \widetilde{C}_u^* \\ \downarrow{\scriptstyle n} & & \downarrow{\scriptstyle n} \\ C_u & \overset{j}{\hookrightarrow} & C_u^* \end{array} \qquad \text{gives rise to the diagram} \qquad \begin{array}{ccc} H^1(\widetilde{C}_u) & \overset{\simeq}{\longleftarrow} & H^1(\widetilde{C}_u^*) \\ \uparrow{\scriptstyle n^*} & & \uparrow{\scriptstyle n^*} \\ H^1(C_u) & \overset{\simeq}{\longleftarrow} & H^1(C_u^*) \end{array}$$

in which the vertical arrows are surjections. Write I_u and $\widetilde{I}_u$ for the intersection forms on C_u and $\widetilde{C}_u$. These are pulled back from the intersection forms on the closed curves C_u^* and $\widetilde{C}_u^*$ by means of the isomorphisms in the preceding diagram. Because $n_* : H_2(\widetilde{C}_u, \partial\widetilde{C}_u) \simeq H_2(C, \partial C)$, it follows by functoriality that

$$\widetilde{I}_u(n^*a, n^*b) = I_u(a, b). \tag{4.1}$$

Note that the form $\widetilde{I}_u$ is non-degenerate.

4.2. de Rham version of I_u

The pairing I_u has the following DE RHAM description. We choose a a pair of collars $U \subset V \subset C_u$ around the boundary ∂C_u and a C^∞ bump-function ρ, equal to 1 on U and 0 outside V. If η is a holomorphic (Kähler) 1-form on C_u, it follows from Stokes theorem that

$$\int_{\partial C} \eta = 0.$$

By integration we can therefore find a holomorphic function α on V with $d\alpha = \eta$ on V. The form η is cohomologous to $\widetilde{\eta} := \eta - d\rho\alpha$ and as $\rho = 1$ on U and there $d\alpha = \eta$, it follows that $\widetilde{\eta}$ is a form with compact support, contained in $C \setminus U$. It is holomorphic and equal to η outside V, but only C^∞ on the annulus $V \setminus U$. One then has, using Stokes theorem

$$I_u([\eta], [\eta']) = I_u([\widetilde{\eta}], [\eta']) = -\int_{\partial C} \alpha\eta'.$$

More details are given in Section 7.

4.3. Extension to $\mathscr{H}^*$ and $\mathscr{H}'$

The pairings I_u on $H^1(C_u)$ combine to give a perfect duality

$$I : \mathscr{H}^* \times \mathscr{H}^* \to \mathcal{O}_S$$

over $S \setminus D$. Because of its topological origin, the intersection form is *horizontal* with respect to the Gauss-Manin connection: for any two sections s_1, s_2 of $\mathscr{H}^*$,

$$d\big(I(s_1, s_2)\big) = I(\nabla s_1, s_2) + I(s_1, \nabla s_2).$$

Using a relative version of the above DE RHAM-description of the intersection pairing one obtains an extension of I, still called I, to $\mathscr{H}'$:

$$I : \mathscr{H}' \times \mathscr{H}' \to \mathcal{O}_S .$$

For two sections η_1, η_2 of $\mathscr{H}'$ one has

$$I(\eta_1, \eta_2)(u) = I_u([\eta_1|C_u], [\eta_2|C_u]). \tag{4.2}$$

4.4. Pulling back the intersection form

Using the period map one can pull-back the intersection form on $H^1(C_u)$ to obtain a 2-form on S. Let us first start with an arbitrary section $s \in \mathscr{H}^*$ over $S \setminus D$. From it we obtain a 2-form

$$\Omega = s^* I \in \Omega^2_{S\setminus D}$$

on $S \setminus D$ by the formula

$$\Omega(\vartheta_1, \vartheta_2) := I(\nabla_{\vartheta_1} s, \nabla_{\vartheta_2} s).$$

Proposition 4.1. *The form Ω is closed.*

Proof. This is 'clear' as we are pulling back the 'constant form I', but here is a nice direct calculation: if a, b and c are germs of commuting vector fields on S, then

$$\begin{aligned} d(s^*I)(a,b,c) &= d\big(I(a,b)\big)(c) - d\big(I(a,c)\big)(b) + d(\big(I(b,c)\big)(a) \\ &= I(\nabla_c\nabla_a s, \nabla_b s) + I(\nabla_a s, \nabla_c\nabla_b s) \\ &\quad - I(\nabla_b\nabla_a s, \nabla_c s) - I(\nabla_a s, \nabla_b\nabla_c s) \\ &\quad + I(\nabla_a\nabla_b s, \nabla_c s) + I(\nabla_b s, \nabla_a\nabla_c s). \end{aligned} \tag{4.3}$$

Because a and b commute and ∇ is flat, $\nabla_a\nabla_b = \nabla_b\nabla_a$, and similarly for $\nabla_a\nabla_c$ and $\nabla_b\nabla_c$. This means that all terms on the right hand side in (4.3) cancel, except the first and last. These cancel because of the anti-symmetry of I. □

Theorem 4.2. ([9]) *If $s = \eta$ is a non-degenerate section of $\mathscr{H}'$, then Ω is itself non-degenerate and hence symplectic, and moreover extends to all of S as a symplectic form.*

5. Results

In [9] one find the formulation of a *principle* that the types of degeneration that occur in the fibres C_u are reflected in the lagrangian properties of the corresponding strata. Our results can be seen as a vindication of this principle in some special cases.

As before, we will consider the versal deformation $\pi : X \longrightarrow S$ of an irreducible curve singularity, a non-degenerate section η of the Brieskorn-module $\mathscr{H}'$ and the resulting symplectic form Ω on S, obtained by pulling back the intersection form on the fibres $H^1(C_u)$.

5.1. The rank of Ω on the logarithmic tangent space

Recall that for a point $u \in S$, the logarithmic tangent space $T_u^{logD} S \subset T_u S$ is the sub-space spanned by the logarithmic vector fields at u.

Theorem 5.1. *The rank of Ω restricted to $T_u^{\log D} S$ is equal to the rank of I_u on $H^1(C_u)$, which is equal to dim $H^1(\widetilde{C}_u) = 2(\delta(C) - \delta(C_u))$.*

Proof. Let $\mathscr{R}(u)$ and $\mathscr{K}(u)$ denote, respectively, the right-equivalence stratum and the $\mathscr{K}$-equivalence stratum containing u. Recall that by 3.3 the period map maps the space $T_u\mathscr{K}(u)$ surjectively to $H^1(C_u)$; its restriction to $T_u\mathscr{R}(u) \subseteq T_u\mathscr{K}(u)$ maps isomorphically to $H^1(C_u)$. From (4.2) it follows that the rank of Ω on $T_u^{\log} D$ at u is equal to the rank of the intersection form I_u on $H^1(C_u)$, which is equal to the rank of $H^1(\widetilde{C}_u)$, and therefore is equal to $\mu(C) - 2\delta(C_u) = 2\delta(C) - 2\delta(C_u)$. □

5.2. Coisotropicity of the Severi strata

Recall that a subspace V of a symplectic vector space $(W, \langle \_, \_ \rangle)$ is *coisotropic* if $V^\perp \subset V$, where $V^\perp = \{w \in W : \langle v, w \rangle = 0 \text{ for all } v \in V\}$. A submanifold X of a symplectic manifold M is coisotropic if for all $x \in X$, $T_x X$ is a coisotropic subspace of $T_x M$. A singular subset X of the symplectic manifold M is coisotropic if X_{reg} is coisotropic.

Theorem 5.2. *All the Severi strata*

$$D(\delta) \subset D(\delta - 1) \subset \cdots \subset D(1) = D$$

are coisotropic with respect to Ω.

Proof. Suppose that u is a regular point of $D(k)$, so C_u has exactly k ordinary double points as singularities. As $\mathscr{R}(u) = \mathscr{K}(u) = D(k)$ near u, the tangent space $T_u D(k)$ is the same as $T_u^{logD} S$. From theorem 5.1 the rank of $\Omega_{|T_u D(k)}$ is equal to $\mu - 2k$, hence dim $\ker \Omega_{|T_u D(k)} = k$. But from the non-degeneracy of Ω it follows that $T_u D(k)^\perp$ has dimension equal to the codimension of $D(k)$, namely k. Thus both sides in the relation

$$T_u D(k)^\perp \supset \ker(\Omega_u|_{T_u D(k)})$$

have dimension k, and are therefore equal. It follows that $T_u D(k)^\perp \subset T_u D(k)$. That is, $D(k)$ is coisotropic. □

The principle mentioned above explains this result by simply saying the near a regular point $u \in D(k)$ there are k mutually non-intersecting cycles vanishing at u, which make up an isotropic subspace of H_1. However, making this into an honest proof is another matter and leads to the considerations outlined above. The form Ω is not unique, and moreover is determined globally rather than locally. One cannot prove anything by using a local normal form $N(\ell) := \{u_1 \cdots u_\ell = 0\}$ for D at a generic point u_0 of a Severi stratum $D(\ell)$, since the symplectic form one picks there will not in general coincide with the pullback of the form Ω by an isomorphism identifying (D, u_0) with $(N(\ell), 0)$.

5.3. Equations for the $D(k)$

Let $\chi_1, \ldots, \chi_\mu$ be a basis for for $\Theta_S(-\log D)$, and let $\omega_1, \ldots, \omega_\mu$ be the dual basis for $\Omega^1_S(\log D)$. Considering Ω as an element of $\Omega^2_S(\log D)$, it can be expressed as the sum

$$\Omega = \sum_{i<j} \Omega(\chi_i, \chi_j)\omega_i \wedge \omega_j.$$

We denote the skew matrix with i, j'th entry $\Omega(\chi_i, \chi_j)$ by $\chi^t\Omega\chi$. Then

$$\wedge^k\Omega = \sum_{1 \leq i_1 < \cdots < i_{2k} \leq \mu} \mathrm{Pf}(\chi^t\Omega\chi(i_1, \ldots, i_{2k}))\omega_{i_1} \wedge \cdots \wedge \omega_{i_{2k}}, \tag{5.1}$$

where $\chi^t\Omega\chi(i_1, \ldots, i_{2k})$ is the submatrix of $\chi^t\Omega\chi$ consisting of rows and columns $i_1, \ldots, i_{2k}$ and Pf denotes its Pfaffian. The ideal generated by the coefficients of $\wedge^k\Omega$ with respect to the basis $\omega_{i_1} \wedge \cdots \wedge \omega_{i_{2k}}$ of $\Omega^{2k}(\log D)$ is the same as the ideal $\mathrm{Pf}_{2k}(\chi\Omega\chi)$ of $2k \times 2k$ Pfaffians of $\chi^t\Omega\chi$.

Theorem 5.3. $D(k) = V\left(Pf_{2(\delta-k+1)}(\chi^t\Omega\chi)\right)$. *In particular, the δ-constant stratum $D(\delta)$ is defined by the entries of $\chi^t\Omega\chi$.*

Proof. Consider an arbitrary $u \in S$. The rank of the matrix $\chi^t\Omega\chi$ at u is the rank of Ω restricted to the space of evaluations at u of the vector fields in $\Theta_S(-\log D)_u$, which is precisely the logarithmic tangent space $T_u^{\log D}S$. Theorem 5.1 states that the rank of Ω on $T_u^{\log}D$ at u is equal to $2\delta(C) - 2\delta(C_u)$. As the rank of a skew-symmetric matrix is always even and equal to the size of the largest non-vanishing Pfaffian, it follows that $D(k)$ is precisely cut out by the Pfaffians of size $2(\delta - k + 1)$ of the matrix $\chi^t\Omega\chi$, i.e., $D(k) = V\left(Pf_{2(\delta-k+1)}(\chi^t\Omega\chi)\right)$. □

A symplectic form Ω on a manifold S gives rise to a Poisson bracket $\{\_,\_\}$ on the sheaf of functions on S, as follows: Ω determines an isomorphism $\Omega^1_S \to \Theta_S$ sending a 1-form α to a vector field $\alpha^\flat$. Then for functions f, g,

$$\{f, g\} = \Omega((df)^\flat, (dg)^\flat).$$

The vector field $\chi_f := (df)^\flat$ is called the *Hamiltonian vector field* associated to f. If $V \subset S$ is a sub-variety and $I(V) \subset \mathcal{O}_S$ the ideal of functions vanishing on V, then it is easy to show that for a regular point $x \in V$ one has

$$T_xV^\perp = \{\chi_f(x) : f \in I(V)_x\}. \tag{5.2}$$

The following is well known:

Proposition 5.4. $V \subset S$ *is coisotropic if and only if the ideal* $I(V)$ *is Poisson-closed:*

$$\{I(V), I(V)\} \subset I(V).$$

For the convenience of the reader we include a proof.

Proof. Let $x \in V$ be a regular point, $v, w \in T_xV^\perp$, and $f, g \in I(V)$ two functions with $\chi_f(x) = v$, $\chi_g(x) = w$ (using (5.2)). Then

$$\Omega(v,w) = \Omega(\chi_f(x), \chi_g(x)) = \{f,g\}(x).$$

From this we see that $\{f,g\}$ vanishes at x if and only if $\Omega(v,w) = 0$, which means that $T_xV^\perp \subset (T_xV^\perp)^\perp = T_xV$, that is, V is coisotropic. □

Thus, for each of the Severi strata $D(k)$, the ideal $I(D(k))$ is involutive. But note that an ideal defining a coisotropic subvariety is not necessarily involutive; the proof only shows that this holds if the ideal is radical.

Conjecture 5.5. *For all* $k = 1, 2, \ldots, \delta$,

(a) $Pf_{2k}(\chi^t\Omega\chi)$ *is involutive;*
(b) $Pf_{2k}(\chi^t\Omega\chi)$ *is radical.*

By the theorem 1.2, (b) $\Longrightarrow$ (a), as vanishing ideals of coisotropic varieties are involutive. Nevertheless, involutivity of the ideals $\mathrm{Pf}_{2k}(\chi^t\Omega\chi)$ may hold even without their being radical.

Problem: How to write the Poisson bracket of two Pfaffians of $\chi^t\Omega\chi$ as a linear combination of Pfaffians? Is there a universal formula?

6. The symplectic form as Extension

The matrix $\chi^t\Omega\chi$ can be considered as an endormorphism of $\mathcal{O}_S^\mu$ and its cokernel N_Ω defines a rank 2 Cohen-Macaulay module on $\mathcal{O}_D$. If the basis χ of $\Theta_S(-\log D)$ is chosen to be *symmetric*, as described in Subsection 2.4, then N_Ω sits in an exact sequence

$$0 \longleftarrow \mathcal{O}_{\tilde{D}} \longleftarrow N_\Omega \longleftarrow \mathcal{O}_{\tilde{D}} \longleftarrow 0\,. \tag{6.1}$$

In fact, we show that the extension (6.1) has a coordinate-independent meaning, depending only on the choice of ω used in the definition of the period map. As such it represents an element in the $\Omega_{\tilde{D}}$-module

$$\operatorname*{Ext}^1_D(\mathcal{O}_{\tilde{D}}, \mathcal{O}_{\tilde{D}})$$

and therefore an infinitesimal deformation of $\mathcal{O}_{\tilde{D}}$ as $\mathcal{O}_D$-module. We refer to N_Ω as the *intersection module.* For a vector field ϑ, let $\vartheta^\#$ denote the contraction of Ω by ϑ. Begin with the exact sequence

$$0 \leftarrow \frac{\Omega^1_S(\log D)}{\Omega^1_S} \leftarrow \frac{\Omega^1_S(\log D)}{\Theta_S(-\log D)^\#} \leftarrow \frac{\Theta_S}{\Theta_S(-\log D)} \leftarrow 0. \tag{6.2}$$

This exists for every divisor D and non-degenerate 2-form Ω. Here the first arrow is induced by contraction with Ω, which maps Θ_S to Ω^1_S and $\Theta_S(-\log D)$ to

$\Theta_S(-\log D)^{\#}$. Then the exact sequence we consider is obtained from (6.2) by composing the last arrow with the isomorphism $k^{-1} \circ \beta : \Omega^1_S(\log D) \to \Theta_S$ described in Subsection 2.4, inducing an isomorphism

$$\frac{\Theta_S}{\Theta_S(-\log D)} \leftarrow \frac{\Omega^1_S(\log D)}{\Omega^1_S}.$$

Since we have a canonical isomorphism $\Theta_S/\Theta_S(-\log D) \to \mathcal{O}_{\widetilde{D}}$ defined by dF, we obtain the exact sequence (6.1). Thus provided the pairing on $\mathcal{O}_\Sigma$ is chosen canonically, the extension class of (6.1) depends only on F and on the symplectic form.

Remark 6.1. *If we apply $k^{-1} \circ \beta$ also to the middle term of the sequence* (6.2) *as well as the third, we obtain* (6.1) *in the slightly different form*

$$0 \leftarrow \frac{\Theta_S}{\Theta_S(-\log D)} \leftarrow \frac{\Theta_S}{k^{-1}\circ\beta(\Theta_S(-\log D)^{\#})} \leftarrow \frac{\Theta_S}{\Theta_S(-\log D)} \leftarrow 0. \tag{6.3}$$

Note that $k^{-1} \circ \beta(\Theta_S(-\log D)^{\#})$ is generated over $\mathcal{O}_S$ by vector fields whose components with respect to the usual basis $\partial/\partial u_1, \ldots, \partial/\partial u_\mu$ are given by the columns of the matrix $\chi\Omega\chi$. It is interesting that in all of the examples where we have made the calculations, $k^{-1} \circ \beta(\Theta_S(-\log D)^{\#}) \subset \Theta_S$ is a Lie sub-algebra, evidently contained in $\Theta_S(-\log D)$. We cannot at present prove this or explain it.

6.1. Calculation of Ext groups

We state without proof the results of some relatively straightforward calculation of Ext groups. Let $\mathscr{C}$ denote the conductor ideal of the projection $n = \pi| : \widetilde{D} \to D$.

Lemma 6.2. (i) *Both $\mathrm{Ext}^1_D(\mathcal{O}_{\widetilde{D}}, \mathcal{O}_{\widetilde{D}})$ and $\mathrm{Ext}^2_D(\mathcal{O}_{\widetilde{D}}, \mathcal{O}_{\widetilde{D}})$ are $\mathcal{O}_{\widetilde{D}}/\mathscr{C}$-modules.*

(ii) $\operatorname{Ext}^1_{\mathcal{O}_D}(\mathcal{O}_{\widetilde{D}}, \mathcal{O}_{\widetilde{D}}) \simeq \dfrac{\{\mathcal{O}_{\widetilde{D}}\text{-syzygies of } g_1, \ldots, g_\mu\}}{\mathcal{O}_{\widetilde{D}} \cdot \{\mathcal{O}_D\text{-syzygies of } g_1, \ldots, g_\mu\}}$;

(iii) $\operatorname{Ext}^2_{\mathcal{O}_D}(\mathcal{O}_{\widetilde{D}}, \mathcal{O}_{\widetilde{D}}) \simeq \mathcal{O}_{\widetilde{D}}/\mathscr{C}$.

Proposition 6.3. *$\mathrm{Ext}^1_D(\mathcal{O}_{\widetilde{D}}, \mathcal{O}_{\widetilde{D}})$ is a maximal Cohen-Macaulay module over $\mathcal{O}_{\widetilde{D}}/\mathscr{C}$ presented by the matrix $\widetilde{\chi}$ obtained from the symmetric matrix χ of the basis for $\Theta_S(-\log D)$ by deleting its last row and column.*

In [1] it is shown that if $n : \widetilde{D} \to D$ has corank 1, then $\mathrm{Coker}\widetilde{\chi} \simeq \pi_* \mathcal{O}_{D^2(n)}$, where, by $D^2(n)$, we mean the double-point scheme of the map n:

$$D^2(n) = \mathrm{closure}\{(x_1, x_2) \in \widetilde{D} \times \widetilde{D} : x_1 \neq x_2, n(x_1) = n(x_2)\}.$$

The isomorphism holds only if n has corank 1. The map $n : \widetilde{D} \to D$, normalising the discriminant in the base of a versal deformation, has corank 1 exactly for the A_μ singularities. Thus, for the A_μ, and only for these, $\mathrm{Ext}^1_D(\mathcal{O}_{\widetilde{D}}, \mathcal{O}_{\widetilde{D}}) \simeq \mathcal{O}_{D^2(n)}$.

7. Computations and Examples

It was described in [8] how the symplectic form Ω can be computed in the case of irreducible quasi-homogeneous curve singularities. The projective closure of such a curve has a unique point at infinity ∞.

Proposition 7.1. *Let C be a curve, $\infty \in C$ a smooth point and ω, η two meromorphic differential form, holomorphic on $C \setminus \{\infty\}$. Then the intersection form of the cohomology classes $[\omega], [\eta] \in H^1(C)$ is*

$$I([\omega], [\eta]) = 2\pi i \operatorname{Res}_{\infty}(\alpha\eta)$$

where α is a meromorphic function in a neighbourhood of ∞ with $d\alpha = \omega$.

Proof. Choose two small open discs $U \subset V \subset C$ around ∞, and a C^∞ bump function ρ on C, equal to 1 on U and 0 outside V. Choose a function α meromorphic on V with $d\alpha = \omega$. Then $\omega - d(\rho\alpha)$ is a C^∞ compactly supported form, cohomologous to $[\omega]$. Using $\omega \wedge \eta = 0$, we find

$$I([\omega], [\eta]) = -\int_C d(\rho\alpha) \wedge \eta = -\int_U d(\rho\alpha \cdot \eta)$$

and by Stokes theorem

$$-\int_U d(\rho\alpha \cdot \eta) = -\int_{\partial U} \alpha\eta$$

which, noticing the reverse of orientation in the boundary, gives the above formula. $\square$

This proposition can be used to calculate intersections using Laurent-series exapansions. If the curve C is given by an affine equation $f(x, y) = 0$ and has a single point at infinity, we can find a Laurent parametrisation of C around ∞

$$x(t), y(t) \in \mathbb{C}[[t]][1/t]$$

If $\omega = A(x, y)dx$ and $\eta = B(x, y)dx$ are the differential forms on C, then by substitution we obtain expansions

$$\omega = a(t)dt, \eta = b(t)dt$$

where $a(t), b(t) \in \mathbb{C}[[t]][1/t]$ are Laurent series. Integrating up one we find

$$\alpha(t) = \int a(t)dt \in \mathbb{C}[[t]][1/t]$$

and we can compute the cohomological intersection as:

$$I([\omega], [\eta]) = Res_0 \alpha(t)b(t)dt.$$

Proposition 7.2. ([9]) *Suppose that f is quasihomogeneous. Then for $\omega = dx \wedge dy$, the period map P_ω is non-degenerate.*

Case A_4

We consider the versal deformation of A_4 given by

$$F(x,a,b,c,d) = y^2 + x^5 + ax^3 + bx^2 + cx + d.$$

We take the symmetric basis for $\Theta_S(-\log D)$ with Saito matrix

$$\chi := \begin{pmatrix} 10a & 15b & 20c & 25d \\ 15b & -6a^2+20c & -4ab+25d & -2ac \\ 20c & -4ab+25d & -6b^2+10ac & -3bc+15ad \\ 25d & -2ac & -3bc+15ad & -4c^2+10bd \end{pmatrix}. \tag{7.1}$$

The symplectic form pulled back by the period mapping induced by the 1-form ydx is

$$\Omega = ada \wedge db + da \wedge dd + 3db \wedge dc. \tag{7.2}$$

Therefore the ideal of entries of the matrix $\chi\Omega\chi$, defining the δ-constant stratum $D(2)$, is generated by

$$a^4 + \frac{27}{4}ab^2 - 9a^2c + 20c^2 - \frac{25}{2}ad, \quad a^3b + \frac{27}{4}b^3 - 9abc - 10a^2d + 50cd, \tag{7.3}$$

and

$$a^3c + \frac{27}{4}b^2c - 4ac^2 - 20abd + \frac{125}{4}d^2.$$

Case A_6

A versal deformation of A_6 is given by

$$F(x,a,b,c,d,e,f) = x^7 + ax^5 + bx^4 + cx^3 + dx^2 + ex.$$

We take the basis of $\Theta_S(-\log D)$ with Saito matrix

$$\begin{pmatrix} 2a & 3b & 4c & 5d & 6e & 7f \\ 3b & -\frac{10}{7}a^2+4c & -\frac{8}{7}ab+5d & -\frac{6}{7}ac+6e & -\frac{4}{7}ad+7f & -\frac{2}{7}ae \\ 4c & -\frac{8}{7}ab+5d & -\frac{12}{7}b^2+2ac+6e & -\frac{9}{7}bc+3ad+7f & -\frac{6}{7}bd+4ae & -\frac{3}{7}be+5af \\ 5d & -\frac{6}{7}ac+6e & -\frac{9}{7}bc+3ad+7f & -\frac{12}{7}c^2+2bd+4ae & -\frac{8}{7}cd+3be+5af & -\frac{4}{7}ce+4bf \\ 6e & -\frac{4}{7}ad+7f & -\frac{6}{7}bd+4ae & -\frac{8}{7}cd+3be+5af & -\frac{10}{7}d^2+2ce+4bf & -\frac{5}{7}de+3cf \\ 7f & -\frac{2}{7}ae & -\frac{3}{7}be+5af & -\frac{4}{7}ce+4bf & -\frac{5}{7}de+3cf & -\frac{6}{7}e^2+2df \end{pmatrix}$$

and symplectic form

$$\Omega = \begin{pmatrix} 0 & -3a^2-c & -6b & 9a & 0 & -3 \\ 3a^2+c & 0 & -5a & 0 & -5 & 0 \\ 6b & 5a & 0 & -15 & 0 & 0 \\ -9a & 0 & 15 & 0 & 0 & 0 \\ 0 & 5 & 0 & 0 & 0 & 0 \\ 3 & 0 & 0 & 0 & 0 & 0 \end{pmatrix}.$$

Each of the ideals $\mathrm{Pf}_{2\ell}$ is Poisson-closed, and defines a Cohen-Macaulay variety of codimension $3-\ell+1$.

Case A_8

For A_8, each of the ideals $\mathrm{Pf}_{2\ell}$ is Poisson-closed, and defines a Cohen-Macaulay variety of codimension $4-\ell+1$.

Case E_6 and E_8

A versal deformation of E_6 is given by

$$F(x,y,a,b,c,d,e,f) = x^3 + y^4 + axy^2 + bxy + cy^2 + dx + ey + f.$$

We take the basis of $\Theta_S(-\log D)$ with symmetric Saito matrix χ equal to

$$\left(\begin{array}{llll}
2a & 5b & 6c & 8d \\
5b & \frac{-a^4}{6} - 4ac + 8d & \frac{a^2b}{2} + 9e & -\frac{a^3b}{12} - \frac{3bc+ae}{2} \\
6c & \frac{a^2b}{2} + 9e & -\frac{5b^2+2a^2c+10ad}{3} & +\frac{7ab^2}{12} - \frac{4a^2d}{3} + 12f \\
8d & -\frac{a^3b}{12} - \frac{3bc+ae}{2} & \frac{7ab^2}{12} - \frac{4a^2d}{3} + 12f & -\frac{a^2b^2}{24} + 4cd - \frac{7be}{2} + 6af \\
9e & \frac{ab^2-a^3c}{6} + \frac{a^2d-9c^2}{3} + 12f & \frac{7abc}{6} - \frac{13bd+4a^2e}{3} & \frac{5b^3-a^2bc}{12} - \frac{7abd}{6} - \frac{3ce}{2} \\
12f & \frac{abd}{6} - \frac{a^3e}{12} - \frac{3ce}{2} & -\frac{8d^2}{3} + \frac{7abe}{12} - 2a^2f & \frac{10b^2d-a^2be}{24} - \frac{4ad^2}{3} - \frac{9e^2}{4} + 6cf
\end{array}\right.$$

$$\left.\begin{array}{rr}
9e & 12f \\
\frac{ab^2-a^3c}{6} + \frac{a^2d-9c^2}{3} + 12f & \frac{abd}{6} - \frac{a^3e}{12} - \frac{3ce}{2} \\
\frac{7abc}{6} - \frac{13bd}{3} - \frac{4a^2}{3}e & -\frac{8d^2}{3} + \frac{7abe}{12} - 2a^2f \\
\frac{5b^3-a^2bc}{12} - \frac{7abd}{6} - \frac{3ce}{2} & \frac{10b^2d-a^2be}{24} - \frac{4ad^2}{3} - \frac{9e^2}{4} + 6cf \\
\frac{4b^2c}{3} - \frac{a^2c^2}{6} + \frac{8acd-8d^2-5abe-6a^2f}{3} & \frac{bcd}{2} + \frac{5b^2e-a^2ce}{12} + \frac{5ade}{6} - 3abf \\
\frac{bcd}{2} + \frac{5b^2e-a^2ce}{12} + \frac{5ade}{6} - 3abf & -\frac{4cd^2}{3} + \frac{11bde}{6} - \frac{a^2e^2}{24} - b^2f - 2adf
\end{array}\right). \tag{7.4}$$

The symplectic form Ω has matrix

$$\begin{pmatrix}
0 & -\frac{1}{15}ab & \frac{1}{5}c & \frac{2}{15}a^2 & 0 & \frac{1}{5} \\
\frac{1}{15}ab & 0 & 0 & 0 & \frac{1}{2} & 0 \\
-\frac{1}{5}c & 0 & 0 & 1 & 0 & 0 \\
-\frac{2}{15}a^2 & 0 & -1 & 0 & 0 & 0 \\
0 & -\frac{1}{2} & 0 & 0 & 0 & 0 \\
-\frac{1}{5} & 0 & 0 & 0 & 0 & 0
\end{pmatrix}. \tag{7.5}$$

The ideal of 2×2 Pfaffians (i.e., the ideal of entries) of $\chi\Omega\chi$, defining the δ-constant stratum, is Cohen-Macaulay of codimension 3, and Poisson-closed. Below we comment on the computations involved in proving Cohen-Macaulayness. The ideal J of 4×4 Pfaffians is also Poisson closed, and has codimension 2 but projective dimension 3.

For both E_6 and E_8 we check the Cohen Macaulay property for the ideal generated by the entries in the matrix $\chi\Omega\chi$ using the *Depth* package of *Macaulay 2*. To show that this ideal is radical, we use the result of [7], that the geometric degree of $D(\delta)$ is equal to the Euler characteristic of the compactified Jacobian. This Euler characteristic is calculated in [18]: for E_6 it is 5 and for E_8 7. Using *Singular* we computed the algebraic degree of $\mathcal{O}_{D(\delta)}$, as defined by the ideal of entries of $\chi\Omega\chi$, and found that it took these values, showing, in view of Cohen-Macaulayness, that this is the reduced structure.

Betti numbers of the Severi strata for A_{2k}

The following table shows the non-zero betti numbers of minimal free resolutions of the ideals of Pfaffians, $\mathrm{Pf}_{2\ell}$, of the matrix $\chi\Omega\chi$ for singularities of type A_{2k} for $1 \leq k \leq 4$.

	A_2	A_4		A_6			A_8			
ℓ	β_0	β_0	β_1	β_0	β_1	β_2	β_0	β_1	β_2	β_3
1	1	3	2	6	8	3	10	20	15	4
2	–	1	–	5	4	–	15	24	10	–
3	–	–	–	1	–	–	7	6	–	–
4	–	–	–	–	–	–	1	–	–	–

(7.6)

Since depth + projective dimension = dimension S and codim $D(j) = j$, it follows from the data in the table that for A_{2k} with $k \leq 4$, each of the rings $\mathcal{O}_S\,/\mathrm{Pf}_{2\ell}$, and therefore each of the Severi strata $D(k-\ell+1) = V(\mathrm{Pf}_{2\ell}) \subset S$, is Cohen-Macaulay.

Conjecture 7.3. *For all ℓ and k with $\ell \leq k$, each of the Severi strata $D(\ell)$ in the base of a miniversal deformation of A_{2k} is Cohen Macaulay.*

References

[1] Ayşe Altıntaş and David Mond. Free resolutions for multiple point spaces. *Geom. Dedicata*, 162:177–190, 2013.

[2] V. I. Arnol′d, S. M. Guseĭn-Zade, and A. N. Varchenko. *Singularities of differentiable maps. Vol. II*, volume 83 of *Monographs in Mathematics*. Birkhäuser Boston, Inc., Boston, MA, 1988. Monodromy and asymptotics of integrals, Translated from the Russian by Hugh Porteous, Translation revised by the authors and James Montaldi.

[3] Egbert Brieskorn. Die Monodromie der isolierten Singularitäten von Hyperflächen. *Manuscripta Math.*, 2:103–161, 1970.

[4] Paul Cadman. Deformations of plane curve singularities and the δ-constant stratum. PhD thesis, University of Warwick, 2011.

[5] Arthur Cayley. On the higher singularities of a plane curve. *Quarterly Journal*, VII:212–222, 1866.

[6] Steven Diaz and Joe Harris. Ideals associated to deformations of singular plane curves. *Trans. Amer. Math. Soc.*, 309(2):433–468, 1988.

[7] B. Fantechi, L. Göttsche, and D. van Straten. Euler number of the compactified Jacobian and multiplicity of rational curves. *J. Algebraic Geom.*, 8(1):115–133, 1999.

[8] A. B. Givental′. Singular Lagrangian manifolds and their Lagrangian mappings. In *Current problems in mathematics. Newest results, Vol. 33 (Russian)*, Itogi Nauki i Tekhniki, pages 55–112, 236. Akad. Nauk SSSR Vsesoyuz. Inst. Nauchn. i Tekhn. Inform., Moscow, 1988. Translated in J. Soviet Math. **52** (1990), no. 4, 3246–3278.

[9] A. B. Givental′ and A. N. Varchenko. The period mapping and the intersection form. *Funktsional. Anal. i Prilozhen.*, 16(2):7–20, 96, 1982.

[10] G.-M. Greuel, C. Lossen and E. Shustin, *Introduction to Singularities and Deformations*, Springer Monographs in Mathematics, Springer-Verlag, 2006.

[11] G.-M. Greuel, Der Gauss-Manin-Zusammenhang isolierter Singularitäten von vollständingen Durchschnitten, *Math. Ann.* 214:235–266, 1975.

[12] Joe Harris. On the severi problem. *Invent. math.*, 84:445–461, 1986.

[13] Claus Hertling. *Frobenius manifolds and moduli spaces for singularities*, volume 151 of *Cambridge Tracts in Mathematics*. Cambridge University Press, Cambridge, 2002.

[14] Claus Hertling. Letter to the authors. 2014.

[15] Valentine S. Kulikov. *Mixed Hodge structures and singularities*, volume 132 of *Cambridge Tracts in Mathematics*. Cambridge University Press, Cambridge, 1998.

[16] E. J. N. Looijenga. *Isolated singular points on complete intersections*, volume 77 of *London Mathematical Society Lecture Note Series*. Cambridge University Press, Cambridge, 1984.

[17] Alexei Oblomkov and Vivek Shende. The Hilbert scheme of a plane curve singularity and the HOMFLY polynomial of its link. *Duke Math. J.*, 161(7):1277–1303, 2012.

[18] Jens Piontkowski. Topology of the compactified Jacobians of singular curves. *Math. Z.*, 255(1):195–226, 2007.

[19] Kyoji Saito. Period mapping associated to a primitive form. *Publ. Res. Inst. Math. Sci.*, 19(3):1231–1264, 1983.

[20] Charlotte Scott. On the higher singularities of plane curves. *Am.J.Math*, 14(4):301–325, 1892.

[21] Francesco Severi. *Vorlesungen über algebraische Geometrie. Geometrie auf einer Kurve, Riemannsche Flächen, Abelsche Integrale.* B.G.Teubner, Leipzig, 1921.

[22] Vivek Shende. Hilbert schemes of points on a locally planar curve and the Severi strata of its versal deformation. *Compos. Math.*, 148(2):531–547, 2012.

[23] Bernard Teissier. Résolution simultanée i:familles des courbes. In *Séminaire sur les singularités des surfaces*, volume 777 of *Lecture Notes in Math.* Springer, Berlin, 1980.

[24] Duco van Straten. Some problems on Lagrangian singularities. In *Singularities and computer algebra*, volume 324 of *London Math. Soc. Lecture Note Ser.*, pages 333–349. Cambridge Univ. Press, Cambridge, 2006.

[25] A. N. Varchenko. Period mapping and discriminant. *Mat. Sb. (N.S.)*, 134(176)(1):66–81, 142, 1987.

Paul Cadman and David Mond
Mathematics Institute, University of Warwick, Coventry CV4 7AL
United Kingdom
e-mail: pcadman@gmail.com
d.m.q.mond@warwick.ac.uk

Duco van Straten
Institut für Mathematik, FB 08 – Physik, Mathematik und Informatik
Johannes Gutenberg-Universität, Staudingerweg 9, 4. OG, 55128 Mainz
Germany
e-mail: straten@mathematik.uni-mainz.de

Classification of Isolated Polar Weighted Homogeneous Singularities

José Luis Cisneros-Molina and Agustín Romano-Velázquez

To Pepe Seade for his 60th birthday

Abstract. Polar weighted homogeneous polynomials are real analytic maps which generalize complex weighted homogeneous polynomials. In this article we give classes of mixed polynomials in three variables which generalize Orlik and Wagreich classes of complex weighted homogeneous polynomials. We give explicit conditions for this classes to be polar weighted homogeneous polynomials with isolated critical point. We prove that under small perturbation of their coefficients they remain with isolated critical point and the diffeomorphism type of their link does not change.

Mathematics Subject Classification (2000). Primary 32C18, 32S50; Secondary 14B05, 57R45.

Keywords. Polar weighted homogeneous polynomials, Orlik and Wagreich classes.

1. Introduction

Let $f\colon \mathbb{C}^3 \to \mathbb{C}$ be a complex weighted homogeneous polynomial with isolated critical point. Let $V = f^{-1}(0)$ be its zero-set and consider its link given by $K = V \cap \mathbb{S}^5$. It is now a classical result by Orlik and Wagreich [9, §3.1] that the link of such polynomial is equivariantly diffeomorphic to the link of a polynomial in one of six classes given explicitly in the aforementioned paper.

In this article we generalize Orlik and Wagreich classes for polar weighted homogeneous polynomials with isolated critical point. These are real analytic maps which generalize complex weighted homogeneous polynomials, they are polynomials in complex variables and their conjugates (mixed functions) and they are

Research supported by projects UNAM-DGAPA-PAPIIT IN106614 and CONACYT 253506. The first author is Regular Associate of the Abdus Salam International Centre for Theoretical Physics, Trieste, Italy.

weighted homogeneous with respect to an $\mathbb{R}^+$-action and also with respect to a $\mathbb{S}^1$-action. The first examples of polar weighted homogeneous polynomials were twisted Brieskorn-Pham polynomials given in [10] by Ruas, Seade and Verjovsky to give explicit examples of real analytic maps f with isolated critical point which Milnor fibration is given by $f/\| f \|$ as for holomorphic maps. Inspired by these examples polar weighted homogeneous polynomials were introduced by Cisneros-Molina in [4] and later they were studied by Oka in [6, 7, 8].

The organization of the article is as follows. In Section 2 we recall some basic facts about mixed functions, in particular, when a mixed polynomial is full and the definition of polar weighted homogeneous polynomials. We also generalize a lemma by Arnold about the existence of certain monomials in mixed polynomials with isolated critical point. In Section 3 we prove that polar weighted homogeneous polynomials with isolated critical point at the origin under small perturbation of their coefficients remain with isolated critical point (Corollary 3.6). In Section 4 we give the classes of mixed polinomials which generalize Orlik and Wagreich classes. In contrast with the complex case, these classes of mixed polynomials are not automatically polar weighted homogeneous, so we compute the explicit conditions for these families to be polar weighted homogeneous with isolated singularity at the origin (Theorem 4.5 and Theorem 4.10). As a result of these computations we list the classes which are full polar weighted homogeneous polynomials (Corollary 4.7). In Section 5 we prove that the diffeomorphism type of the link of a polar weighted homogeneous polynomial with isolated singularity at the origen does not change under small perturbation of the coefficients of the polynomial (Theorem 5.3).

2. Mixed functions

Consider $\mathbb{C}^n$ with coordinates $z_1, \dots, z_n$. Let $\bar{z}_j$ be the complex conjugate of z_j. We will write $z_j = x_j + iy_j$ with $x_j, y_j \in \mathbb{R}$. To simplify notation we shall write $\mathbf{z} = (z_1, \dots, z_n)$, $\bar{\mathbf{z}} = (\bar{z}_1, \dots, \bar{z}_n)$, $\mathbf{x} = (x_1, \dots, x_n)$ and $\mathbf{y} = (y_1, \dots, y_n)$. We also denote by $\mathbf{0}$ the origin in $\mathbb{C}^n$, by $\mathbb{C}^*$ the non-zero complex numbers and by $\mathbb{R}^+$ the positive real numbers.

Let $\mu = (\mu_1, \dots, \mu_n)$ and $\nu = (\nu_1, \dots, \nu_n)$ with $\mu_j, \nu_j \in \mathbb{N} \cup \{0\}$, set $\mathbf{z}^\mu = z_1^{\mu_1} \dots z_n^{\mu_n}$ and $\bar{\mathbf{z}}^\nu = \bar{z}_1^{\nu_1} \dots \bar{z}_n^{\nu_n}$. Consider a complex valued function $f\colon \mathbb{C}^n \to \mathbb{C}$ expanded in a convergent power series of variables $\mathbf{z}$ and $\bar{\mathbf{z}}$,

$$f(\mathbf{z}) = \sum_{\mu,\nu} c_{\mu,\nu} \mathbf{z}^\mu \bar{\mathbf{z}}^\nu \,.$$

We call f a *mixed analytic function* (or a *mixed polynomial*, if f is a polynomial).

We consider f as a function $f\colon \mathbb{R}^{2n} \to \mathbb{R}^2$ in the $2n$ real variables $(\mathbf{x}, \mathbf{y})$ writing $f(\mathbf{z}) = g(\mathbf{x}, \mathbf{y}) + ih(\mathbf{x}, \mathbf{y})$, taking $z_j = x_j + iy_j$ where $g, h\colon \mathbb{C}^n \cong \mathbb{R}^{2n} \to \mathbb{R}$ are real analytic functions. Recall that for any real analytic function $k\colon \mathbb{R}^{2n} \to \mathbb{R}$ we have

$$\frac{\partial k}{\partial z_j} = \frac{1}{2}\left(\frac{\partial k}{\partial x_j} - i\frac{\partial k}{\partial y_j}\right), \qquad \frac{\partial k}{\partial \bar{z}_j} = \frac{1}{2}\left(\frac{\partial k}{\partial x_j} + i\frac{\partial k}{\partial y_j}\right).$$

So we have

$$\frac{\partial f}{\partial z_j} = \frac{\partial g}{\partial z_j} + i\frac{\partial h}{\partial z_j}\,, \qquad \frac{\partial f}{\partial \bar{z}_j} = \frac{\partial g}{\partial \bar{z}_j} + i\frac{\partial g}{\partial \bar{z}_j}\,.$$

As usual, we define the real gradients of g and h by

$$\mathrm{d}_{\mathbb{R}}g(\mathbf{x},\mathbf{y}) = \left(\frac{\partial g}{\partial x_1},\dots,\frac{\partial g}{\partial x_n},\frac{\partial g}{\partial y_1},\dots,\frac{\partial g}{\partial y_n}\right)\,,$$

$$\mathrm{d}_{\mathbb{R}}h(\mathbf{x},\mathbf{y}) = \left(\frac{\partial h}{\partial x_1},\dots,\frac{\partial h}{\partial x_n},\frac{\partial h}{\partial y_1},\dots,\frac{\partial h}{\partial y_n}\right)\,.$$

Following Oka [6] set

$$\mathrm{d}f(\mathbf{z}) = \left(\frac{\partial f(\mathbf{z})}{\partial z_1},\dots,\frac{\partial f(\mathbf{z})}{\partial z_n}\right)\,, \qquad \bar{\mathrm{d}}f(\mathbf{z}) = \left(\frac{\partial f(\mathbf{z})}{\partial \bar{z}_1},\dots,\frac{\partial f(\mathbf{z})}{\partial \bar{z}_n}\right)\,.$$

The following proposition is an useful criterium to determine whether a point $\mathbf{z} \in \mathbb{C}^n$ is a critical point of a mixed function f.

Proposition 2.1 (Oka's Criterium [6, Proposition 1]**).** *Let $\mathbf{z} \in \mathbb{C}^n$. The following two conditions are equivalent,*

1. *The vectors $\mathrm{d}_{\mathbb{R}}g(\mathbf{z})$ and $\mathrm{d}_{\mathbb{R}}h(\mathbf{z})$ are linearly dependent over $\mathbb{R}$.*
2. *There exists a complex number $\alpha \in \mathbb{S}^1$ such that $\overline{\mathrm{d}f(\mathbf{z})} = \alpha\bar{\mathrm{d}}f(\mathbf{z})$.*

We need the following condition which will be automatically satisfied by the family of polar weighted homogeneous polynomials that we will consider later.

Condition 2.2. *If the monomial z_j appears in f, then the monomial $\overline{z}_j$ does not appear in f.*

The following lemma is a generalization of [1, Proposition 11.1] by Arnold.

Lemma 2.3. *Fix $i \in \{1,\dots,n\}$. If f is a mixed polynomial with isolated singularity at the origin of $\mathbb{C}^n$ satisfying Condition 2.2, then there exist $a, b \in \mathbb{N}\cup\{0\}$ with $a+b \neq 0$, such that the monomial $z_i^a\bar{z}_i^b x$ appears in f, with $x \in \{z_1,\bar{z}_1,\dots,z_n,\bar{z}_n\}$.*

Proof. Assume that for all $a \geq 0$, $b \geq 0$ there are no monomial $z_i^a\bar{z}_i^b x$. By Condition 2.2, f does not have a linear term, if so, $f = 0$ has no singularity at the origin. Consider $\mathrm{d}f$ and $\bar{\mathrm{d}}f$ on the axis $z_1 = \cdots = z_{i-1} = z_{i+1} = \cdots = z_n = 0$. This axis is a subset of $f^{-1}(0)$ and we have that both gradient vectors vanish simultaneously. This means that the axis is included in the singular locus, which contradicts the fact that f has an isolated singularity at the origin. □

Following Oka [6, §2.3] we have the following definition.

Definition 2.4. Let $\mu_j = (\mu_{j,1},\dots,\mu_{j,n})$ and $\nu_j = (\nu_{j,1},\dots,\nu_{j,n})$ be multi-indices and let $f\colon \mathbb{C}^n \to \mathbb{C}$ be a mixed polynomial written as

$$f(\mathbf{z}) = \sum_{j=1}^{m} c_j \mathbf{z}^{\mu_j}\bar{\mathbf{z}}^{\nu_j}\,,$$

where $c_1, \dots, c_m$ are non-zero. Consider the following matrices

$$P = \begin{pmatrix} \mu_{1,1}+\nu_{1,1} & \cdots & \mu_{1,n}+\nu_{1,n} \\ \vdots & \vdots & \vdots \\ \mu_{m,1}+\nu_{m,1} & \cdots & \mu_{m,n}+\nu_{m,n} \end{pmatrix}, \qquad Q = \begin{pmatrix} \mu_{1,1}-\nu_{1,1} & \cdots & \mu_{1,n}-\nu_{1,n} \\ \vdots & \vdots & \vdots \\ \mu_{m,1}-\nu_{m,1} & \cdots & \mu_{m,n}-\nu_{m,n} \end{pmatrix}.$$

We say that f is *radial full* (respectivelly *angular full*) if $n = m$ and P (respectively, Q) has rank n. If f is radial and angular full, then we say that f is *full*. We call the matrix P the *radial matrix* and Q the *angular matrix* of f.

Define the *associated Laurent polynomial* $\hat{f}\colon \mathbb{C}^{*n} \to \mathbb{C}$ by

$$\hat{f}(\mathbf{w}) = \sum_{j=1}^{m} c_j \mathbf{w}^{\mu_j - \nu_j} .$$

Theorem 2.5 ([6, Theorem 10]). *Let $f(\mathbf{z})$ be a full mixed polynomial and let $\hat{f}(\mathbf{w})$ be its associated Laurent polynomial. Then there exists a diffeomorphism $\phi\colon \mathbb{C}^{*n} \to \mathbb{C}^{*n}$ such that $\hat{f} \circ \phi = f|_{\mathbb{C}^{*n}}$.*

Corollary 2.6. *The associated Laurant polynomial $\hat{f}\colon \mathbb{C}^{*n} \to \mathbb{C}$ has no critical points.*

Proof. Let Q be the angular matrix. As in [5, page 68] define the map $\psi_Q\colon \mathbb{C}^{*n} \to \mathbb{C}^{*n}$ by

$$\psi_Q(\mathbf{w}) = (w_1^{\mu_{1,1}-\nu_{1,1}} \dots w_n^{\mu_{1,n}-\nu_{1,n}}, \dots, w_1^{\mu_{m,1}-\nu_{m,1}} \dots w_n^{\mu_{m,n}-\nu_{m,n}}),$$

and define $h\colon \mathbb{C}^{*n} \to \mathbb{C}$ by $h(\mathbf{w}) = c_1 w_1 + \cdots + c_m w_m$. Then we have that $\hat{f}(\mathbf{w}) = h(\psi_Q(\mathbf{w}))$. By [5, Assertion (1.3.2), page 109] ψ_Q is a $\det(Q)$-fold covering and clearly h has no critical points. □

An useful property of a radial full or angular full polynomial is that we can have more control on the coefficients c_j.

Lemma 2.7. *Let f be a mixed polynomial and suppose that k rows of the radial matrix P are linearly independent. Then under a change of coordinates we can assume that k coefficients are on $\mathbb{S}^1$.*

Lemma 2.8. *Let f be a mixed polynomial and suppose that k rows of the angular matrix Q are linearly independent. Then under a change of coordinates we can assume that k coefficients are on $\mathbb{R}^+$.*

Corollary 2.9 ([8, Lemma 8]). *If f is full, then under a change of coordinates we can assume that all the coefficients are* 1.

We are just going to prove Lemma 2.7 (actually it is just an adaptation of the proof of [8, Lemma 8]).

Proof of Lemma 2.7. We have $f(\mathbf{z}) = \sum_{j=1}^{m} c_j \mathbf{z}^{\mu_j} \bar{\mathbf{z}}^{\nu_j}$. We can apply a change of coordinates $z_j \to z_{\sigma(j)}$ with σ a permutation of $\{1, \ldots, n\}$ so that the matrix

$$P' = \begin{pmatrix} \mu_{1,1} + \nu_{1,1} & \ldots & \mu_{1,k} + \nu_{1,k} \\ \vdots & \vdots & \vdots \\ \mu_{k,1} + \nu_{k,1} & \ldots & \mu_{k,k} + \nu_{k,k} \end{pmatrix}$$

is invertible.

We are going to construct a change of coordinates of the form $z_j \to e^{t_j} z_j$ where $t_j \in \mathbb{R}$ with $j = 1, \ldots, k$. Write $c_j = e^{a_j}\theta_j$ and notice that we want some numbers $e^{t_j} \in \mathbb{R}^+$ such that

$$(\mu_{j,1} + \nu_{j,1})t_1 + \cdots + (\mu_{j,k} + \nu_{j,k})t_k = -a_j$$

then we have the system

$$P'(t_1, \ldots, t_k)^\top = (-a_1, \ldots, -a_k)^\top .$$

Since P' is invertible we can solve this system. □

2.1. Polar weighted homogeneous polynomials

Let $p_1, \ldots, p_n$ and $q_1, \ldots, q_n$ be non-zero integers such that $\gcd(p_1, \ldots, p_n) = 1$ and $\gcd(q_1, \ldots, q_n) = 1$. Let $w \in \mathbb{C}^*$ written in its polar form $w = t\tau$, with $t \in \mathbb{R}^+$ and $\tau \in \mathbb{S}^1$. A *polar* $\mathbb{C}^*$*-action* on $\mathbb{C}^n$ with *radial weights* $(p_1, \ldots, p_n)$ and *angular weights* $(q_1, \ldots, q_n)$ is given by:

$$t\tau \bullet \mathbf{z} = (t^{p_1}\tau^{q_1} z_1, \ldots, t^{p_n}\tau^{q_n} z_n) . \tag{2.1}$$

Definition 2.10. A mixed function $f\colon \mathbb{C}^n \to \mathbb{C}$ is *polar weighted homogeneous* if there exists $p_1, \ldots, p_n$ positive integers, $q_1, \ldots, q_n$ non-zero integers, a, c positive integers, and a polar $\mathbb{C}^*$-action given by (2.1) such that f satisfies the following functional equation:

$$f(t\tau \bullet \mathbf{z}) = t^a \tau^c f(\mathbf{z}) . \tag{2.2}$$

We say that the polar weighted homogeneous function f has *radial weight type* $(p_1, \ldots, p_n; a)$ and *angular weight type* $(q_1, \ldots, q_n; c)$.

Sometimes it is more convenient to consider the *normalized radial weights* $(p'_1, \ldots, p'_n)$ given by $p'_i = \frac{p_i}{a}$ and the *normalized angular weights* $(q_1, \ldots, q_n)$ given by $q'_i = \frac{q_i}{c}$.

We will say that f is *generalized polar weighted homogeneous* if it satisfies (2.2) with $p_1, \ldots, p_n$ and $q_1, \ldots, q_n$ integers, i.e. some p_j or q_j can be zero or negative.

Remark 2.11. The definition of polar weighted homogeneous functions follows the original definition given in [4] but allowing the q_i's to be negative. Other authors (for instance [6, 3]) call polar weighted homogeneous functions to more general notions allowing the p_i's or q_i's to be zero; we call this more general definition generalized polar weighted homogeneous functions to emphazise the difference. Originally, the *angular weights* were called *polar weights* and this has caused some confusion in the literature because some authors (for instance [7, 3]) call *polar weighted homogeneous* to mixed functions which are weighted homogeneous with respect to the *angular weights* and not to both radial and angular weights. To avoid this ambiguity in [2] the authors propose to use the term mixed weighted homogeneous instead of what we call polar weighted homogeneous. We think it is better to keep the term polar weighted homogeneous for the original definition given in [4] and use the term *angular weights* instead of *polar weights* and respectively *angular weighted homogeneous*; the reason is that the polar coordinate system on the plane consists of two coordinates: the radial coordinate and the angular coordinate, and *polar* $\mathbb{C}^*$*-actions* are defined writing the acting element $w \in \mathbb{C}^*$ in its *polar form*.

Remark 2.12. Notice that a complex weighted homogeneous polynomial is a particular case of a polar weighted homogeneous polynomial with $p_j = q_j$.

Notice that given a polar $\mathbb{C}^*$-action on $\mathbb{C}^n$, we get a radial $\mathbb{R}^+$-action on $\mathbb{C}^n$ given by

$$t * \mathbf{z} := (t^{p_1} z_1, \dots, t^{p_n} z_n).$$

Sometimes we will be interested in the general case of real analytic maps $f\colon \mathbb{R}^n \to \mathbb{R}^k$, so we also consider the following definition.

Definition 2.13. Let $p_1, \dots, p_n$ be integers with $\gcd(p_1, \dots, p_n) = 1$. Let $f\colon \mathbb{R}^n \to \mathbb{R}^m$ be an analytic map and consider an $\mathbb{R}^+$-action on $\mathbb{R}^n$ given by

$$t * \mathbf{x} := (t^{p_1} x_1, \dots, t^{p_n} x_n)\ .$$

Let a be a positive integer. We call f a *radial weighted homogeneous map of type* $(p_1, \dots, p_n; a)$ if

$$f(t * \mathbf{x}) = t^a f(\mathbf{x})\ ,$$

where p_j is a positive integer for $j = 1, \dots, n$. We say that f is a *generalized radial weighted homogeneous* if $p_1, \dots, p_n$ are arbitrary integers.

Proposition 2.14 ([4, §3],[6, §2]). *Let $f(\mathbf{z})$ be a generalized polar weighted homogeneous function with radial weight type $(p_1, \dots, p_n; a)$ and angular weight type $(q_1, \dots, q_n; c)$. Then it satisfies the following properties:*

1. *Euler identities:*

$$af(\mathbf{z}) = \sum_{j=1}^{n} p_j z_j \frac{\partial f}{\partial z_j}(\mathbf{z}) + \sum_{j=1}^{n} p_j \bar{z}_j \frac{\partial f}{\partial \bar{z}_j}(\mathbf{z})\ ,$$

$$cf(\mathbf{z}) = \sum_{j=1}^{n} q_j z_j \frac{\partial f}{\partial z_j}(\mathbf{z}) - \sum_{j=1}^{n} q_j \bar{z}_j \frac{\partial f}{\partial \bar{z}_j}(\mathbf{z})\ .$$

2. *The maps $\frac{\partial f}{\partial z_j}$ and $\frac{\partial f}{\partial \bar{z}_j}$ are also generalized weighted homogeneous.*
3. *The only critical value of f is* 0.
4. *The fiber $F_\alpha := f^{-1}(\alpha)$ is a manifold of real dimension $2(n-1)$ and it is canonical diffeomorphic to $F_1 = f^{-1}(1)$.*

5. *If the weights* $p_1, \ldots, p_n$ *are positive, then:*
   (a) *The function* f *is indeed a polynomial.*
   (b) *The zero-set* $V = f^{-1}(0)$ *is contractible to the origin.*
   (c) *The restriction* $f\colon (\mathbb{C}^n \setminus V) \to \mathbb{C}^*$ *is a locally trivial fibration.*
   (d) *The map*

$$\phi = \frac{f}{|f|} : \mathbb{S}_\epsilon^{2n-1} \setminus K_\epsilon \to \mathbb{S}_1 \tag{2.3}$$

   is a fiber bundle, for any $\epsilon > 0$.
   (e) *The fibration* $f_{\mathbb{S}^1} := f|_{f^{-1}(\mathbb{S}^1)} \colon f^{-1}(\mathbb{S}^1) \to \mathbb{S}^1$ *is equivalent to the fibration* (2.3).

   Furthermore, if the origin is an isolated singularity of V
   (f) $V \setminus \{\mathbf{0}\}$ *is smooth.*
   (g) *The sphere* $\mathbb{S}_\epsilon^{2n-1}$ *of radius* ϵ *around* $\mathbf{0}$ *is transverse to* V *for any* $\epsilon > 0$.
   (h) *Let* $K_\epsilon := V \cap \mathbb{S}_\epsilon^{2n-1}$. *Then for any* $\epsilon', \epsilon > 0$, $K_{\epsilon'}$ *and* K_ϵ *are* $\mathbb{S}^1$*-equivariantly diffeomorphic (compare with* [9, Proposition 3.1.3]*).*

Remark 2.15. If $f\colon \mathbb{R}^n \to \mathbb{R}^m$ is a radial weighted homogeneous map analogous to item 2.14 of Proposition 2.14 we have that $\frac{\partial f}{\partial x_i}$ is also radial weighted homogeneous.

In this article we restrict to the case when the radial weights are positive and non-zero angular weights, that is, we are only interested in polar weighted homogeneous polynomials.

Lemma 2.16. *If* $f\colon \mathbb{C}^n \to \mathbb{C}$ *is polar weighted homogeneous and* $\mathbf{z}$ *is a critical point of* f*, then* $t\lambda \bullet \mathbf{z}$ *is a critical point for all* $t\lambda \in \mathbb{C}^*$.

The analogous statement is true for a radial weighted map.

Proof. We will prove it for polar weighted homogeneous polynomials, the other case is analogous. Suppose f has radial weight type $(p_1, \ldots, p_n; a)$ and angular weight type $(q_1, \ldots, q_n; c)$.

Since $\mathbf{z}$ is a critical point of f, by Lemma 2.1 there exists $\alpha \in \mathbb{S}^1$ such that

$$\overline{\frac{\partial f(\mathbf{z})}{\partial z_j}} = \alpha \frac{\partial f(\mathbf{z})}{\partial \bar{z}_j}, \qquad j \in \{1, \ldots, n\}\,.$$

Since $\frac{\partial f}{\partial z_j}$ and $\frac{\partial f}{\partial \bar{z}_j}$ are also polar weighted homogeneous, for any $t \in \mathbb{R}^+$ and $\lambda \in \mathbb{S}^1$

$$\overline{\frac{\partial f(t\lambda \bullet \mathbf{z})}{\partial z_j}} = t^{a-p_j} \lambda^{-c+q_j} \overline{\frac{\partial f(\mathbf{z})}{\partial z_j}} = t^{a-p_j} \lambda^{c+q_j} \frac{\alpha}{\lambda^{2c}} \frac{\partial f(\mathbf{z})}{\partial \bar{z}_j} = \frac{\alpha}{\lambda^{2c}} \frac{\partial f(t\lambda \bullet \mathbf{z})}{\partial \bar{z}_j}\,,$$

therefore $t\lambda \bullet \mathbf{z}$ is a critical point of f. □

3. Isolated critical point under perturbation of coefficients

The aim of this section is to prove that given a polar weighted homogeneous polynomial with isolated critical point, with a small perturbation of its coefficients it still has isolated critical point.

Definition 3.1. Let $f = (f_1, \dots, f_m) \colon (\mathbb{R}^n, \mathbf{0}) \to (\mathbb{R}^m, \mathbf{0})$ be a map where $f_j \colon \mathbb{R}^n \to \mathbb{R}$ is a polynomial for $j \in \{1, \dots, m\}$. Suppose that

$$f_j(\mathbf{x}) = \sum_{l=1}^{k_j} c_{j,l} P_{j,l}(\mathbf{x}) ,$$

where $c_{j,l} \in \mathbb{R}^*$ and $P_{j,l}$ are monomials with coefficient 1.

We can identify the set of coefficients $c_{j,l}$ of f (up to a permutation) with a point in $\mathbb{R}^{k_1+\cdots+k_m}$. Let $\epsilon > 0$ and $\mathbb{B}(\mathbf{0}, \epsilon)$ be the open ball in $\mathbb{R}^{k_1+\cdots+k_m}$ centered at the origin with radius ϵ and let $p \in \mathbb{B}(\mathbf{0}, \epsilon)$ with coordinates $p = (p_{j,l})$, $j = 1, \dots, m$ and $l = 1, \dots, k_j$.

We can consider the polynomials

$$f_{j,p}(\mathbf{x}) = \sum_{l=1}^{k_j} (c_{j,l} + p_{j,l}) P_{j,l}(\mathbf{x}) ,$$

and the map

$$f_p = (f_{1,p}, \dots, f_{k,p}) \colon \mathbb{R}^n \to \mathbb{R}^m . \tag{3.1}$$

Suppose that f has an isolated critical point at the origin. We say that f is *stable under a small perturbation of its coefficients*, if there exist $\epsilon > 0$ small enough such that f_p has an isolated critical point at the origin for all $p \in \mathbb{B}(\mathbf{0}, \epsilon)$.

Remark 3.2. Suppose that f has an isolated critical point at the origin, let $\mathbf{x} \in \mathbb{R}^n$ be a regular point of f and let $M_1, \dots, M_k$ be all the minors of size $m \times m$.

Each minor M_j is a polynomial on the variables $x_1, \dots, x_n$ and if we fix the variables and allow to change the coefficients of f, we have that M_j is also a polynomial on the coefficients $c_{1,1}, \dots, c_{m,k_m}$.

Therefore we think M_j as a polynomial

$$M_j \colon \mathbb{R}^n \times \mathbb{R}^{k_1+\cdots+k_m} \to \mathbb{R} . \tag{3.2}$$

That f is stable under a small perturbation of its coefficient is equivalent to say that there exist $\epsilon > 0$ such that for any $\mathbf{x} \in \mathbb{R}^n \setminus \{\mathbf{0}\}$ we have $M_j(\mathbf{x}, p) \neq 0$ for some $j \in \{1, \dots, k\}$ and every $p \in \mathbb{B}(\mathbf{0}, \epsilon) \subset \mathbb{R}^{k_1+\cdots+k_m}$.

The following lemma is a direct consequence of Lemma 2.16.

Lemma 3.3. *Let f be a polar weighted homogeneous polynomial and $\mathbf{z}_0 \in \mathbb{C}$ a regular point of f. Let $M(\mathbf{z})$ be a 2×2 minor of the Jacobian matrix of f, seen as a real analytic map, such that $M(\mathbf{z}_0) \neq 0$. Then $M(t \bullet \mathbf{z}_0) \neq 0$ for all $t \in \mathbb{R}^+$.*

The analogous statement is true for a radial weighted homogeneous map.

Proof. Suppose that

$$M(\mathbf{z}) = \frac{\partial g(\mathbf{z})}{\partial x_j}\frac{\partial h(\mathbf{z})}{\partial y_k} - \frac{\partial g(\mathbf{z})}{\partial x_k}\frac{\partial h(\mathbf{z})}{\partial y_j}.$$

Since the partials derivatives are polar weighted homogeneous (in particular radial weighted homogeneous) then

$$M(t \bullet \mathbf{z}) = t^{2a-p_j-p_k} \left(\frac{\partial g(\mathbf{z})}{\partial x_j} \frac{\partial h(\mathbf{z})}{\partial y_k} - \frac{\partial g(\mathbf{z})}{\partial x_k} \frac{\partial h(\mathbf{z})}{\partial y_j} \right) ,$$

therefore $M(t \bullet \mathbf{z}_0) = t^{2a-p_j-p_k} M(\mathbf{z}_0) \neq 0$ for all $t \in \mathbb{R}^+$. □

Proposition 3.4. *Let $f\colon \mathbb{R}^n \to \mathbb{R}^m$ be a radial weighted homogeneous map. If f has an isolated critical point at the origin, then f is stable under a small perturbation of its coefficients.*

Proof. Let $\mathbf{x} \in \mathbb{S}_1^{n-1}$. By Lemma 2.16 the origin is the only critical point of f, therefore $\mathbf{x}$ is a regular point of f and there exist a minor $M_{\mathbf{x}}$ of size $m \times m$ with $M_{\mathbf{x}}(\mathbf{x}) \neq 0$ and an open set $U_{\mathbf{x}} \subset \mathbb{R}^n$ such that $\mathbf{x} \in U_{\mathbf{x}}$ and

$$|M_{\mathbf{x}}(\mathbf{x}) - M_{\mathbf{x}}(\mathbf{y})| < \frac{|M_{\mathbf{x}}(\mathbf{x})|}{2}, \quad \text{for every } \mathbf{y} \in U_{\mathbf{x}}.$$

Therefore we have a cover of $\mathbb{S}_1^{n-1}$ consisting of $\{U_{\mathbf{x}}\}_{\mathbf{x}\in\mathbb{S}_1^{n-1}}$ and since it is compact, we have a finite subcover $\{U_j\}$. Denote by M_j the minor corresponding to the open set U_j and consider M_j as in (3.2). Now consider the following function $D_j\colon \overline{U}_j \to \mathbb{R}^+$

$$D_j(\mathbf{y}) = \sup\{\epsilon > 0 \,|\, M_j(\mathbf{y}, p) \neq 0 \text{ for every } p \in \mathbb{B}(\mathbf{0}, \epsilon) \subset \mathbb{R}^{k_1+\cdots+k_m}\} ,$$

where $\overline{U}_j$ is the closure of U_j and take $\epsilon_j = \min\{ D_j(\mathbf{y}) \mid \mathbf{y} \in \overline{U}_j \}$. Consider $\epsilon = \min\{\epsilon_1, \ldots, \epsilon_j\}$, therefore for every $0 < \epsilon' \leq \epsilon$, $\mathbf{y} \in \mathbb{S}_1^{n-1}$ and $p \in \mathbb{B}(\mathbf{0}, \epsilon')$ we have $M_j(\mathbf{y}, p) \neq 0$, for some minor M_j.

Now using Lemma 3.3 we have that the same holds for any $\mathbf{y} \in \mathbb{R}^n \setminus \{\mathbf{0}\}$ and any $p \in \mathbb{B}(\mathbf{0}, \epsilon')$ for every $0 < \epsilon' \leq \epsilon$. □

Corollary 3.5. *Let $f_p = (f_{1,p}, \ldots, f_{k,p})\colon \mathbb{R}^n \to \mathbb{R}^m$ be a family of radial weighted homogeneous maps as in* (3.1). *Then the subspace U of $\mathbb{R}^{k_1+\cdots+k_m}$ of parameters p for which f_p has an isolated singularity is an open set.*

Corollary 3.6. *If f is a polar weighted homogeneous polynomial with isolated singularity, then f is stable under a small perturbation of its coefficients.*

4. Classification of polar weighted homogeneous polynomials in $\mathbb{C}^3$

In this section we want to study polar weighted homogeneous polynomials $f\colon \mathbb{C}^3 \to \mathbb{C}$ with isolated singularity at the origin. We do it in four steps:

First step. We define families of mixed polinomials which contain terms which are necessary in order to have isolated singularity.

Second step. We give conditions on the exponents of the elements of these families to be polar weighted homogeneous polinomials.

Third step. Under a suitable change of coordinates we simplify the coefficient of these families taking them to a *special form.*

Fourth step. In the *special form* we give conditions to have isolated singularity.

Following Orlik and Wagreich [9, §3.1] we have the following definition.

Definition 4.1. A mixed function $f(\mathbf{z})$ is said to be of class **I** (respectively **II**,..., **V**) if there is a permutation σ of the set $\{1,2,3\}$ and non-zero complex numbers $\alpha_1, \alpha_2, \alpha_3$ such that $f(z_{\sigma(1)}, z_{\sigma(2)}, z_{\sigma(3)})$ is equal to

I. $\alpha_1 z_1^{a_1}\bar{z}_1^{b_1} + \alpha_2 z_2^{a_2}\bar{z}_2^{b_2} + \alpha_3 z_3^{a_3}\bar{z}_3^{b_3}$,
II. $\alpha_1 z_1^{a_1}\bar{z}_1^{b_1} + \alpha_2 z_2^{a_2}\bar{z}_2^{b_2} + \alpha_3 z_3^{a_3}\bar{z}_3^{b_3} g_2$,
III. $\alpha_1 z_1^{a_1}\bar{z}_1^{b_1} + \alpha_2 z_2^{a_2}\bar{z}_2^{b_2} g_3 + \alpha_3 z_3^{a_3}\bar{z}_3^{b_3} g_2$,
IV. $\alpha_1 z_1^{a_1}\bar{z}_1^{b_1} + \alpha_2 z_2^{a_2}\bar{z}_2^{b_2} g_1 + \alpha_3 z_3^{a_3}\bar{z}_3^{b_3} g_2$,
V. $\alpha_1 z_1^{a_1}\bar{z}_1^{b_1} g_2 + \alpha_2 z_2^{a_2}\bar{z}_2^{b_2} g_3 + \alpha_3 z_3^{a_3}\bar{z}_3^{b_3} g_1$,

where $g_j \in \{z_j, \bar{z}_j\}$.

Remark 4.2. Notice that if a mixed polynomial does not satisfies Condition 2.2 it cannot be weighted homogeneous with respect to the $\mathbb{S}^1$-action, therefore polar weighted homogeneous polynomials satisfy Condition 2.2. Definition 4.1 lists all possible polynomials that one can get applying Lemma 2.3 to a mixed function with isolated singularity. Notice that taking $b_i = 0$ and $g_i = z_i$ for $i = 1,2,3$ in classes **I** to **V** we recover Orlik and Wagreich classes **I** to **V** of irreducible complex weighted homogeneous polynomials, their class **VI** corresponds to class **III** taking $a_2 = a_3 = 1$.

In contrast with Orlik and Wagreich the classes in Definition 4.1 are not necessarily polar weighted homogeneous, for example,

$$f(\mathbf{z}) = |z_1|^2 + |z_2|^2 + |z_3|^2$$

is a mixed function of class **I** but it is not polar weighted homogeneous. For this reason one has to find the conditions that the a_i and b_i should satisfy in order to get a polar weighted homogeneous polynomial.

Remark 4.3. Using the change of coordinates $z_i \mapsto \bar{z}_i$ we can always assume that $g_i = z_i$ but in this case $a_i - b_i$ can be positive, negative or zero. Hereafter we assume that $g_j = z_j$.

We will frequently use the following basic lemma.

Lemma 4.4. *Let $z \in \mathbb{C}$ and $t \in \mathbb{R}$. If $z + t\bar{z}$ belongs to $\mathbb{R}$, then $t = 1$ or $z \in \mathbb{R}$.*

Theorem 4.5. *Let f be a mixed function of one of the classes of Definition 4.1. Then the following conditions must be satisfied in order to f be polar weighted homogeneous:*

Class I. $a_j - b_j \neq 0$ *with* $j = 1,2,3$.
Class II. *One of the following conditions is satisfied:*
- **a)** $a_j - b_j \neq 0$ *with* $j = 1,2,3$ *and* $a_2 \pm b_2 \neq 1$.
- **b)** $a_1 - b_1 \neq 0$, $a_2 - b_2 = 1$, $b_2 \neq 0$ *and* $a_3 = b_3$.

Class III. $a_1 - b_1 \neq 0$ *and* $a_2 - b_2$, $a_3 - b_3$ *are not both* -1. *Also one of the following conditions is satisfied:*

a) *$a_2 \pm b_2$ and $a_3 \pm b_3$ are not* 1.
b) $(a_2+b_2)(a_3+b_3) > 1$, $a_2 - b_2 = 1$ *and* $a_3 - b_3 = 1$.
c) $a_2 = a_3 = 1$ *and* $b_2 = b_3 = 0$.

Class IV. *One of the following conditions is satisfied:*
a) $a_i - b_i \neq 0$ *for* $i = 1,2,3$, $a_1 \pm b_1 \neq 1$ *and* $(a_1, a_2) \neq (b_1 - 1, b_2 + 2)$.
b) $a_2 = b_2$, $a_1 - b_1 = 1$ *and* $b_1 \neq 0$.
c) $a_3 = b_3$, $a_1 - b_1 \neq 0$, $a_1 + b_1 > 1$ *and* $(a_1, a_2) = (b_1 - 1, b_2 + 2)$.

Class V. $(a_1 - b_1)(a_2 - b_2)(a_3 - b_3) \neq -1$ *and*

$$\begin{cases} (a_{i-1}, a_{i+1}) \neq (b_{i-1}+1, b_{i+1}), \\ (a_{i-1}, a_{i+1}) \neq (b_{i-1}-1, b_{i+1}+2), \end{cases} \qquad i = 1,2,3.$$

Proof. **Class I.** Suppose that

$$f(\mathbf{z}) = \alpha_1 z_1^{a_1} \bar{z}_1^{b_1} + \alpha_2 z_2^{a_2} \bar{z}_2^{b_2} + \alpha_3 z_3^{a_3} \bar{z}_3^{b_3}\,.$$

In this case the radial and angular matrices are

$$P = \begin{pmatrix} a_1+b_1 & 0 & 0 \\ 0 & a_2+b_2 & 0 \\ 0 & 0 & a_3+b_3 \end{pmatrix}, \quad Q = \begin{pmatrix} a_1-b_1 & 0 & 0 \\ 0 & a_2-b_2 & 0 \\ 0 & 0 & a_3-b_3 \end{pmatrix}.$$

In order to f be polar weighted homogeneous we want to find solutions to the system

$$P(p_1, p_2, p_3)^\top = (1,1,1)^\top, \qquad Q(q_1, q_2, q_3)^\top = (1,1,1)^\top,$$

with $p_j \in \mathbb{Q}^+$ and $q_j \in \mathbb{Q} \setminus \{0\}$.

If some $a_j - b_j = 0$, then we can not solve the system, therefore $a_j - b_j \neq 0$ for $j = 1,2,3$.

The solution of the system give us the normalized radial and angular weights so we need to take m, m', M and M' to get the weights.

Class II. Suppose that

$$f(\mathbf{z}) = \alpha_1 z_1^{a_1} \bar{z}_1^{b_1} + \alpha_2 z_2^{a_2} \bar{z}_2^{b_2} + \alpha_3 z_3^{a_3} \bar{z}_3^{b_3} z_2\,.$$

In this case the radial and angular matrices are

$$P = \begin{pmatrix} a_1+b_1 & 0 & 0 \\ 0 & a_2+b_2 & 0 \\ 0 & 1 & a_3+b_3 \end{pmatrix}, \quad Q = \begin{pmatrix} a_1-b_1 & 0 & 0 \\ 0 & a_2-b_2 & 0 \\ 0 & 1 & a_3-b_3 \end{pmatrix}.$$

We have two cases:

a) The easiest case is that P and Q are invertible. In this case we have that

$$a_j - b_j \neq 0\,, \qquad j = 1,2,3\,.$$

The weights are just the solution of the system given by P and Q as in the previous case.

b) By construction, P is always invertible, so suppose that Q is not invertible but the system has solution. In this case we have that $a_1 - b_1$, $a_2 - b_2 \neq 0$ and $a_3 - b_3 = 0$. Therefore we have

$$Q = \begin{pmatrix} a_1-b_1 & 0 & 0 \\ 0 & a_2-b_2 & 0 \\ 0 & 1 & 0 \end{pmatrix},$$

and since we want $Q(q_1, q_2, q_3)^\top = (1,1,1)^\top$, therefore we have $q_2 = 1$ and $a_2 - b_2 = 1$. Under this assumptions we have

$$P = \begin{pmatrix} a_1+b_1 & 0 & 0 \\ 0 & 2a_2-1 & 0 \\ 0 & 1 & 2a_3 \end{pmatrix},$$

although the matrix is invertible, if $b_2 = 0$ (i.e., $a_2 = 1$), then $p_2 = 1$, so p_3 must be 0 but we do not allow this kind of solutions. Therefore $b_2 \neq 0$. If f satisfies the aforementioned conditions, then it is polar weighted homogeneous and the weights are just the solutions to the systems given by P and Q.

Class III. Suppose that

$$f(\mathbf{z}) = \alpha_1 z_1^{a_1} \bar{z}_1^{b_1} + \alpha_2 z_2^{a_2} \bar{z}_2^{b_2} z_3 + \alpha_3 z_3^{a_3} \bar{z}_3^{b_3} z_2 .$$

In this case the radial and angular matrices are

$$P = \begin{pmatrix} a_1+b_1 & 0 & 0 \\ 0 & a_2+b_2 & 1 \\ 0 & 1 & a_3+b_3 \end{pmatrix}, \quad Q = \begin{pmatrix} a_1-b_1 & 0 & 0 \\ 0 & a_2-b_2 & 1 \\ 0 & 1 & a_3-b_3 \end{pmatrix}.$$

We have basically two cases:

a) If P and Q are invertible, then we have

$$\det P = (a_1+b_1)\left((a_2+b_2)(a_3+b_3) - 1\right) \neq 0 ,$$
$$\det Q = (a_1-b_1)\left((a_2-b_2)(a_3-b_3) - 1\right) \neq 0 ,$$

and since we always have $a_1 + b_1 \neq 0$, then the conditions are

$$(a_2+b_2)(a_3+b_3) \neq 1 ,$$
$$(a_1-b_1)\left((a_2-b_2)(a_3-b_3) - 1\right) \neq 0 .$$

b) Suppose Q is not invertible but P it is invertible. Since $a_1 - b_1$ must be different from 0, then suppose that $(a_2 - b_2)(a_3 - b_3) = 1$.
We have two cases, the first one is $a_2 - b_2 = a_3 - b_3 = -1$, then

$$Q = \begin{pmatrix} a_1-b_1 & 0 & 0 \\ 0 & -1 & 1 \\ 0 & 1 & -1 \end{pmatrix},$$

but in this case we can not find q_2 and q_3 such that $q_2 - q_3 = 1$ and $q_3 - q_2 = 1$, so we do not have to consider this case.
The second case is $a_2 - b_2 = a_3 - b_3 = 1$. Under this assumption, we have

$$Q = \begin{pmatrix} a_1-b_1 & 0 & 0 \\ 0 & 1 & 1 \\ 0 & 1 & 1 \end{pmatrix}.$$

The angular weights are deduced from this matrix and the additional condition comes from the requirement to have non-zero integers.

c) Suppose P is not invertible but Q it is invertible, therefore $a_2 + b_2 = a_3 + b_3 = 1$ and the only solutions to this equation are:

$a_2 = a_3 = 1$ and $b_2 = b_3 = 0$,
$a_2 = b_3 = 1$ and $b_2 = a_3 = 0$,
$a_3 = b_2 = 1$ and $a_2 = b_3 = 0$,
$a_2 = a_3 = 0$ and $b_2 = b_3 = 1$,

and since $(a_2 - b_2)(a_3 - b_3) \neq 1$, then the solutions are
$a_2 = b_3 = 1$ and $b_2 = a_3 = 0$,
$a_3 = b_2 = 1$ and $a_2 = b_3 = 0$.
Consider the equations

$$a_2 = b_3 = 1 \text{ and } b_2 = a_3 = 0,$$

therefore

$$P = \begin{pmatrix} a_1+b_1 & 0 & 0 \\ 0 & 1 & 1 \\ 0 & 1 & 1 \end{pmatrix}, \quad Q = \begin{pmatrix} a_1-b_1 & 0 & 0 \\ 0 & 1 & 1 \\ 0 & 1 & -1 \end{pmatrix},$$

but in this case $q_2 = 1$ and $q_3 = 0$ but all the angular weights must be non-zero rational numbers, therefore this case does not happen. The case $a_3 = b_2 = 1$ and $a_2 = b_3 = 0$ does not happen by an analogous argument.

d) Suppose P and Q are not invertible. Using the last ideas we have that the only solution (up to a change of coordinates) is

$$a_2 = a_3 = 1 \text{ and } b_2 = b_3 = 0,$$

therefore

$$P = \begin{pmatrix} a_1+b_1 & 0 & 0 \\ 0 & 1 & 1 \\ 0 & 1 & 1 \end{pmatrix}, \quad Q = \begin{pmatrix} a_1-b_1 & 0 & 0 \\ 0 & 1 & 1 \\ 0 & 1 & 1 \end{pmatrix},$$

therefore f is polar weighted homogeneous.

Class IV. Suppose that

$$f(\mathbf{z}) = \alpha_1 z_1^{a_1} \bar{z}_1^{b_1} + \alpha_2 z_2^{a_2} \bar{z}_2^{b_2} z_1 + \alpha_3 z_3^{a_3} \bar{z}_3^{b_3} z_2 .$$

In this case the radial and angular matrices are

$$P = \begin{pmatrix} a_1+b_1 & 0 & 0 \\ 1 & a_2+b_2 & 0 \\ 0 & 1 & a_3+b_3 \end{pmatrix}, \quad Q = \begin{pmatrix} a_1-b_1 & 0 & 0 \\ 1 & a_2-b_2 & 0 \\ 0 & 1 & a_3-b_3 \end{pmatrix}.$$

a) Suppose P and Q are invertible, then $a_j - b_j \neq 0$.
Now if $a_1 + b_1 = 1$, then $p_1 = 1$ but this gives us that $p_2 = 0$ and this can not happen. Using the same idea we can check that $a_1 - b_1 \neq 1$. Consider the system $Q(q_1, q_2, q_3)^\top = (1, 1, 1)^\top$, solving this system we get that

$$q_3 = \frac{(a_1 - b_1)(a_2 - b_2 - 1) + 1}{(a_1 - b_1)(a_2 - b_2)(a_3 - b_3)},$$

and since q_3 can not be 0, then $(a_1 - b_1)(a_2 - b_2 - 1) \neq -1$. Since P is always invertible, hence we just have to consider the following cases:

b) Suppose $a_2 = b_2$, then the angular matrix is

$$Q = \begin{pmatrix} a_1-b_1 & 0 & 0 \\ 1 & 0 & 0 \\ 0 & 1 & a_3-b_3 \end{pmatrix},$$

hence $a_1 - b_1$ must be 1. Notice that if $b_1 = 0$, then the radial matrix is

$$P = \begin{pmatrix} 1 & 0 & 0 \\ 1 & 2a_2 & 0 \\ 0 & 1 & a_3+b_3 \end{pmatrix},$$

but this implies that $p_2 = 0$ but this can not happen.

c) Suppose $a_2 \neq b_2$ and $a_3 = b_3$. Then the angular matrix is

$$Q = \begin{pmatrix} a_1-b_1 & 0 & 0 \\ 1 & a_2-b_2 & 0 \\ 0 & 1 & 0 \end{pmatrix} ,$$

therefore $q_2 = 1$ and since $q_1 = \frac{1}{a_1-b_1}$, then we have $1 + (a_2 - b_2)(a_1 - b_1) = (a_1 - b_1)$.
The radial matrix is

$$P = \begin{pmatrix} a_1+b_1 & 0 & 0 \\ 1 & a_2+b_2 & 0 \\ 0 & 1 & 2a_3 \end{pmatrix} ,$$

notice that if $a_1 + b_1 = 1$, then again $p_2 = 0$ but this can not happen.

Class V. Suppose that

$$f(\mathbf{z}) = \alpha_1 z_1^{a_1} \bar{z}_1^{b_1} z_2 + \alpha_2 z_2^{a_2} \bar{z}_2^{b_2} z_3 + \alpha_3 z_3^{a_3} \bar{z}_3^{b_3} z_1 .$$

In this case the radial and angular matrices are

$$P = \begin{pmatrix} a_1+b_1 & 1 & 0 \\ 0 & a_2+b_2 & 1 \\ 1 & 0 & a_3+b_3 \end{pmatrix} , \quad Q = \begin{pmatrix} a_1-b_1 & 1 & 0 \\ 0 & a_2-b_2 & 1 \\ 1 & 0 & a_3-b_3 \end{pmatrix} .$$

We have that

$$\det P = (a_1 + b_1)(a_2 + b_2)(a_3 + b_3) + 1 ,$$
$$\det Q = (a_1 - b_1)(a_2 - b_2)(a_3 - b_3) + 1 ,$$

hence P is always invertible. Suppose that Q is invertible, therefore

$$P^{-1} = \frac{1}{\det P} \begin{pmatrix} (a_2+b_2)(a_3+b_3) & -(a_3+b_3) & 1 \\ 1 & (a_1+b_1)(a_3+b_3) & -(a_1+b_1) \\ -(a_2+b_2) & 1 & (a_1+b_1)(a_2+b_2) \end{pmatrix} ,$$
$$Q^{-1} = \frac{1}{\det Q} \begin{pmatrix} (a_2-b_2)(a_3-b_3) & -(a_3-b_3) & 1 \\ 1 & (a_1-b_1)(a_3-b_3) & -(a_1-b_1) \\ -(a_2-b_2) & 1 & (a_1-b_1)(a_2-b_2) \end{pmatrix} ,$$

and since the normalized weights satisfy

$$(p_1, p_2, p_3)^\top = P^{-1}(1, 1, 1)^\top,$$
$$(q_1, q_2, q_3)^\top = Q^{-1}(1, 1, 1)^\top,$$

hence

$$p_j = \frac{(a_{j-1} + b_{j-1})(a_{j+1} + b_{j+1} - 1) + 1}{\det P} ,$$
$$q_j = \frac{(a_{j-1} - b_{j-1})(a_{j+1} - b_{j+1} - 1) + 1}{\det Q} ,$$

for $j = 1, 2, 3 \mod 3$.

Since q_j must be non zero, then

$$(a_{j-1} - b_{j-1})(a_{j+1} - b_{j+1} - 1) \neq -1 ,$$

for $j = 1, 2, 3 \mod 3$.

If $\det Q = 0$, then the only solutions (up to a change of coordinates) are

$$a_1 - b_1 = -1 \text{ and } a_2 - b_2 = a_3 - b_3 = 1\ ,$$
$$a_1 - b_1 = a_2 - b_2 = a_3 - b_3 = -1\ .$$

In both cases the system given by Q has no solutions. □

Corollary 4.6. *Let $f\colon \mathbb{C}^3 \to \mathbb{C}$ be a polar weighted homogeneous polynomial belonging to some class of Theorem 4.5 and $n \in \mathbb{Q} \setminus \{0\}$. Then its normalized radial and angular weights are given by:*

Class I. $p_i' := \frac{1}{a_i+b_i}$, $\quad q_i' := \frac{1}{a_i-b_i}$.

Class II.a.

$$p_1' := \frac{1}{a_1+b_1}\ , \qquad p_2' := \frac{1}{a_2+b_2}\ , \qquad p_3' := \frac{a_2+b_2-1}{(a_2+b_2)(a_3+b_3)}\ ,$$

$$q_1' := \frac{1}{a_1-b_1}\ , \qquad q_2' := \frac{1}{a_2-b_2}\ , \qquad q_3' := \frac{a_2-b_2-1}{(a_2-b_2)(a_3-b_3)}\ .$$

Class II.b.

$$p_1' := \frac{1}{a_1+b_1}\ , \qquad p_2' := \frac{1}{1+2b_2}\ , \qquad p_3' := \frac{b_2}{(1+2b_2)b_3}\ ,$$

$$q_1' := \frac{1}{a_1-b_1}\ , \qquad q_2' := 1\ , \qquad q_3' := \frac{n}{a_1-b_1}.$$

Class III.a.

$$p_1' := \frac{1}{a_1+b_1}\ , \qquad p_2' := \frac{1-(a_3+b_3)}{(1-(a_2+b_2)(a_3+b_3))}\ , \qquad p_3' := \frac{1-(a_2+b_2)}{(1-(a_2+b_2)(a_3+b_3))}\ ,$$
$$q_1' := \frac{1}{a_1-b_1}\ , \qquad p_2' := \frac{1-(a_3-b_3)}{(1-(a_2-b_2)(a_3-b_3))}\ , \qquad p_3' := \frac{1-(a_2-b_2)}{(1-(a_2-b_2)(a_3-b_3))}\ .$$

Class III.b. *The radial weights are the same as in* **Class III.a** *but*

$$q_1' := \frac{1}{a_1-b_1}, \quad q_2' := \frac{a_1-b_1-n}{a_1-b_1}, \quad q_3' := \frac{n}{a_1-b_1}, \quad n \neq a_1 - b_1.$$

Class III.c. *The angular weights are the same as in* **Class III.a** *but*

$$p_1' := \frac{1}{a_1+b_1}, \quad p_2' := \frac{a_1+b_1-n}{a_1+b_1}, \quad p_3' := \frac{n}{a_1+b_1}, \quad 1 \leq n \leq a_1+b_1-1.$$

Class IV.a.

$$p_1' := \frac{1}{a_1+b_1}\ , \qquad p_2' := \frac{a_1+b_1-1}{(a_1+b_1)(a_2+b_2)}\ , \qquad p_3' := \frac{(a_1+b_1)(a_2+b_2-1)+1}{(a_1+b_1)(a_2+b_2)(a_3+b_3)}\ ,$$

$$q_1' := \frac{1}{a_1-b_1}\ , \qquad q_2' := \frac{a_1-b_1-1}{(a_1-b_1)(a_2-b_2)}\ , \qquad q_3' := \frac{(a_1-b_1)(a_2-b_2-1)+1}{(a_1-b_1)(a_2-b_2)(a_3-b_3)}\ .$$

Class IV.b.

$$p_1' := \frac{1}{1+2b_1}\ , \quad p_2' := \frac{b_1}{(1+2b_1)a_2}\ , \quad p_3' := \frac{a_2(1+2b_1)-b_1}{(1+2b_1)a_2(a_3+b_3)}\ ,$$

$q_1' := 1\ ,\quad q_2' := 1 - q_3'(a_3-b_3)\ ,\quad q_3' \in \mathbb{Q}$ *with* $q_3' \neq 0$ *and* $q_3'(a_3-b_3) \neq 1$.

Class IV.c.

$$p'_1 := \frac{1}{a_1+b_1}\,, \qquad p'_2 := \frac{a_1+b_1-1}{(a_1+b_1)(a_2+b_2)}\,, \qquad p'_3 := \frac{(a_1+b_1)(a_2+b_2-1)+1}{(a_1+b_1)(a_2+b_2)2b_3}\,,$$

$$q'_1 := \frac{1}{a_1-b_1}\,, \qquad q'_2 := 1\,, \qquad q'_3 := \frac{n}{a_1-b_1}.$$

Class V. *Define*

$$r_i := (1 + (a_i + b_i)(a_{i+1} + b_{i+1}) - (a_{i-1} + b_{i-1})),$$
$$s_i := (1 + (a_i - b_i)(a_{i+1} - b_{i+1}) - (a_{i-1} - b_{i-1})),$$

for $i = 1, 2, 3$. *Then*

$$p'_i := \frac{r_i}{1+(a_1+b_1)(a_2+b_2)(a_3+b_3)}, \qquad q'_i := \frac{s_i}{1+(a_1-b_1)(a_2-b_2)(a_3-b_3)}\,.$$

Proof. Solving the systems appearing in the proof of Theorem 4.5 for each of the classes. □

Corollary 4.7. *Let* $f\colon \mathbb{C}^3 \to \mathbb{C}$ *be a polar weighted homogeneous polynomial belonging to some class of Theorem* 4.5. *Then*

1. f *is full if it is of one of the classes:* **I**, **II.a**, **III.a**, **IV.a**, **V**.
2. f *is radial full if it is of one of the classes:* **II.b**, **III.b**, **IV.b**, **IV.c**.

Corollary 4.8. *Let* $f\colon \mathbb{C}^3 \to \mathbb{C}$ *be a polar weighted homogeneous polynomial belonging to some class of Theorem* 4.5. *Then there exists a change of coordinates such that we get:*

Class I.

$$z_1^{a_1}\bar{z}_1^{b_1} + z_2^{a_2}\bar{z}_2^{b_2} + z_3^{a_3}\bar{z}_3^{b_3}.$$

Class II.a.

$$z_1^{a_1}\bar{z}_1^{b_1} + z_2^{a_2}\bar{z}_2^{b_2} + z_3^{a_3}\bar{z}_3^{b_3}z_2.$$

Class II.b.

$$z_1^{a_1}\bar{z}_1^{b_1} + z_2^{a_2}\bar{z}_2^{b_2} + \tau z_3^{a_3}\bar{z}_3^{b_3}z_2, \quad \tau \in \mathbb{S}^1.$$

Class III.a.

$$z_1^{a_1}\bar{z}_1^{b_1} + z_2^{a_2}\bar{z}_2^{b_2}z_3 + z_3^{a_3}\bar{z}_3^{b_3}z_2.$$

Class III.b.

$$z_1^{a_1}\bar{z}_1^{b_1} + z_2^{a_2}\bar{z}_2^{b_2}z_3 + \tau z_3^{a_3}\bar{z}_3^{b_3}z_2, \quad \tau \in \mathbb{S}^1.$$

Class III.c.

$$z_1^{a_1}\bar{z}_1^{b_1} + z_2z_3.$$

Class IV.a.

$$z_1^{a_1}\bar{z}_1^{b_1} + z_2^{a_2}\bar{z}_2^{b_2}z_1 + z_3^{a_3}\bar{z}_3^{b_3}z_2.$$

Class IV.b.

$$z_1^{a_1}\bar{z}_1^{b_1} + \tau z_2^{a_2}\bar{z}_2^{b_2}z_1 + z_3^{a_3}\bar{z}_3^{b_3}z_2, \quad \tau \in \mathbb{S}^1.$$

Class IV.c.

$$z_1^{a_1}\bar{z}_1^{b_1} + \tau z_2^{a_2}\bar{z}_2^{b_2}z_1 + z_3^{a_3}\bar{z}_3^{b_3}z_2, \quad \tau \in \mathbb{S}^1$$

Class V.

$$z_1^{a_1}\bar{z}_1^{b_1}z_2 + z_2^{a_2}\bar{z}_2^{b_2}z_3 + z_3^{a_3}\bar{z}_3^{b_3}z_1.$$

Proof. By Corollary 4.7 the polynomials in classes **I**, **II.a**, **III.a**, **IV.a**, **V** are full, so just apply Lemma 2.7 and Lemma 2.8. Also by Corollary 4.7 the polynomials in classes **II.b**, **III.b**, **IV.b**, **IV.c** are only radial full. Applying Lemma 2.7 we can assume that all the coefficients are in $\mathbb{S}^1$. Now, consider for instance Class **IV.c**. Its angular matrix has rows pairwise linearly independent, hence, applying Lemma 2.8 we can make any two coefficients equal to 1. Suppose $\alpha_1 = \alpha_2 = 1$, taking the change of coordinates $(z_1, z_2, z_3) \mapsto (z_1, \bar{\alpha}_3 z_2, z_3)$ we get

$$z_1^{a_1}\bar{z}_1^{b_1} + \alpha_3^{b_2-a_2} z_2^{a_2}\bar{z}_2^{b_2}z_1 + z_3^{a_3}\bar{z}_3^{b_3}z_2.$$

For $\alpha_2 = \alpha_3 = 1$, let $\alpha_1 = e^{i\theta}$ and let $\tau = e^{i\frac{\theta}{a_1-b_1}}$. Taking the change of coordinates $(z_1, z_2, z_3) \mapsto (\tau z_1, z_2, z_3)$ we get the expresion we want.

The other classes which are radial full are analogous. □

Definition 4.9. Each of the polynomials given in Corollary 4.8 are called the *special representative* of its corresponding subclass.

Theorem 4.10. *Let $f\colon \mathbb{C}^3 \to \mathbb{C}$ be the special representative of some subclass. Then*

1. *If f is of one the classes* **I**, **II.a**, **III.a**, **III.c**, **IV.a** *or* **V**, *then f has an unique singularity at the origin.*
2. *If f is of one of the classes* **II.b**, **III.b** *or* **IV.b**, *then f has an unique singularity at the origin if and only if $\tau \neq -1$.*
3. *If f is of the classe* **IV.c** *then f has an unique singularity if and only if $\tau \neq 1$.*

Proof. **Class I.** We have

$$\begin{aligned} \overline{df(\mathbf{z})} &= \left(a_1\bar{z}_1^{a_1-1}z_1^{b_1}, a_2\bar{z}_2^{a_2-1}z_2^{b_2}, a_3\bar{z}_3^{a_3-1}z_3^{b_3}\right), \\ \bar{d}f(\mathbf{z}) &= \left(b_1 z_1^{a_1}\bar{z}_1^{b_1-1}, b_2 z_2^{a_2}\bar{z}_2^{b_2-1}, b_3 z_3^{a_3}\bar{z}_3^{b_3-1}\right). \end{aligned} \tag{4.1}$$

Suppose that (z_1, z_2, z_3) is a critical point of f, then by Proposition 2.1 there exist $\alpha \in \mathbb{S}^1$ such that

$$\overline{df(\mathbf{z})} = \alpha\bar{d}f(\mathbf{z}).$$

The previous equality and (4.1) give us the following system

$$\begin{aligned} a_1\bar{z}_1^{a_1-1}z_1^{b_1} &= \alpha b_1 z_1^{a_1}\bar{z}_1^{b_1-1}, \\ a_2\bar{z}_2^{a_2-1}z_2^{b_2} &= \alpha b_2 z_2^{a_2}\bar{z}_2^{b_2-1}, \\ a_3\bar{z}_3^{a_3-1}z_3^{b_3} &= \alpha b_3 z_3^{a_3}\bar{z}_3^{b_3-1}. \end{aligned} \tag{4.2}$$

Suppose $z_j \neq 0$ for some $j \in \{1, 2, 3\}$, then by (4.2) we have

$$a_j\bar{z}_j^{a_j-1}z_j^{b_j} = \alpha b_j z_j^{a_j}\bar{z}_j^{b_j-1},$$

computing the norm we have

$$a_j|z_j|^{a_j+b_j-1} = b_j|z_j|^{a_j+b_j-1},$$

so $a_j = b_j$ which can not occur. Then $\mathbf{0}$ is the only critical point.

Class II.a. We have that $a_j - b_j \neq 0$ with $j = 1,2,3$ and $a_2 \pm b_2 \neq 1$. Again, computing $\overline{df(\mathbf{z})}$, $\bar{d}f(\mathbf{z})$ and using Proposition 2.1, there exist $\alpha \in \mathbb{S}^1$ which gives the following system

$$a_1\bar{z}_1^{a_1-1}z_1^{b_1} = \alpha b_1 z_1^{a_1}\bar{z}_1^{b_1-1}\,, \tag{4.3}$$

$$a_2\bar{z}_2^{a_2-1}z_2^{b_2} + \bar{z}_3^{a_3}z_3^{b_3} = \alpha b_2 z_2^{a_2}\bar{z}_2^{b_2-1}\,, \tag{4.4}$$

$$a_3\bar{z}_3^{a_3-1}z_3^{b_3}\bar{z}_2 = \alpha b_3 z_3^{a_3}\bar{z}_3^{b_3-1}z_2\,. \tag{4.5}$$

Using the ideas of the previous case, it is clear that $z_1 = 0$. Now suppose that $z_2 \neq 0$, we have two cases:

1. If $z_3 \neq 0$, considering equation (4.5) and using the norm we get that $a_3 = b_3$ which can not occur.
2. If $z_3 = 0$, considering equation (4.4) and again using the norm we get $a_2 = b_2$.

We conclude that $z_2 = 0$ and using equation (4.4) we get that $z_3 = 0$. Therefore the only critical point is $\mathbf{0}$.

Class II.b. We have that $a_1 \neq b_1$, $a_2 - b_2 = 1$, $a_3 = b_3$ and $b_2 \neq 0$. By Proposition 2.1, there exist $\alpha \in \mathbb{S}^1$ such that

$$a_1\bar{z}_1^{a_1-1}z_1^{b_1} = \alpha b_1 z_1^{a_1}\bar{z}_1^{b_1-1}\,, \tag{4.6}$$

$$a_2\bar{z}_2^{a_2-1}z_2^{b_2} + \bar{\tau}\bar{z}_3^{a_3}z_3^{b_3} = \alpha b_2 z_2^{a_2}\bar{z}_2^{b_2-1}\,, \tag{4.7}$$

$$\bar{\tau}a_3\bar{z}_3^{a_3-1}z_3^{b_3}\bar{z}_2 = \alpha\tau b_3 z_3^{a_3}\bar{z}_3^{b_3-1}z_2\,. \tag{4.8}$$

Again, we have that $z_1 = 0$ and we can simplify the equations to get

$$\begin{aligned} a_2|z_2|^{2(a_2-1)} + \bar{\tau}|z_3|^{2a_3} &= \alpha(a_2-1)z_2^{a_2}\bar{z}_2^{a_2-2}\,,\\ \bar{\tau}a_3|z_3|^{2(a_3-1)}z_3\bar{z}_2 &= \alpha\tau a_3 z_3|z_3|^{2(a_3-1)}z_2\,. \end{aligned} \tag{4.9}$$

If $z_3 = 0$, then

$$a_2|z_2|^{2(a_2-1)} = \alpha(a_2-1)z_2^{a_2}\bar{z}_2^{a_2-2}\,,$$

therefore z_2 must be 0.

If $z_2 = 0$, then

$$\bar{\tau}|z_3|^{2a_3} = 0\,,$$

therefore z_3 must be 0.

Now suppose $z_2, z_3 \neq 0$. We can simplify equations (4.9) to the following equations

$$\begin{aligned} a_2|z_2|^{2(a_2-1)} + \bar{\tau}|z_3|^{2a_3} &= \alpha(a_2-1)z_2^{a_2}\bar{z}_2^{a_2-2}\,,\\ \bar{\tau}\bar{z}_2 &= \alpha\tau z_2\,, \end{aligned}$$

so we get

$$\tau a_2|z_2|^{2(a_2-1)} + \tau\bar{\tau}|z_3|^{2a_3} = \bar{\tau}(a_2-1)|z_2|^{2(a_2-1)}\,,$$

therefore

$$|z_2|^{2(a_2-1)}(\tau a_2 - \bar{\tau}(a_2-1)) + |z_3|^{2a_3} = 0\,, \tag{4.10}$$

in particular $\tau a_2 - \bar{\tau}(a_2-1)$ must be a real number and by Lemma 4.4 we have that $\tau \in \mathbb{R}$, hence $\tau = \pm 1$.

We can simplify equation (4.10) to

$$|z_2|^{2(a_2-1)}\tau + |z_3|^{2a_3} = 0\,, \tag{4.11}$$

so we have that:

1. If $\tau \neq -1$, then the only critical point is the origin.
2. If $\tau = -1$, then f does not have isolated singularity, for instance the point $(0,1,1)$ satisfies (4.11) and by Lemma 2.16 all the points of the form $(0, t^{p_2}, t^{p_3})$ for $t \in \mathbb{R}^+$ are singular.

Class III.a. Suppose that (z_1, z_2, z_3) is a critical point of f. We have that $a_1 - b_1 \neq 0$ and $a_2 - b_2$, $a_3 - b_3$ are not both -1, also $a_2 \pm b_2$ and $a_3 \pm b_3$ are not 1. By Proposition 2.1, there exist $\alpha \in \mathbb{S}^1$ such that

$$\begin{aligned} a_1 \bar{z}_1^{a_1-1} z_1^{b_1} &= \alpha b_1 z_1^{a_1} \bar{z}_1^{b_1-1}\,,\\ a_2 \bar{z}_2^{a_2-1} z_2^{b_2} \bar{z}_3 + \bar{z}_3^{a_3} z_3^{b_3} &= \alpha b_2 z_2^{a_2} \bar{z}_2^{b_2-1} z_3\,,\\ \bar{z}_2^{a_2} z_2^{b_2} + a_3 \bar{z}_3^{a_3-1} z_3^{b_3} \bar{z}_2 &= \alpha b_3 z_3^{a_3} \bar{z}_3^{b_3-1} z_2\,. \end{aligned} \tag{4.12}$$

As before, we have that $z_1 = 0$. Since $a_2 + b_2$ and $a_3 + b_3$ are not 1, then $z_2 = 0$ if and only if $z_3 = 0$.

For $z_1 = 0$ and $z_2 z_3 \neq 0$ by Theorem 2.5 and Corollary 2.6 the point $(0, z_2, z_3)$ is not a critical point of $z_2^{a_2} \bar{z}_2^{b_2} z_3 + z_3^{a_3} \bar{z}_3^{b_3} z_2$.

Class III.b. We have $a_1 - b_1 \neq 0$ and $a_2 - b_2$, $a_3 - b_3$ are not both -1, also $a_2 + b_2$, $a_3 + b_3$ are not 1 and $a_2 - b_2 = a_3 - b_3 = 1$. The set of equations given by Proposition 2.1 is also given by (4.12). As before, the first equation implies that $z_1 = 0$.

Suppose that $(0, z_2, z_3)$ is a critical point of f, we can simply equations (4.12) to get

$$a_2 \bar{z}_2^{a_2-1} z_2^{a_2-1} \bar{z}_3 + \bar{z}_3^{a_3} z_3^{a_3-1} = \alpha(a_2-1) z_2^{a_2} \bar{z}_2^{a_2-2} z_3\,, \tag{4.13}$$

$$\bar{z}_2^{a_2} z_2^{a_2-1} + a_3 \bar{z}_3^{a_3-1} z_3^{a_3-1} \bar{z}_2 = \alpha(a_3-1) z_3^{a_3} \bar{z}_3^{a_3-2} z_2\,. \tag{4.14}$$

It is clear that $z_2 = 0$ if and only if $z_3 = 0$.

Suppose $z_2, z_3 \neq 0$. Now we have that

$$\begin{aligned} f(0, z_2, z_3) &= z_2 z_3 (z_2^{a_2-1} \bar{z}_2^{a_2-1} + \tau z_3^{a_3-1} \bar{z}_3^{a_3-1})\\ &= z_2 z_3 (|z_2|^{2(a_2-1)} + \alpha |z_3|^{2(a_3-1)}) = 0\,. \end{aligned}$$

Just as for the class **II.b** we have that if $\alpha = -1$, then $(0, t^{p_2}, t^{p_3})$ are singular points of f. Therefore f has an isolated singularity if and only if $\tau \neq -1$.

Class III.c. It is immediate.

Class IV.a. We have that $a_j - b_j \neq 0$, $a_1 \pm b_1 \neq 1$ and $(a_1-b_1)(a_2-b_2-1) \neq -1$. We have the following equations

$$a_1 \bar{z}_1^{a_1-1} z_1^{b_1} + \bar{z}_2^{a_2} z_2^{b_2} = \alpha b_1 z_1^{a_1} \bar{z}_1^{b_1-1}\,, \tag{4.15}$$
$$a_2 \bar{z}_2^{a_2-1} z_2^{b_2} \bar{z}_1 + \bar{z}_3^{a_3} z_3^{b_3} = \alpha b_2 z_2^{a_2} \bar{z}_2^{b_2-1} z_1\,, \tag{4.16}$$
$$a_3 \bar{z}_3^{a_3-1} z_3^{b_3} \bar{z}_2 = \alpha b_3 z_3^{a_3} \bar{z}_3^{b_3-1} z_2\,. \tag{4.17}$$

If $z_1 = 0$, then by (4.15) and (4.16)

$$\bar{z}_2^{a_2} z_2^{b_2} = 0\,,$$
$$\bar{z}_3^{a_3} z_3^{b_3} = 0\,,$$

therefore $z_2 = z_3 = 0$.

If $z_2 = 0$, then by (4.15)

$$a_1 \bar{z}_1^{a_1-1} z_1^{b_1} = \alpha b_1 z_1^{a_1} \bar{z}_1^{b_1-1}\,,$$

which implies

$$a_1 |z_1|^{a_1+b_1-1} = b_1 |z_1|^{a_1+b_1-1}\,,$$

and this only happens if $z_1 = 0$.

If $z_3 = 0$, then by (4.16)

$$a_2 \bar{z}_2^{a_2-1} z_2^{b_2} \bar{z}_1 = \alpha b_2 z_2^{a_2} \bar{z}_2^{b_2-1} z_1\,,$$

therefore

$$a_2 |z_2|^{a_2+b_2-1} |z_1| = b_2 |z_2|^{a_2+b_2-1} |z_1|\,,$$

so $z_1 = 0$ or $z_2 = 0$.

Finally suppose z_1, z_2, z_3 are not 0, then by (4.17)

$$a_3 |z_3|^{a_3+b_3-1} |z_2| = b_3 |z_3|^{a_3+b_3-1} |z_2|\,,$$

but this implies $a_3 = b_3$.

Therefore f has an isolated singularity at the origin.

Class IV.b. We have that $a_2 = b_2$, $a_3 - b_3 \neq 0$, $a_1 - b_1 = 1$ and $b_1 \neq 0$. Using the ideas of the previous cases, we have

$$a_1 \bar{z}_1^{a_1-1} z_1^{b_1} + \bar{\tau} \bar{z}_2^{a_2} z_2^{b_2} = \alpha b_1 z_1^{a_1} \bar{z}_1^{b_1-1}\,, \tag{4.18}$$
$$\bar{\tau} a_2 \bar{z}_2^{a_2-1} z_2^{b_2} \bar{z}_1 + \bar{z}_3^{a_3} z_3^{b_3} = \alpha \tau b_2 z_2^{a_2} \bar{z}_2^{b_2-1} z_1\,, \tag{4.19}$$
$$a_3 \bar{z}_3^{a_3-1} z_3^{b_3} \bar{z}_2 = \alpha b_3 z_3^{a_3} \bar{z}_3^{b_3-1} z_2\,. \tag{4.20}$$

Let (z_1, z_2, z_3) be a critical point of f.

If z_1, z_2, z_3 are not 0, then by (4.20)

$$a_3 |z_3|^{a_3+b_3-1} |z_2| = b_3 |z_3|^{a_3+b_3-1} |z_2|\,,$$

so $a_3 = b_3$ but this can not happen.

Suppose $z_1 = 0$, then by (4.18)

$$\bar{\tau} \bar{z}_2^{a_2} z_2^{b_2} = 0\,,$$

so $z_2 = 0$ and by (4.19)

$$\bar{z}_3^{a_3} z_3^{b_3} = 0 ,$$

therefore $z_3 = 0$.

If $z_2 = 0$, then by (4.18)

$$a_1 \bar{z}_1^{a_1-1} z_1^{b_1} = \alpha b_1 z_1^{a_1} \bar{z}_1^{b_1-1} ,$$

hence

$$a_1 |z_1|^{a_1+b_1-1} = b_1 |z_1|^{a_1+b_1-1} ,$$

but this only happen if $z_1 = 0$ and therefore $z_3 = 0$.

If $z_3 = 0$ and $z_1, z_2 \neq 0$, then by (4.18) and (4.19)

$$a_1 |z_1|^{2(a_1-1)} + \bar{\tau} |z_2|^{a_2+b_2} = \alpha b_1 z_1^2 |z_1|^{2(a_1-2)} , \quad (4.21)$$

$$\bar{\tau} z_2 |z_2|^{2(a_2-1)} \bar{z}_1 = \alpha \tau z_2 |z_2|^{2(a_2-1)} z_1 . \quad (4.22)$$

We have using (4.22)

$$\bar{\tau} \bar{z}_1 = \alpha \tau z_1 ,$$

therefore by (4.21) and since $a_1 - b_1 = 1$,

$$|z_1|^{2(a_1-1)} (a_1 \tau - a_1 \bar{\tau} + \bar{\tau}) + |z_2|^{2a_2} = 0$$

and by Lemma 4.4 this only happen if $\tau = \pm 1$.

If $\tau = 1$, then

$$|z_1|^{2(a_1-1)} + |z_2|^{2a_2} = 0$$

but this can not happen.

If $\tau = -1$, then all the points of the form $(t^{p_1}, t^{p_2}, 0)$ are singular points.

Therefore f has an isolated singularity if and only if $\tau \neq -1$.

Class IV.c. We have that $a_3 = b_3$, $1 + (a_2 - b_2)(a_1 - b_1) = a_1 - b_1$, $a_1 + b_1 \neq 1$ and $a_1 - b_1, a_2 - b_2$ are not 0. The system is

$$a_1 \bar{z}_1^{a_1-1} z_1^{b_1} + \bar{\tau} \bar{z}_2^{a_2} z_2^{b_2} = \alpha b_1 z_1^{a_1} \bar{z}_1^{b_1-1} , \quad (4.23)$$

$$\bar{\tau} a_2 \bar{z}_2^{a_2-1} z_2^{b_2} \bar{z}_1 + \bar{z}_3^{a_3} z_3^{b_3} = \alpha \tau b_2 z_2^{a_2} \bar{z}_2^{b_2-1} z_1 , \quad (4.24)$$

$$a_3 \bar{z}_3^{a_3-1} z_3^{b_3} \bar{z}_2 = \alpha b_3 z_3^{a_3} \bar{z}_3^{b_3-1} z_2 . \quad (4.25)$$

If $z_1 = 0$, then by (4.23)

$$\bar{\tau} \bar{z}_2^{a_2} z_2^{b_2} = 0 ,$$

therefore $z_2 = 0$ and by (4.24)

$$\bar{z}_3^{a_3} z_3^{b_3} = 0 ,$$

so $z_3 = 0$.

If $z_2 = 0$, then by (4.23)

$$a_1 \bar{z}_1^{a_1-1} z_1^{b_1} = \alpha b_1 z_1^{a_1} \bar{z}_1^{b_1-1} ,$$

computing the norm

$$a_1 |z_1|^{a_1+b_1-1} = b_1 |z_1|^{a_1+b_1-1} ,$$

this only happen if $z_1 = 0$.

If $z_3 = 0$, then by (4.24)

$$\bar{\tau} a_2 \bar{z}_2^{a_2-1} z_2^{b_2} \bar{z}_1 = \alpha \tau b_2 z_2^{a_2} \bar{z}_2^{b_2-1} z_1 ,$$

using the norm

$$a_2 |z_2|^{a_2+b_2-1} |z_1| = b_2 |z_2|^{a_2+b_2-1} |z_1| ,$$

hence if z_2, z_3 are not 0, then $a_2 = b_2$ but this can not happen, therefore $z_1 = 0$ or $z_2 = 0$.

If z_1, z_2, z_3 are not 0, then the equation (4.25)

$$a_3 \bar{z}_3^{a_3-1} z_3^{b_3} \bar{z}_2 = \alpha b_3 z_3^{a_3} \bar{z}_3^{b_3-1} z_2 ,$$

give us

$$\bar{z}_2 = \alpha z_2 .$$

Using the polar action we can assume that $z_2 \in \mathbb{R}^+$, therefore the last equality give us $\alpha = 1$ and we can simplify equation (4.24) to

$$a_2 w + r = b_2 \bar{w} , \tag{4.26}$$

where $w = \bar{\tau} z_2^{a_2+b_2-1} \bar{z}_1$ and $r = |z_3|^{2a_3}$.

Therefore $a_2 w - b_2 \bar{w}$ must be a real and by Lemma 4.4 this only happen if $w \in \mathbb{R}$, so $\tau z_1 \in \mathbb{R}$. We can simplify the equation (4.26) to

$$w(a_2 - b_2) + r = 0 . \tag{4.27}$$

Notice that $1 + (a_2 - b_2)(a_1 - b_1) = a_1 - b_1$ only has the solutions

$$a_2 = b_2 , \qquad a_1 - b_1 = 1 , \quad \text{or}$$
$$a_2 - b_2 = 2 , \qquad a_1 - b_1 = -1 ,$$

but $a_2 \neq b_2$, therefore $a_2 - b_2 = 2$ and $a_1 - b_1 = -1$.

So we can simplify (4.27) to

$$2w + r = 0 ,$$

this only happen if $\tau z_1 \in \mathbb{R}^-$.

Multiplying equation (4.23) by $\bar{z}_1$ we get

$$a_1 \bar{z}_1^{a_1} z_1^{b_1} + \bar{\tau} \bar{z}_2^{a_2} z_2^{b_2} \bar{z}_1 = b_1 z_1^{a_1} \bar{z}_1^{b_1} ,$$

since $w = \bar{\tau} z_2^{a_2+b_2-1} \bar{z}_1$,

$$a_1 \bar{z}_1^{a_1} z_1^{b_1} + z_2 w = b_1 z_1^{a_1} \bar{z}_1^{b_1} ,$$

and $b_1 = a_1 + 1$, therefore

$$a_1 \bar{z}_1^{a_1} z_1^{a_1+1} + z_2 w = (a_1 + 1) z_1^{a_1} \bar{z}_1^{a_1+1} ,$$

so

$$|z_1|^{2a_1} (a_1 z_1 - (a_1 + 1)\bar{z}_1) + z_2 w = 0 , \tag{4.28}$$

but $|z_1|, z_2 w \in \mathbb{R}$, so by Lemma 4.4 the only solution is $z_1 \in \mathbb{R} \setminus \{0\}$ and since $w = \bar{\tau} z_2^{a_2+b_2-1} \bar{z}_1 \in \mathbb{R}^-$, then τ must be ± 1.

If $\tau = -1$, equation (4.28) becomes

$$-|z_1|^{2a_1} z_1 - z_2^{a_2+b_2} z_1 = 0 \, ,$$

hence

$$z_1^{2a_1} + z_2^{a_2+b_2} = 0 \, ,$$

but this can not happen since z_1 and z_2 are real numbers different from 0.

If $\tau = 1$, equation (4.28) becomes

$$-z_1^{2a_1} + z_2^{a_2+b_2} = 0 \, ,$$

therefore $(-1, 1, 2^{\frac{1}{2a_3}})$ is a singular point.

Then f has an isolated singularity if and only if $\tau \neq 1$.

Class V. We have $1 + (a_1 - b_1)(a_2 - b_2)(a_3 - b_3) \neq 0$. The system is

$$a_1 \bar{z}_1^{a_1-1} z_1^{b_1} \bar{z}_2 + \bar{z}_3^{a_3} z_3^{b_3} = \alpha b_1 z_1^{a_1} \bar{z}_1^{b_1-1} z_2 \, , \tag{4.29}$$

$$\bar{z}_1^{a_1} z_1^{b_1} + a_2 \bar{z}_2^{a_2-1} z_2^{b_2} \bar{z}_3 = \alpha b_2 z_2^{a_2} \bar{z}_2^{b_2-1} z_3 \, , \tag{4.30}$$

$$\bar{z}_2^{a_2} z_2^{b_2} + a_3 \bar{z}_3^{a_3-1} z_3^{b_3} \bar{z}_1 = \alpha b_3 z_3^{a_3} \bar{z}_3^{b_3-1} z_1 \, . \tag{4.31}$$

If $z_1 = 0$, then by (4.31)

$$\bar{z}_2^{a_2} z_2^{b_2} = 0 \, ,$$

so $z_2 = 0$ and by (4.29)

$$\bar{z}_3^{a_3} z_3^{b_3} = 0 \, ,$$

therefore $z_3 = 0$.

If $z_2 = 0$, then $z_3 = 0$ and by (4.30)

$$\bar{z}_1^{a_1} z_1^{b_1} = 0,$$

so $z_1 = 0$.

For $z_1 z_2 z_3 \neq 0$ by Theorem 2.5 and Corollary 2.6 the point (z_1, z_2, z_3) is not a critical point of f.

Therefore the origin is the only singularity. □

5. Diffeomorphism type of the link under perturbation of the coefficients

In Section 3 we proved that given a polar weighted homogeneous polynomial with isolated critical point, with a small perturbation of its coefficients it still has isolated critical point. In this section we prove that under such perturbation the diffeomorphism type of the link does not change. We follow the proof of [9, Theorem 3.1.4] by Orlik and Wagreich.

Proposition 5.1. *Let $f \colon \mathbb{C}^3 \to \mathbb{C}$ be a polar weighted homogeneous polynomial of radial weight type $(p_1, p_2, p_3; a)$ and angular weight type $(q_1, q_2, q_3; b)$ with isolated singularity at the origin. Then f can be written as*

$$f(\mathbf{z}) = h(\mathbf{z}) + g(\mathbf{z}),$$

where h belongs to one of the classes of Definition 4.1, h and g have no monomials in common and both are also polar weighted homogeneous polynomials of radial weight type $(p_1, p_2, p_3; a)$ and angular weight type $(q_1, q_2, q_3; b)$.

Proof. Applying Lemma 2.3 several times we obtain that f must contain a polynomial h in one of the classes. □

Definition 5.2. Let $f\colon \mathbb{C}^3 \to \mathbb{C}$ be a polar weighted homogeneous polynomial. By Proposition 5.1 it can be written in the form $f(\mathbf{z}) = h(\mathbf{z}) + g(\mathbf{z})$ where h belongs to one of the classes of Definition 4.1. We say that f *corresponds* to that class.

Let $f\colon \mathbb{C}^3 \to \mathbb{C}$ be a polar weighted homogeneous polynomial with isolated singularity at the origin. Let $V = f^{-1}(0)$ and $K = V \cap \mathbb{S}^5$. By Proposition 5.1 we can write $f(\mathbf{z}) = h(\mathbf{z}) + g(\mathbf{z})$. Let

$$f = \sum_{j=1}^{r} \alpha_j M_j \ ,$$

where M_j is a monomial on the variables $z_i, \bar{z}_i$ for $i = 1, 2, 3$ and

$$h = \sum_{j=1}^{3} \alpha_j M_j \ , \qquad g = \sum_{j=4}^{r} \alpha_j M_j \ .$$

Given $\mathbf{w} = (w_1, \dots, w_r) \in \mathbb{C}^r$ consider the mixed function

$$f_{\mathbf{w}}(\mathbf{z}) = \sum_{j=1}^{r} w_j M_j(\mathbf{z})$$

and let $V_{\mathbf{w}} = f_{\mathbf{w}}^{-1}(0) \subset \mathbb{C}^3$ be its zero-locus and $K_{\mathbf{w}} = V_{\mathbf{w}} \cap \mathbb{S}^5$ its link. Notice that for $\alpha = (\alpha_1, \dots, \alpha_r) \in \mathbb{C}^r$ we have $f_\alpha = f$, $V_\alpha = V$ and therefore $K = K_\alpha$. Hence we have a family of polar weighted homogeneous polynomials $f_{\mathbf{w}}\colon \mathbb{C}^n \to \mathbb{C}$ where the parameter $\mathbf{w}$ belongs to the parameter space $\mathbb{C}^r$.

We want to construct a manifold M with a $\mathbb{S}^1$-action, an open set $U \subset \mathbb{C}^r$ and a map $\phi\colon M \to U$, such that the action leaves $\phi^{-1}(\mathbf{w})$ invariant for all $\mathbf{w} \in U$, $\phi^{-1}(\mathbf{w}) \cong K_{\mathbf{w}}$ equivariantly and ϕ is a locally trivial fibration.

Consider the function $k\colon \mathbb{C}^{r+3} \to \mathbb{C}$ given by

$$k(\mathbf{z}, w_1, \dots, w_r) = \sum_{j=1}^{r} w_j M_j(\mathbf{z}) \ ,$$

let

$$N := k^{-1}(0) = \bigsqcup_{\mathbf{w} \in \mathbb{C}^r} V_{\mathbf{w}} \times \{\mathbf{w}\} \subset \mathbb{C}^{r+3} \ ,$$

$$C := \mathbb{S}^5 \times \mathbb{C}^r,$$

let $\phi_0\colon \mathbb{C}^{r+3} \to \mathbb{C}^r$ be the projection onto the last r coordinates and set $\phi_1 := \phi_0|_N$.

Define

$$U := \{\mathbf{w} \in \mathbb{C}^r \mid \phi_1^{-1}(\mathbf{w}) = V_{\mathbf{w}} \times \{\mathbf{w}\} \text{ has an isolated singularity at } 0\},$$
$$M := C \cap \phi_1^{-1}(U) = \bigsqcup_{\mathbf{w} \in U} K_{\mathbf{w}} \times \{\mathbf{w}\} \subset \mathbb{C}^{r+3},$$
$$\phi := \phi_1|_M \colon M \to U.$$

Notice that for any $\mathbf{w} \in U$ we have that $\phi^{-1}(\mathbf{w}) = K_{\mathbf{w}} \times \{\mathbf{w}\}$ and by Corollary 3.5 and Corollary 3.6 U is an open set. If $f_{\mathbf{w}}$ is a family of polar weighted homogeneous polynomials, another way to see that U is open (pointed out to us by the referee) is the following: the singular locus

$$W = \left\{ (\mathbf{z}, \mathbf{w}) \in \mathbb{S}^{2n+1} \times \mathbb{C}^r \;\middle|\; \begin{array}{l} f_{\mathbf{w}}(\mathbf{z}) = 0 \text{ is singular at } (\mathbf{z}, \mathbf{w}) \\ \text{as a mixed variety in } \{p\} \times \mathbb{C}^n \end{array} \right\}$$

is a real algebraic set as it is defined by the vanishing of 2×2 minors of the Jacobian matrices of the real and imaginary part of $f_p(\mathbf{z})$. In particular it is a closed set. So the projection of the complement of W onto $\mathbb{C}^r$ is open and it is precisely U. We have that

$$\tau \circ (z_1, z_2, z_3, w_1, \dots, w_r) = (\tau^{q_1} z_1, \tau^{q_2} z_2, \tau^{q_3} z_3, w_1, \dots, w_r), \qquad \tau \in \mathbb{S}^1$$

is the required $\mathbb{S}^1$-action on M.

Theorem 5.3. *The map $\phi \colon M \to U$ is a locally trivial fibration.*

Proof. The proof is a generalization of the proof of [9, Theorem 3.1.4] by Orlik and Wagreich.

Step 1: The map $\phi \colon M \to U$ has no critical points.

Let $m = (\mathbf{z}, w_1, \dots, w_r) \in M$ and $\mathbf{w} = (w_1, \dots, w_r)$. Let $k_1, k_2 \colon \mathbb{C}^{r+3} \cong \mathbb{R}^{2(r+3)} \to \mathbb{R}$ be the real and imaginary parts of k respectively. Consider the matrix of partial derivatives at m

$$A = \begin{pmatrix} \frac{\partial k_1}{\partial x_1}(m) & \frac{\partial k_1}{\partial y_1}(m) & \dots & \frac{\partial k_1}{\partial y_{r+3}}(m) \\ \frac{\partial k_2}{\partial x_1}(m) & \frac{\partial k_2}{\partial y_1}(m) & \dots & \frac{\partial k_2}{\partial y_{r+3}}(m) \end{pmatrix},$$

we are taking coordinates $z_j = x_j + iy_j$ for $j = 1, 2, 3$ and $w_{j-3} = x_j + iy_j$ for $j = 4, \dots, r$.

Since $m \in \phi_1^{-1}(U)$, the point $\mathbf{z} \in \mathbb{C}^3$ is a regular point of $f_{\mathbf{w}}$ and the six first columns of A are precisely the Jacobian of $f_{\mathbf{w}}$ at $\mathbf{z} \in \mathbb{C}^3$ therefore the rank of A is 2 and m is a regular point of k.

Let $T_m N$ and $T_m C$ denote the tangent spaces at m to N and C respectively. We know that $T_m C$ is the real hyperplane orthogonal to $(\mathbf{z}, 0, \dots, 0)$.

Using the radial action given by f, we have an action on $\mathbb{C}^{r+3}$ given by

$$t * (\mathbf{z}, w_1, \dots, w_r) := (t \bullet \mathbf{z}, w_1, \dots, w_r)$$

for any $t \in \mathbb{R}^+$.

With this, we have that $k(t * (\mathbf{z}, w_1, \ldots, w_r)) = t^a k(\mathbf{z}, w_1, \ldots, w_r)$, therefore if we denote by

$$v := \frac{d}{dt}(t * m)|_{t=1} = (p_1 z_1, p_2 z_2, p_3 z_3, 0, \ldots, 0)$$

then $v \in T_m N$ and $v \notin T_m C$, therefore $T_m N$ and $T_m C$ intersect transversely at m.

So we can denote by $T_m M$ the tangent space at m to M and we have that $T_m M = T_m N \cap T_m C$.

We need to prove that

$$\ker \phi_0 + T_m M = \mathbb{C}^{r+3} .$$

Since $T_m C = \{(v_1, v_2, v_3) \in \mathbb{C}^3 \mid \Re\langle \mathbf{z}, (v_1, v_2, v_3)\rangle = 0\} \times \mathbb{C}^r$ and $T_m M = T_m N \cap T_m C$, then it is enough to prove that

$$\ker \phi_0 + T_m N = \mathbb{C}^{r+3}.$$

Denote by $\{e_1, \ldots, e_{2(r+3)}\}$ the canonical basis of $\mathbb{R}^{2(r+3)} \cong \mathbb{C}^{r+3}$, so we have that $e_j \in \ker \phi_0$ for $j = 1, \ldots, 6$.

Notice that

$$Ae_{2j-1}^{\top} = \left(\frac{\partial k_1}{\partial x_j}(m), \frac{\partial k_2}{\partial x_j}(m)\right)^{\top}, \quad Ae_{2j}^{\top} = \left(\frac{\partial k_1}{\partial y_j}(m), \frac{\partial k_2}{\partial y_j}(m)\right)^{\top}$$

for $j = 1, 2, 3$.

Since $m \in \phi_1^{-1}(U)$, then there exists two vectors e_{j_1}, e_{j_2} such that $Ae_{j_1}^{\top} \neq 0$, $Ae_{j_2}^{\top} \neq 0$ and $Ae_{j_1}^{\top} + tAe_{j_2}^{\top} \neq 0$ for all $t \in \mathbb{R}$. Therefore we have two vectors $e_{j_1}, e_{j_2} \in \ker \phi_0$ such that $e_{j_1}, e_{j_2} \notin T_m N$ and

$$\operatorname{span}\{e_{j_1}, e_{j_2}\} \cap T_m N = \mathbf{0} ,$$

therefore the intersection is transversal.

Step 2: The map $\phi \colon M \to U$ is proper.

Let L be a compact subset of $U \subset \mathbb{C}^r$. We have that

$$M = (\mathbb{S}^5 \times \mathbb{C}^r) \cap N \cap (\mathbb{C}^3 \times U) ,$$

and

$$\phi^{-1}(L) = (\mathbb{S}^5 \times \mathbb{C}^r) \cap N \cap (\mathbb{C}^3 \times L) = (\mathbb{S}^5 \times L) \cap N \subset \mathbb{C}^{3+r} .$$

Since $\mathbb{S}^5 \times L$ is close and bounded and N is closed in $\mathbb{C}^{3+r}$, hence $\phi^{-1}(L)$ is compact in $\mathbb{C}^{3+r}$, therefore $\phi^{-1}(L)$ it is also compact in M with the subspace topology. This proves that ϕ is proper.

Since $\phi \colon M \to U$ is a proper submersion, by Ehresmann Fibration Theorem it is a smooth fibre bundle over it image. □

Corollary 5.4. *Let $\phi \colon M \to U$ be as in Theorem 5.3. Let $\tilde{\mathbf{w}} \in U$ and consider the polar weighted homogeneous polynomial with isolated critical point $f_{\tilde{\mathbf{w}}}$. Then there exist a ball $\mathbb{B}(\tilde{\mathbf{w}}, \epsilon)$ centred at $\tilde{\mathbf{w}}$ such that for any $\mathbf{w} \in \mathbb{B}(\tilde{\mathbf{w}}, \epsilon)$ the link $K_{\mathbf{w}}$ of $f_{\mathbf{w}}$ is diffeomorphic to the link $K_{\tilde{\mathbf{w}}}$ of $f_{\tilde{\mathbf{w}}}$.*

Remark 5.5. Recall that we considered $f: \mathbb{C}^3 \to \mathbb{C}$ be a polar weighted homogeneous polynomial with isolated singularity at the origin written as $f(\mathbf{z}) = h(\mathbf{z}) + g(\mathbf{z})$ by Proposition 5.1 and such that

$$f = \sum_{j=1}^{r} \alpha_j M_j ,$$

where M_j is a monomial on the variables $z_i, \bar{z}_i$ for $i = 1, 2, 3$ and

$$h = \sum_{j=1}^{3} \alpha_j M_j , \qquad g = \sum_{j=4}^{r} \alpha_j M_j .$$

By Theorem 4.10 the vector $\mathbf{w}_0 = (1, 1, 1, 0, \ldots, 0) \in U$ (except for the class **IV.c** for which we take $\mathbf{w}_0 = (1, -1, 1, 0, \ldots, 0) \in U$), and also $\alpha \in U$. It may happen that the set $U \subset \mathbb{C}^r$ is not connected, in this case we cannot conclude that $K_{\mathbf{w}_0} = \phi^{-1}(\mathbf{w}_0)$ is diffeomorphic to $K_\alpha = \phi^{-1}(\alpha)$ as in the complex case. If this is the case, one has to study the connected components of U.

This phenomenon is shown in two variables in an example given by Oka in [7, Example 59] or [2, §3.2].

Corollary 5.6. *Let f be a polar weighted homogeneous polynomial with isolated singularity at the origin and let K be its link.*

- *In the classes* **II.b**, **III.b IV.b** *or* **IV.d**, *the diffeomorphism type of the link K of f is the same for any $\tau \neq -1$. In particular, we can take $\tau = 1$.*
- *In the classe* **IV.c**, *the diffeomorphism type of the link K of f is the same for any $\tau \neq 1$. In particular we can take $\tau = -1$.*

Proof. Let f be a special representative of each of the aforementioned classes. If we vary τ in $\mathbb{C}$ by Theorem 4.10 for classes **II.b**, **III.b IV.b** or **IV.d** the locus where f has non-isolated critical point is the non-positive real ray. Also by Theorem 4.10 for class **IV.c** the locus where f has non-isolated critical point is the non-negative real ray. Therefore in these cases U is connected. □

The word classification in the title is meant in a coarse sense: by Proposition 5.1 every polar weighted homogeneous polynomial with isolated singularity corresponds to one of the subclasses of Corollary 4.8. By Remark 5.5 the parameter space can have several connected components, so the natural step to follow is to study the topology of the Milnor fibre (Milnor number, characteristic polynomial, etc.) for the special representatives of each subclass and then study how this topology change when we change connected component. This will appear in a future work.

Acknowledgment

We thank Professor Mutsuo Oka for pointing out to us his example and for fruitful conversations. We thank the referee for his/her corrections and suggestions which greatly improve and simplified the presentation of this article.

References

[1] V. I. Arnold. Normal forms of functions in the neighbourhood of degenerate critical points. *Russian Mathematical Surveys*, 29(2):19–48, 1974.

[2] Vincent Blanlœil and Mutsuo Oka. Topology of strongly polar weighted homogeneous links. *SUT J. Math.*, (1):119–128, 2015.

[3] Ying Chen. Milnor fibration at infinity for mixed polynomials. *Cent. Eur. J. Math.*, 12(1):28–38, 2014.

[4] José Luis Cisneros-Molina. Join theorem for polar weighted homogeneous singularities. In *Singularities II*, volume 475 of *Contemp. Math.*, pages 43–59. Amer. Math. Soc., Providence, RI, 2008.

[5] Mutsuo Oka. *Non-degenerate complete intersection singularity*. Actualités Mathématiques. [Current Mathematical Topics]. Hermann, Paris, 1997.

[6] Mutsuo Oka. Topology of polar weighted homogeneous hypersurfaces. *Kodai Math. J.*, 31(2):163–182, 2008.

[7] Mutsuo Oka. Non-degenerate mixed functions. *Kodai Math. J.*, 33(1):1–62, 2010.

[8] Mutsuo Oka. On mixed Brieskorn variety. In *Topology of algebraic varieties and singularities*, volume 538 of *Contemp. Math.*, pages 389–399. Amer. Math. Soc., Providence, RI, 2011.

[9] Peter Orlik and Philip Wagreich. Isolated singularities of algebraic surfaces with $\mathbb{C}^*$ action. *Ann. of Math. (2)*, 93:205–228, 1971.

[10] Maria Aparecida Soares Ruas, José Seade, and Alberto Verjovsky. On real singularities with a Milnor fibration. In A. Libgober and M. Tibar, editors, *Trends in singularities*, Trends Math., pages 191–213. Birkhäuser, Basel, 2002.

José Luis Cisneros-Molina and Agustín Romano-Velázquez
Instituto de Matemáticas, Unidad Cuernavaca
Universidad Nacional Autónoma de México
Avenida Universidad s/n
Colonia Lomas de Chamilpa
CP62210, Cuernavaca, Morelos Mexico
e-mail: jlcm@matcuer.unam.mx
agustin.romano@im.unam.mx

Rational and Iterated Maps, Degeneracy Loci, and the Generalized Riemann-Hurwitz Formula

James F. Glazebrook and Alberto Verjovsky

Dedicated to Professor José Seade on his 60th birthday

Abstract. We consider a generalized Riemann-Hurwitz formula as it may be applied to rational maps between projective varieties having an indeterminacy set and fold-like singularities. The case of a holomorphic branched covering map is recalled. Then we see how the formula can be applied to iterated maps having branch-like singularities, degree lowering curves, and holomorphic maps having a fixed point set. Separately, we consider a further application involving the Chern classes of determinantal varieties when the latter are realized as the degeneracy loci of certain vector bundle morphisms.

Mathematics Subject Classification (2000). Primary 57M12, 32C10; Secondary 57R19, 32H50.

Keywords. Riemann-Hurwitz formula, rational maps, iterated maps, degeneracy locus, determinantal variety.

1. Introduction

This paper has to main aims: Firstly we consider (generalized) higher dimensional versions of the classical Riemann-Hurwitz formula as initially applied to rational maps of complex projective varieties $f : X \longrightarrow Y$, where X and Y have the same complex dimension. The main results presented here (Theorem 2.1 and Theorem 2.2) are derived from the general setting of [12] formulated mainly in the category of CW-complexes, and then applying the basics of the topological theory of characteristic classes of G-bundles (for suitable groups G) and associated vector bundles. Other versions of a generalized Riemann-Hurwitz formula, such as in the differentiable category, had previously been obtained in [6, 28, 25]. In this first part we will be applying the main results to operations involving rational maps, as for

This work was partially supported by PAPIIT (Universidad Nacional Autónoma de México) #IN103914.

instance, realized in various algebraic-geometric and complex-dynamical constructions (beyond, that is, the familiar case of holomorphic branched covering maps). Applications and examples include:

(1) iterates of rational maps $f : X \longrightarrow Y$ [4](cf. [10]);
(2) rational self-maps $f : X \longrightarrow X$, with respect to their fixed point sets [1], and degree lowering curves [10].

In the second part, we will apply those same main results and the allied constructions to the study of determinantal varieties when the latter are realized as the degeneracy loci of morphisms $\psi : E \longrightarrow F$, of complex vector bundles over X. From a general formula established in §5.1, we pay attention to two particular cases: general symmetric bundle maps [19] and flagged bundles [11].

Throughout, the formulas are established in terms of Chern classes, as by now the traditional method for studying invariants in the algebraic-geometric category. Specifically, such formulas regulate the necessary topological conditions for the existence of a given class of rational maps (or morphisms), just as the classical Riemann-Hurwitz formula applies when studying holomorphic maps of algebraic curves (viz. compact Riemann surfaces).

The construction and main results of [12] were adapted in [14] to cover the case of generalized monoidal transformations. A further work [15] is intended to further elaborate on this construction, as well as to bring into focus a number of results obtained by other authors concerning the blowing-up process of (singular) Chern classes following the original study undertaken by I. R. Porteous [26, 27] and R. Thom [29] (see also [2, 13]).

2. The topological background

Given a topological group G, we start by recalling from [12] a general result valid in the characteristic ring of G-bundles when defined initially within the category of CW-complexes.

2.1. Adapted pairs

Following [12], let Λ be a given (commutative) coefficient ring and (M, M_1) a pair of CW-complexes, with M of dimension n and M_1 a subcomplex of codimension $r \geq 2$, so that $H_q(M, \Lambda) = 0$ for $q > n$, and $H_q(M_1, \Lambda) = 0$ for $q \geq n - 1$. Then the pair (M, M_1) is called (n, Λ)-*adapted* if $H_n(M, \Lambda) \cong \Lambda$, and the following condition holds: there exists a neighborhood $N(M_1)$ of M_1, such that M_1 is a deformation retract of the interior $N^0(M_1)$ of $N(M_1)$, and the inclusion map $p : M \longrightarrow (M, M - N^0(M_1))$ induces an isomorphism

$$p_* : H_n(M) \xrightarrow{\cong} H_n(M, M - N^0(M_1)). \tag{2.1}$$

Having defined an adapted pair we next form the subspace $\mathcal{K}$ of $M \times I$, where

$$\mathcal{K} = (M \times \partial I) \cup ((M, M - N^0(M_1)) \times I), \tag{2.2}$$

together with the double $S(M_1) \subset \mathcal{K}$, given by

$$S(M_1) = (\partial N(M_1) \times I) \cup (N(M_1) \times \partial I). \tag{2.3}$$

Now let

$$\mathcal{K}_1 = (M \times \{0\}) \cup ((M, M - N^0(M_1)) \times [0, \frac{3}{4}],$$

and

$$\mathcal{K}_2 = (M \times \{1\}) \cup ((M, M - N^0(M_1)) \times [0, \frac{1}{4}],$$

and let $S(M_1)_i = S(M_1) \cap \mathcal{K}_i$, for $i = 1, 2$.

By this construction, the spaces $\mathcal{K}_i$ are homotopically equivalent to M_1 and the spaces $\mathcal{K}/\mathcal{K}_1$ and $S(M_1)/S(M_1)_1$, are both homotopically equivalent to the generalized Thom space $M/(M - N^0(M_1))$.

It follows from the cofibration

$$S(M_1)_1 \longrightarrow S(M_1) \longrightarrow S(M_1)/S(M_1)_1 \tag{2.4}$$

and the above definition that

$$H_n(S(M_1)_1, \Lambda) \cong H_n(S(M_1)/S(M_1)_1) \cong \Lambda. \tag{2.5}$$

A choice of generators for $H_n(M, \Lambda)$ and $H_n(S(M_1), \Lambda)$ will be called an orientation or a fundamental class $[M]$ of M and $[S(M_1)]$ of $S(M_1)$, respectively.

Remark 2.1. Observe that the conditions defining an (n, Λ)-adapted pair above are immediately satisfied when M is a closed (compact without boundary) connected orientable n-manifold, and M_1 is a closed connected and orientable submanifold of codimension $r \geq 2$, with Λ any coefficient ring. This example also applies to topological, PL, as well as smooth (sub)manifolds, with $S(M_1)$ a corresponding normal sphere bundle (also closed, connected and orientable for $H_n(S(M_1), \Lambda) \cong \Lambda$, given the above topological type of M_1).

More generally, M could be considered as an orientable simple n-circuit (or 'triangulated pseudomanifold' in the sense of [16]; see also [9, 23]), and M_1 an arbitrary subcomplex of codimension $r \geq 2$, such that (M, M_1) satisfies the conditions of an (n, Λ)-adapted pair.

Let BG denote the classifying space of the topological group G. If $P \in H^q(BG, \Lambda)$ is a cohomology class, and $E \longrightarrow X$ is a G-bundle over a space X, then we shall denote by $P(E) \in H^q(X, \Lambda)$ the class defined by the characteristic class $P(E) = \Phi_E^*(P)$ where $\Phi_E : X \longrightarrow BG$ is the classifying map (for the basic details see, e.g., [5, 7, 21]).

If $\tau \in H_q(X, \Lambda)$, then we denote the Kronecker pairing by $\langle P(E), \tau \rangle$, which by definition is zero in the case where the cohomological degree of $P(E)$ is different from the homological degree (or dimension) of the cycle τ.

We have then the general result from [12, Theorem 1.1]:

Theorem 2.1. *Suppose (M, M_1) is an (n, Λ)-adapted pair and E, F are G-bundles over M, such that on $M - M_1$ there exists a homotopy*

$$\theta : \Phi_E|_{M-M_1} \sim \Phi_F|_{M-M_1}. \tag{2.6}$$

Then there exists a G-bundle $\xi_\theta \longrightarrow S(M_1)$ and orientations $[M]$ and $[S(M_1)]$, such that for any class $P \in H^n(BG, \Lambda)$, we have the following equality of Kronecker pairings as established in [12, Th. (1.1)]*:*

$$\langle P(E) - P(F), [M] \rangle = \langle P(\xi_\theta), [S(M_1)] \rangle. \tag{2.7}$$

Remark 2.2. An analogous result in the context of generalized monoidal transformations was given in [14].

2.2. The clutching construction

We now give a more explicit construction of the G-bundle $\xi_\theta \longrightarrow S(M_1)$ which can be used for the applications which are to follow.

Suppose now that the n-adapted pair (M, M_1) is a pair of closed and connected orientable manifolds of dimension n, respectively of codimension $r \geq 2$ (as in Remark 2.1; again, this could be taken in the topological, PL, or smooth context). Also, without much loss of generality, we suppose that G is one of the Lie groups $\mathrm{SU}(\ell), \mathrm{SO}(\ell)$ or $\mathrm{Sp}(\ell)$, corresponding to a complex, real oriented or symplectic vector bundle of rank ℓ, respectively. We may also consider $G = \mathrm{O}(\ell)$ and real vector bundles of rank ℓ when $\Lambda = \mathbb{Z}_2$. For ease of notation, we retain E and F to denote the associated vector bundles to those G-bundles as previously.

Suppose now we have a homomorphism $\psi : E \longrightarrow F$, such that

i) $\psi : E|_{M-M_1} \xrightarrow{\cong} F|_{M-M_1}$;
ii) $\psi|_{M_1}$ has constant rank.

Here ψ is viewed as a 'clutching function' [3, 22] used to clutch E and F over $S(M_1)$ (cf. [25]). With these assumptions, we have then the following exact sequence of vector bundles on M_1:

$$0 \longrightarrow K_1 \longrightarrow E|_{M_1} \xrightarrow{\psi} F|_{M_1} \longrightarrow K_2 \longrightarrow 0, \tag{2.8}$$

where $K_1 \cong \ker \psi$, and $K_2 \cong \operatorname{coker} \psi$. Further, let

$$L := \psi(E|_{M_1}) \subset F|_{M_1}. \tag{2.9}$$

From this we may exhibit on $S(M_1)$ vector bundle isomorphisms for the clutched bundle as given by

$$E|_{M_1} \cong K_1 \oplus L, \text{ and } F|_{M_1} \cong K_2 \oplus L. \tag{2.10}$$

In the development of ideas that follow, the vector bundle L in (2.9) along with its characteristic classes will be essential objects for producing formulas that will specialize the righthand side of (2.7).

In this setting, a geometric realization of $S(M_1)$ can be given as follows. Let $B(M_1)$ be a closed tubular neighborhood of M_1. For $i = 1, 2$, let $B_i(M_1)$ be two distinct copies of $B(M_1)$. Identifying $B_1(M_1)$ and $B_2(M_1)$ along their common boundary $\partial B(M_1) = \partial B_1(M_1) = \partial B_2(M_1)$, then $S(M_1)$ may be realized as the 'double' $S(M_1)$ obtained by setting

$$S(M_1) = B_1(M_1) \cup_{\partial B(M_1)} B_2(M_1). \tag{2.11}$$

This induces a closed S^r-fibration $q : S(M_1) \longrightarrow M_1$, with restriction maps

$$q_i = q|_{B_i(M_1)} : B_i(M_1) \longrightarrow M_1, \tag{2.12}$$

that can be seen to be deformation retracts (see, e.g., [3, 22]).

Observe that $\psi : E \longrightarrow F$ is a isomorphism when restricted to the common boundary $\partial B_1(M_1) = \partial B_2(M_1)$.

As in [25], we may exhibit a vector bundle isomorphism

$$\xi = (E, \psi, F) \cong (q_1^* K_1, \eta, q_2^* K_2) \oplus q^* L, \tag{2.13}$$

where η is the appropriate clutching function. For ease of notation, let us set the clutched bundle $(q_1^* K_1, \eta, q_2^* K_2) = K$, so that ξ is expressed as the direct sum

$$\xi = K \oplus q^* L. \tag{2.14}$$

2.3. The result in the characteristic ring

Observe that given an oriented sphere bundle $q : S(M_1) \longrightarrow M_1$ with fibre S^r as above, we have the associated (long exact) Gysin sequence (see, e.g., [5, 21])

$$\ldots H^i(S(M_1)) \xrightarrow{q_*} H^{i-r}(M_1) \xrightarrow{\cup e} H^{i+1}(M_1) \xrightarrow{q^*} H^{i+1}(S(M_1)) \ldots \tag{2.15}$$

in which the maps q_*, $\cup e$, and q^* are integration along the fibre, the product with the Euler class, and the natural pull-back, respectively.

For a given vector bundle E of rank ℓ over M and with the corresponding characteristic class $P = P_E \in H^*(BG, \Lambda)$, for $G = \mathcal{U}(\ell), \mathrm{SO}(\ell)$ or $\mathrm{Sp}(\ell)$, typically one considers the image $\Phi_E^*(P)$ in $H^*(M, \Lambda)$. Recall (from, e.g., [5, 21]) that $H^*(BG, \Lambda)$ is a polynomial ring in the Chern classes (for complex vector bundles, with $\Lambda = \mathbb{Z}$); in the Pontrjagin classes and Euler class (for real oriented vector bundles, with $\Lambda = \mathbb{Z}[\frac{1}{2}]$); or the corresponding Pontrjagin classes (for symplectic vector bundles, with $\Lambda = \mathbb{Z}$), and in the Stiefel-Whitney classes (for real vector bundles, with $\Lambda = \mathbb{Z}_2$).

A differentiable version of the following result was established [25] and stated explicitly in terms of the Chern forms of a principal $\mathrm{U}(q)$-bundle (for some q).

Theorem 2.2. *Let K and L be as in (2.14), with $q : S(M_1) \longrightarrow M_1$ the r-sphere bundle as above. With respect to the clutched bundle $\xi = (E, \psi, F)$ in (2.14) and the classifying map Φ, suppose*

1) *there exists a splitting $\Phi_\xi^*(P) = \Phi_K^*(P) \cup \Phi_{q^*L}^*(P)$, and*
2) *the cohomological degree of $P_K \leq \mathrm{rank}(K) = r$.*

Then we have the following equality in characteristic numbers,

$$\langle \Phi_\xi^*(P), [S(M_1)] \rangle = \langle \Phi_K^*(P) \cup \Phi_{q^*L}^*(P), [S(M_1)] \rangle = k \langle \Phi_L^*(P), [M_1] \rangle, \tag{2.16}$$

and hence

$$\langle P(E) - P(F), [M] \rangle = \langle P(\xi), [S(M_1)] \rangle = k \langle P(L), [M_1] \rangle, \tag{2.17}$$

for some constant $k \in \Lambda$.

Proof. Let $\alpha \in \Phi^*_L(P)$, and $\beta \in \Phi^*_K(P)$. Then if q_* and q^* are the maps in (2.15), we have via fibre-integration along $S(M_1)|_{x\in M_1}$, the equality $q_*(q^*(\alpha) \cup \beta) = \alpha \cup q_*(\beta)$, which on integrating over M_1, yields

$$\langle \Phi^*_L(P)\langle \Phi^*_K(P), [S(M_1)_x]\rangle, [M_1]\rangle = \langle \Phi^*_K(P), [S(M_1)_x]\rangle\langle \Phi^*_L(P), [M_1]\rangle. \tag{2.18}$$

Now $\langle \Phi^*_K(P), [S(M_1)_x]\rangle = \langle \Phi^*_{K_x}(P), [S(M_1)_x]\rangle$, where

$$K_x = (q_1^* K_1|_{B_1(M_1)_x}, \eta_x, q_2^* K_2|_{B_2(M_1)_x}), \tag{2.19}$$

the vector bundle over $S(M_1)|_{x\in M_1}$ constructed via the transition function η_x seen as the restriction of η to $\partial B(M_1)_x$. Accordingly, we have an isomorphism

$$\eta_x : q_1^* K_1|_{\partial B(M_1)_x} \xrightarrow{\cong} q_2^* K_2|_{\partial B(M_1)_x}. \tag{2.20}$$

If $c(x, y)$ is a curve in M_1 joining two points $x, y \in M_1$, then the restrictions $K_1|_{c(x,y)}$, and $K_2|_{c(x,y)}$ are trivial. We have then the following diagram in which the vertical maps are isomorphisms

$$\begin{array}{ccc} q_1^* K_1|_{\partial B(M_1)_x} & \xrightarrow{\eta_x} & q_2^* K_2|_{\partial B(M_1)_x} \\ \cong\big\downarrow & & \big\downarrow\cong \\ q_1^* K_1|_{\partial B(M_1)_y} & \xrightarrow{\eta_y} & q_2^* K_2|_{\partial B(M_1)_y} \end{array} \tag{2.21}$$

and modulo these isomorphisms, η_x, and η_y are homotopic. Thus K_x and K_y regarded as bundles on $S^r \cong S(M_1)_x \cong S(M_1)_y$, are isomorphic. Since M_1 is connected, this implies $\langle \Phi^*_{K_x}(P), [S(M_1)_x]\rangle$ is a constant, k, say, independent of x from which (2.16) follows. Then (2.17) follows by (2.7). □

In particular, if $\Phi^* P(\ ) = e(\ )$ is the Euler class of a real oriented vector bundle, assumptions 1) and 2) in Theorem 2.2 are satisfied, and we obtain as in [25]:

Corollary 2.1. *For ξ as defined above, we have in terms of Euler classes*

$$\langle e(E) - e(F), [M]\rangle = \langle e(\xi), [S(M_1)]\rangle = k\langle e(L), [M_1]\rangle, \tag{2.22}$$

where $k \in \mathbb{Z}$ is a constant.

Proof. It is instructive to include the straightforward proof from [12, Lemma1.2] which implements the maps in (2.15) (cf. [25]). Starting from (2.14), we have

$$\begin{aligned} \langle e(\xi), [S(M_1)]\rangle &= \langle e(q^* L) \cup e(K), [S(M_1)]\rangle \\ &= \langle q^* e(L), e(K) \cap [S(M_1)]\rangle \\ &= \langle e(L), q_*(e(K) \cap [S(M_1)])\rangle \\ &= k\langle e(L), [M_1]\rangle. \end{aligned}$$

□

Remark 2.3. The applications in the following sections of this paper mainly involve complex vector bundles (so that $\Lambda = \mathbb{Z}$), with P corresponding to the total (or top) Chern class c_* (or c_{top}), so that both assumptions 1) and 2) in Theorem 2.2

are satisfied (with r the rank of L as a complex vector bundle), so that M_1 is of real codimension $2r$, with cohomological degree $c_*(L) \leq 2r$.

3. Rational maps of projective varieties

In the following, we shall be applying the general construction and results of §2 in the category of complex manifolds with morphisms the meromorphic maps. Here M and M_1 are assumed to be closed connected and oriented smooth manifolds. In this case there will be a slight adjustment in the roles played by M and M_1 as result of redefining certain terms. When the context is clear, it is assumed that complexified tangent bundles are taken in each case.

The natural examples in this context include rational maps of (algebraic) projective varieties, and this fully enriched situation is the one to which we pay some attention. But we will point out now (as the astute minded reader can see) that the development of ideas, and constructions, etc. apply equally well if the spaces in question are just taken to be compact complex manifolds. But restricting to the algebraic case affords us some access to using significant numerical data, which otherwise might not necessarily be the case in the more general setting.

3.1. Application of the general result

Let X and Y be compact projective manifolds, $\dim_{\mathbb{C}} X = \dim_{\mathbb{C}} Y = n$, and let $f : X \supset U \longrightarrow Y$ be a rational map. In general, such a map will have a closed algebraic *indeterminacy set* I_f, namely the locus of points in X for which f fails to be defined, with open complement $U = X - I_f$.

Let the locally closed complex algebraic subset

$$X(s) := \{x \in X : \operatorname{rank}_{\mathbb{C}}(df(x)) \leq s < n\} \subset X, \tag{3.1}$$

be such that $f^{-1}(f(X(s)) = X(s)$.

Letting Z be the complex algebraic subset $Z = I_f \cup X(n-1)$, we consider (X, Z), or more generally $(M, M \cap Z)$, as an $(n, \mathbb{Z})$ adapted pair (with $r = \operatorname{codim}_{\mathbb{C}} Z \geq 1$, respectively $2r = \operatorname{codim}_{\mathbb{R}}(M \cap Z)$), with $M \subset X$ a closed and connected smooth oriented submanifold transversal to Z.

More specifically, M does not intersect the singular locus Z_{sing} of Z, and is transversal to the regular part Z_{reg} of Z, with $M_1 = M \cap Z$ a closed and connected smooth oriented submanifold, $\operatorname{codim}_{\mathbb{R}} M_1 = 2r \geq 2$. In particular, when $M = X$ is connected of real dimension $2n$, the space Z_{sing} is assumed empty, with Z connected.

Remark 3.1. The point of taking $s = n - 1$ in the definition of Z, is because at a later stage we will need $\psi = df$ to be an isomorphism outside of Z, and $\partial B_1(N) = \partial B_2(N)$.

Remark 3.2. We recall from, e.g., [18] that such a rational map $f : X \longrightarrow Y$ can be specified by a holomorphic map $\tilde{f} : X - I_f \longrightarrow Y$, for which $\operatorname{codim}_{\mathbb{C}} I_f \geq 2$. Thus for now, we are motivated to take $r \geq 2$, and view Z as a 'singular projective variety'. Also, in cases where $X = Y$, for instance, we might replace $X(s)$ above, by the fixed point set $\mathrm{Fix}(f)$ of f, in the case of a self-map $f : X \longrightarrow X$ (see, e.g., §4.4).

3.2. A certain $2p$-cycle and application of Theorem 2.1

We fix the smooth submanifold $M \subset X$, with $\dim_{\mathbb{R}} M = 2p$ (for a fixed p with $1 \leq p \leq n$) that intersects Z transversally, and set

$$N = M \cap Z = M \cap (X(n-1) \cup I_f), \tag{3.2}$$

so that $\dim_{\mathbb{C}} N = p - r$, for $p \geq r$. On applying our general result, we note that the isomorphisms in question are simply *topological* unless otherwise stated.

To this extent, we take (M, N) to be a $(2p, \mathbb{Z})$-adapted pair, and take $B(N)$ to be a tubular neighborhood of N in M. Thus, N now plays the role of M_1 in §2.1. With this slight modification in mind, we construct as in §2.2, the smooth double $S(N)$ producing the S^{2r}-fibration $q : S(N) \longrightarrow N$, along with projections $q_i : B_i(N) \longrightarrow N$ (for $i = 1, 2$) as in (2.12). Note this produces a $2p$-cycle $[S(N)]$.

Here we will set

$$E = TX|_{B_1(N)}, \text{ and } F = f^*TY|_{B_2(N)}, \tag{3.3}$$

where, as before, $\psi = df : TX \longrightarrow f^*TY$ is a isomorphism when restricted to $\partial B_1(N) = \partial B_2(N)$.

Following Theorem 2.1, and from the construction of §2.2 in the context of complex vector bundles with structure group $\mathrm{U}(q)$, for some q, we straightaway obtain

$$\langle \Phi^*_E(P) - \Phi^*_F(P), [M] \rangle = \langle \Phi^*_\xi(P), [S(N)] \rangle, \tag{3.4}$$

in terms of Chern polynomials $\Phi_\diamond(P)$.

On applying (3.2) together with (3.3), then (3.4) with $\Phi^*_\diamond(P) = c_p(\diamond)$ reduces to the following form (cf. [6, 25]):

$$\langle c_p(X) - f^*c_p(Y), [M] \rangle = \langle c_p(\xi), [S(N)] \rangle, \tag{3.5}$$

where $\xi = (E, \psi, F)$ is given by §2.2.

3.3. The righthand side of (3.5)

In order to deal with enumerating the righthand side of (3.5), we return to the setting and conditions of §2.2. Here we take ψ to have constant rank $n - r$ along N, and following (2.9) we have the isomorphism

$$TZ|_N \cong L = \psi(TZ)|_N, \tag{3.6}$$

so that $\mathrm{rank}_{\mathbb{C}} L = n - r$. We also recall from §2.2, the relations

$$\begin{aligned} E|_N \cong K_1 \oplus L,\ F|_N \cong K_2 \oplus L, \\ E \cong q_1^*(K_1) \oplus q_1^*(L),\ F \cong q_2^*(K_2) \oplus q_2^*(L), \end{aligned} \tag{3.7}$$

while noting that N is a deformation retract of $B_i(N)$, for $i = 1, 2$. Hence on $S(N)$ we have the isomorphism $\xi \cong (q_1^*K_1, \eta, q_2^*K_2) \oplus q^*L = K \oplus q^*L$, as in (2.14), with $\mathrm{rank}_{\mathbb{C}}\xi = n$, from which we deduce $\mathrm{rank}_{\mathbb{C}} K = r$ (note that we have identified

K_1 with the restriction to N of a complex rank r vector bundle normal to TZ in $TX|Z$).

As deduced from the total Chern classes of K and $q^*(L)$, it is straightforward to show that

$$\begin{aligned} c_k(K \oplus q^*L) &= \sum_{\nu=1}^{r} c_\nu(K) \cup q^* c_{k-\nu}(L) + q^* c_k(L), \ 1 \leq k \leq n-1, \\ c_n(K \oplus q^*L) &= \sum_{\nu=1}^{r} c_\nu(K) \cup q^* c_{n-\nu}(L). \end{aligned} \tag{3.8}$$

Remark 3.3. Following from [6, §4](cf. [28, pp. 408-409]), the existence on $S(N)$ of ℓ linearly independent trivializing sections of $q^*(L)$, with $\ell = (n-r)-(k-r) = n-k$, leads to the result

$$\langle q^* c_k(L), [S(N)] \rangle = 0, \ \text{if } 1 \leq k \leq n-1. \tag{3.9}$$

Theorem 3.1. *With regards to* (3.2) *and* 3.6, *we have*

$$\langle c_p(TX|M) - c_p(f^*TY|M), [M] \rangle = k \langle c_{p-r}(L), [N] \rangle, \tag{3.10}$$

for some constant $k \in \mathbb{Z}$, *with* $\dim_{\mathbb{R}} M = 2p$ *and* $\dim_{\mathbb{R}} N = 2(p-r)$.

Proof. In view of Theorem 2.2, we will here set $E = TX|M$, $F = f^*TY|M$, and consider the clutched bundle $\xi = (E, \psi, F)$ over M. Note that on recalling Remark 2.3, the hypotheses 1) and 2) of Theorem 2.2 are satisfied in this case. To proceed, it suffices to consider total Chern classes c_*, which are multiplicative, and then commence by substituting this data into the left hand side of (2.17).

We have $\text{rank}_{\mathbb{C}} K = r$, $\text{rank}_{\mathbb{C}} L = n - r$, and the total Chern class $c_*(K)$ is of cohomological degree $\leq 2r = \text{codim}_{\mathbb{R}} N$, so Theorem 2.2, in particular (2.17), can be applied. Finally, we have $\langle c_*(L), [N] \rangle = \langle c_{p-r}(L), [N] \rangle$, since $\dim_{\mathbb{R}} N = 2(p-r)$. □

3.4. Interpreting Theorem §3.1

In view of applying the clutching construction of §2.2, Theorem §3.1 produces a significantly general formula that can be observed when regulating the topology of a rational map $f : X \longrightarrow Y$ of compact projective varieties of equal (complex) dimension in terms of the cycles $[M]$ and $[N]$ as defined. Note that $[N]$ is a cycle which contains part of the (possibly singular) variety $Z = X(n-1) \cup I_f$, once Z is intersected by the $2p$-cycle $[M]$ as in (3.2).

A working principle is to 'resolve' the indeterminacy set I_f, for instance by blowing up along $f(I_f)$, and then reduce matters to considering a holomorphic map $\hat{f} : X \longrightarrow Y$.

For instance, if $M = X$ (so $p = n$), then (3.10) in this case reduces to:

$$\langle c_n(X) - \hat{f}^* c_n(Y), [X] \rangle = k \langle c_{n-r}(X(n-r)), [X(n-r)] \rangle, \tag{3.11}$$

for a constant $k \in \mathbb{Z}$, with $X(n-r) = \{x \in X : \mathrm{rank}_{\mathbb{C}}(d\hat{f}(x)) \leq n-r\}$ smooth and connected, $\mathrm{codim}_{\mathbb{C}} X(n-r) = r \geq 1$, and $d\hat{f}$ of constant (complex) rank $n-r$ along $Z = N = X(n-r)$ (so that $L \cong TZ \cong TX(n-r)$).

More specifically, given $f : X \longrightarrow Y$, one may pass to a proper modification $\hat{f} : \widehat{X} \longrightarrow Y$, and apply the formula in Theorem 3.1 for a holomorphic map, provided $\widehat{X}$ is a compact complex manifold.

Remark 3.4. We recall from [17] that a *proper modification* $\hat{f} : \widehat{X} \longrightarrow Y$ means that $\hat{f}$ is a proper surjective holomorphic map, such that there exists nowhere dense analytic subsets $X' \subset X$ and $Y' \subset Y$, such that: i) $\hat{f}(X') \subset Y'$, ii) $\hat{f} : \widehat{X} - X' \longrightarrow Y - Y'$ is a biholomorphism, and iii) each fiber $\hat{f}^{-1}(y)$, for $y \in Y'$, consists of more than one point. The set $X' = \hat{f}^{-1}(y)$ is called the *exceptional set.* Note that $\widehat{X}$ may not necessarily be an algebraic variety in general, even if X has this property [20].

3.5. The case of a holomorphic ramified covering map

Consider the case where $s = n-1$ is constant, and let $f : X \longrightarrow Y$ be a holomorphic branched covering map for which $X_1 := X(n-1) \subset X$ is the ramification divisor on which $\mathrm{rank}_{\mathbb{C}} f|_{X_1} = n-1$, with $r = \mathrm{codim}_{\mathbb{C}} X_1 = 1$. As before, let $M \subset X$ be any compact oriented smooth submanifold with $\dim_{\mathbb{R}} M = 2p$ that meets X_1 transversally (with $1 \leq p \leq n$), with $N = M \cap X_1$. From (3.6), we have $L = TX_1|_N$. This leads to a version of the higher dimensional Riemann-Hurwitz formula as given in [6, Proposition 2] (see also [28, p. 409]):

$$\langle f^* c_p(Y) - c_p(X), [M] \rangle = (\mu - 1)\langle c_{p-1}(X_1), [N] \rangle, \tag{3.12}$$

where $\mu = \deg(f|_{X(1)}) \in \mathbb{Z}$ is the local topological degree of f along X_1. Note that in general, the global degree $\deg(f) := \delta \neq \mu$.

Together with (3.10), this case reveals the interest in enumerating the right-hand side of (2.7) and (2.16) in general.

Example 3.1. Let $n = 2$, where X is now a compact complex surface, and let $Y = \mathbb{C}P^2$. Consider a holomorphic map $f : X \longrightarrow \mathbb{C}P^2$ which is branched over a curve C of genus g with normal crossings. Thus $f : X - X_1 \longrightarrow \mathbb{C}P^2 - C$ is an unramified covering map of degree $\deg f = \delta$ say, where the branch set $X_1 = f^{-1}(C)$.

As an example of showing consistency in the data, suppose in this case X is a K-3 surface. One can construct a map with $\delta = \mu = 4$, branched over a quartic curve C (see, e.g., [24]). In this top dimension, we have $\langle c_2(X), [X] \rangle = \chi(X) = 24$, and $\langle f^* c_2(\mathbb{C}P^2), [X] \rangle = 4(\chi(\mathbb{C}P^2) = 4(3) = 12$. From (3.12), with $n = p = 2$, and the adjunction formula (e.g., [18, p.221]), it is straightforward to see that $\langle c_1(X_1), [X_1] \rangle = \chi(C) = -4$, and therefore $g = 3$.

4. Rational self-maps

4.1. Iterated self-maps of projective varieties

Now we consider the case where $X = Y$, and $f : X \longrightarrow X$ is a rational map with k-th iterate denoted f^k. Let

$$X(k,s) := \{x \in X : \mathrm{rank}_{\mathbb{C}}(df^k(x)) \leq s < n\}, \tag{4.1}$$

and set $Z = X(k,n-1) \cup I_{f^k}$, with $\mathrm{codim}_{\mathbb{C}} Z = r_k$. Let $\Delta_k = \deg(f^k)$, and note that in general, $\Delta_k \neq \delta^k = \deg(f)^k$.

We recall from §3.1, making slight modifications, that $M \subset X$ is a closed and connected smooth oriented submanifold, $\dim_{\mathbb{R}} M = 2p$, transversal to $Z = X(k,n-1) \cup I_{f^k}$. Again, M does not meet Z_{sing}, and is transversal to Z_{reg}, with $M_1 = M \cap Z$ a closed and connected smooth oriented submanifold, $\mathrm{codim}_{\mathbb{R}} M_1 = 2r_k \geq 2$. When $M = X$ is connected, $\dim_{\mathbb{R}} M = 2n$, then $Z_{\mathrm{sing}} = \emptyset$, and Z is connected.

Also, we take $E = TX|_{B_1(N)}$, and $F = (f^k)^* TX|_{B_2(N)}$. These are only the essential differences, otherwise the basic construction leading to the various formulas remains the same. In particular, Theorem 3.1 holds with $X(n-1)$ replaced by $X(k,n-1)$ in (4.1).

4.2. Holomorphic and analytically stable maps

Let us deal first with a holomorphic map $f : X \longrightarrow X$, where the k-th iterate f^k is a holomorphic branched covering map with ramification divisor $X_{1,k} := X(k,n-1)$, with $r_k = 1$. Then (3.12) reads as

$$\langle (f^k)^* c_p(X) - c_p(X), [M] \rangle = (\mu_k - 1)\langle c_{p-1}(X_{1,k}), [N] \rangle, \tag{4.2}$$

where $\mu_k \in \mathbb{Z}$ is the local topological degree of f^k along $X_{1,k}$. We have $\Delta_k \neq \mu_k$, in general.

However, in this case where f is holomorphic and $X = \mathbb{C}P^n$, we do have $\Delta_k = \delta^k$ [10]. On the other hand, it is clear for $M = \mathbb{C}P^n$, and $p = n$, that such maps are thus regulated by the expression derived from (4.2):

$$(n+1)(\Delta_k - 1) = (\mu_k - 1)\langle c_{n-1}(X_{1,k}), [X_{1,k}] \rangle. \tag{4.3}$$

For instance:

Proposition 4.1. *Let $f : \mathbb{C}P^2 \longrightarrow \mathbb{C}P^2$ be a holomorphic map. Suppose that the k-th iterate $f^k : \mathbb{C}P^2 \longrightarrow \mathbb{C}P^2$ is a holomorphic branched covering map over a curve of genus g. Then necessarily the global degree $\Delta_k \neq \mu_k$ (the local degree).*

Proof. A straightforward enumeration of (4.3) leads to

$$3(\Delta_k - 1) = 3(\delta^k - 1) = (\mu_k - 1)(2 - 2g), \tag{4.4}$$

which is meaningless if $\Delta_k = \mu_k$. □

Taking $n = 2$, a bimeromorphic map $f : X \longrightarrow X$ is said to be *analytically stable* if: i) for all $k \geq 0$, we have $(f^k)^* = (f^*)^k$, and ii) for each curve C in X, $f^k(C) \notin I_f$ (see [10]). Recall that the blow-up of Y at y, is the proper modification $\hat{f} : \widehat{X} \longrightarrow Y$ which replaces y with the exceptional curve $\pi^{-1}(y) \cong \mathbb{C}P^1$, the set of holomorphic tangent directions at y, and $\hat{f}$ is a biholomorphism elsewhere. In fact, in this instance, any proper modification $\hat{f} : \widehat{X} \longrightarrow Y$ arises as a composition of finitely many point blow ups (see, e.g., [8, Th. 1.1]). Following [8, Th. 01], if $f : X \longrightarrow X$ is a bimeromorphic map, then there always exists a proper modification $\hat{f} : \widehat{X} \longrightarrow X$ that lifts f to an analytically stable map.

Remark 4.1. In [4] it is shown, that for $n = 2$, there are countably many sequences $\{d_\ell\} \subset \mathbb{N}$ for which a rational map $f : \mathbb{C}P^2 \longrightarrow \mathbb{C}P^2$ exists, satisfying $\Delta_\ell = d_\ell$, for all $\ell \in \mathbb{N}$.

It would be interesting to see how the topology of these procedures can be regulated by results of the type Theorem 3.1, and by the known results for the Chern classes of the blowing-up process (e.g., [2, 26]).

4.3. Degree lowering curves

For a general rational map $f : \mathbb{C}P^n \longrightarrow \mathbb{C}P^n$, there may be any amount of peculiar behavior. For instance, it is possible that an iteration f^k may, for some k, map an (irreducible) curve C into the indeterminacy set I_{f^k} (thus f cannot be analytically stable). In this case one sees that $\Delta_k < \delta^k$ [10], and so enumerating (3.10), for $p = n$, can easily be seen to give

$$\Delta_k = 1 - \frac{1}{n+1}(k\langle c_{n-r}(L), [N]\rangle) < \delta^k, \tag{4.5}$$

where N is given by (3.2), $\mathrm{codim}_{\mathbb{R}} N = 2r$, and L is as in (3.6).

4.4. Holomorphic maps with fixed point set

Next we will consider a holomorphic map $f : X \longrightarrow X$, with a (possibly singular) fixed point set $S = \mathrm{Fix}(f)$, with $r = \mathrm{codim}_{\mathbb{C}} S$. In the general setting of §2, we will regard (X, S) or $(M, M \cap S)$ as an $(n, \mathbb{Z})$-adapted pair with S playing the role of Z as in §3.2, and $M_1 = N = M \cap S$. Again, the transversality and other conditions relating to M, Z_{reg}, and Z_{sing} as made in §3.1 and §4, also apply here.

This application is partially motivated by certain constructions in [1] (where X can be taken more generally to be a complex manifold) to which we will apply the general results of Theorem 2.1 and Theorem 2.2. But this will necessitate assigning some different data compared to that of the previous sections and redefining terms accordingly.

Specifically, let us start by letting Q_S denote the normal bundle to S in X. Then let

$$E = (Q_S)^{\otimes \nu_f}, \text{ and } F = TX, \tag{4.6}$$

where $\nu_f \in \mathbb{N}$. We take up the hypotheses in §2.2 in terms of a homomorphism $\psi : (Q_S)^{\otimes \nu_f} \longrightarrow TX$.

In particular, as in [1], the restricted morphism $\psi|_S : (Q_S)^{\otimes \nu_f} \longrightarrow TX|_S$, is identified with a holomorphic section of $TX|_S \otimes (Q_S^*)^{\otimes \nu_f}$. The bundles K and L are taken to be as in §2.2, with

$$L = \mathrm{Im}(\psi|_S) \subset TX|_S. \tag{4.7}$$

Let M be a $2p$-dimensional submanifold $M \subseteq X$ intersecting S transversally. As before, we set $N = M \cap S$, with $\dim_{\mathbb{C}} N = p - r$ (for $p \geq r$).

Having made these adjustments for the bundles E and F, etc., we apply Theorem 2.1 together with essentially the same proof as that used in proving Theorem 3.1, to obtain

$$\langle c_p((Q_S)^{\otimes \nu_f}) - c_p(X), [M] \rangle = k \langle c_{p-r}(L), [N] \rangle. \tag{4.8}$$

Example 4.1. Consider the case $r = 1$, and S is a (possibly singular) hypersurface regarded as an oriented $(n-1)$-circuit. Then we have $\mathrm{rank}_{\mathbb{C}} Q_S = 1$, and $\mathrm{rank}_{\mathbb{C}} L = n - 1$, following which (4.8) gives

$$\langle c_p((Q_S)^{\otimes \nu_f}) - c_p(X), [M] \rangle = k \langle c_{p-1}(L), [N] \rangle, \tag{4.9}$$

with $c_1((Q_S)^{\otimes \nu_f}) = \nu_f c_1(\mathcal{O}_S)$, and $c_p((Q_S)^{\otimes \nu_f}) = 0$ for $p \geq 2$ (since Q_S is a line bundle). Note that applying (4.9) for $M = X$ ($p = n$), we have $N = S$.

Remark 4.2. In the setting of [1], the quantity ν_f is considered as a measure of 'order of contact' between the map f and S. For the case of such a hypersurface S there are the connected components Λ_α of the union of singular sets $\mathrm{Sing}(\mathfrak{X}_f) \cup \mathrm{Sing}(S)$, where $\mathrm{Sing}(\mathfrak{X}_f)$ is the set of zeros of a vector field $\mathfrak{X}_f$ associated to f that induces a (generally singular) holomorphic foliation. This leads to a residue formula $\sum_\alpha \mathrm{Res}(\mathfrak{X}_f, S, \Lambda_\alpha) = \langle c_1^{n-1}(S), [S] \rangle$ as in [1, Th. 01]. Our approach leads to somewhat different formulas as seen above. Though enumerating (4.9) for the case $n = p = 2$, reveals the right-hand (up to a constant) to be such a residual quantity.

5. Determinantal varieties

5.1. The degeneracy locus

In this section we commence the second part of the paper by turning to a related, but essentially more general setting. In the previous sections we considered applying the general result of §2 to rational maps of projective varieties. But now we tweak the setting of those sections somewhat with several terms redefined for the sake of replacing maps of projective varieties by vector bundle morphisms over a compact complex manifold X.

More specifically, consider a morphism $\psi : E \longrightarrow F$ of complex (smooth) vector bundles of the same complex rank $\ell \geq 1$, over a projective variety X (where $\dim_{\mathbb{C}} X = n$). For some given $s \in \mathbb{N}$, we have the *degeneracy locus* of ψ, as defined by

$$\Omega(s) := \{x \in X : \mathrm{rank}_{\mathbb{C}} \psi(x) \leq s < \ell\}. \tag{5.1}$$

As in §3.2, we fix a smooth submanifold $M \subset X$, with $\dim_{\mathbb{R}} M = 2p$ $(1 \leq p \leq n)$ that intersects $\Omega(\ell-1)$ transversally (as previously $s = \ell - 1$ is taken since we require ψ to be an isomorphism outside of $\Omega(\ell-1)$).

Once again we will apply the general setting of §2, where $(X, \Omega(\ell-1))$ (respectively, $(M, M \cap \Omega(\ell-1))$) is regarded as an n-adapted pair, so that $\Omega(\ell-1)$ (respectively, $M \cap \Omega(\ell-1)$) plays the role of M_1 in §2), with $r = \operatorname{codim}_{\mathbb{C}} \Omega(\ell-1) \geq 1$.

Theorem 2.1, in particular, (2.7) immediately applies to give the general statement

$$\langle P(E) - P(F), [X] \rangle = \langle P(\xi_\theta), [S(\Omega(\ell-1))] \rangle. \tag{5.2}$$

Remark 5.1. The cohomology class $\{\Omega(s)\}$ of $\Omega(s)$ in X can be determined by polynomials in the Chern classes of E and F, and in certain cases the codimension of $\Omega(s)$ can be determined (see [11, 19] which also cover a historical background to the general problem in the algebraic geometric context). Note that [19] deals initially with results in the differentiable category, thus (5.2) applies in that case as well.

Example 5.1. Let V be a vector space and V^* its dual vector space. A linear map $\psi : V \longrightarrow V^*$ is said to be *symmetric* if $(\psi(x), y) = (\psi(y), x)$, for all $x, y \in V$, where $(\ ,\ )$ is the dual pairing between V^* and V. Equivalently, ψ is symmetric if $\psi = \psi^T$. The precise meaning of 'general' is explained in [19, Note 2, p.72]. Likewise, ψ is *skew-symmetric* if $\psi = -\psi^T$.

If $E \longrightarrow X$ is a complex vector bundle, $\operatorname{rank}_{\mathbb{C}} E = \ell$, $F = E^*$, and $\psi : E \longrightarrow E^*$ is a general symmetric bundle map, then following [19, Th. 1], the cohomology class $\{\Omega(s)\}$ is given by a polynomial $P_s(c_1(E^*), \ldots, c_\ell(E^*))$. This latter polynomial has an explicit expression given in terms of the determinant

$$2^{\ell-s} \begin{bmatrix} c_{\ell-s} & c_{\ell-s+1} & c_{\ell-s+2} & \\ c_{\ell-s-2} & c_{\ell-s-1} & c_{\ell-s} & \\ \vdots & & c_{\ell-s-2} & \\ \vdots & \cdots & & \ddots c_1 \end{bmatrix} = \{\Omega(s)\}, \tag{5.3}$$

where $c_i = c_i(E^*)$, and further, $\Omega(s)$ has codimension $r = \binom{\ell-s+1}{2}$ (for $s < \ell$).

In keeping with the previous sections, we will be interested in finding expressions linking the Chern classes of some degree (p, say) of the spaces in question.

Let $N = M \cap \Omega(\ell-1)$, with $\dim_{\mathbb{C}} N = p - r$, for $p \geq r$ (again, the case $r \geq p$ can be treated likewise). We recall the tubular neighborhoods $B_i(N)$ of N (for $i = 1, 2$), and consider E, F as restricted to $B_1(N)$ and $B_2(N)$ respectively, so that in accordance with §2.2, we have $E|\partial B_1(N) \cong F|\partial B_2(N)$. Note that we do not yet assume that ψ has constant rank on $\Omega(\ell-1)$, since we are still in the context of Theorem 2.1. We also recall from §2.2 the S^{2r}-fibration $q : S(N) \longrightarrow N$.

We shall be applying the same basic strategy as in §3 (and in §4). Thus (5.2) reduces to a statement that is more general than (3.5):

$$\langle c_p(E) - c_p(F), [M] \rangle = \langle P(\xi_\theta)^{[2p]}, [S(N)] \rangle, \tag{5.4}$$

where $P(\xi_\theta) \in H^*(BU(\ell))$ is a characteristic class whose component $P(\xi_\theta)^{[2p]} \in H^{2p}(BU(\ell))$ corresponds to the p-th Chern class c_p (e.g., P corresponds as before to the total Chern class c_*). Again, it is interesting to enumerate the right-hand side of (5.4) once the cohomology class $\{\Omega(\ell-1)\}$ has been determined.

Remark 5.2. In view of this last comment, if the cohomology class $\{N\}$ of N in M happens to be cohomologous to $\{\Omega(\ell-1)\}^{[2p]}$ in $H^*(\Omega(\ell-1),\mathbb{Z})$, then the class $P(\xi_\theta)^{[2p]}$ is expressible in the form

$$P(\xi_\theta)^{[2p]} = q^*\{\Omega(\ell-1)\}^{[2p]} \cup \gamma, \tag{5.5}$$

for some $\gamma \in H^{2(p-r)}(S(N),\mathbb{Z})$.

The following observations summarized as a proposition shows that, in the context of the symmetric bundle map of Example 5.1, there is indeed a restriction on components of the class $P(\xi_\theta)$.

Proposition 5.1. *With regards to the context of Example* 5.1, *we have the following relationships:*

(1) *For p odd,*

$$2\langle c_p(E),[M]\rangle = \langle P(\xi_\theta)^{[2p]},[S(N)]\rangle. \tag{5.6}$$

(2) *For p even, the class $P(\xi_\theta)^{[2p]}$ is trivial in $H^{2p}(S(N),\mathbb{Z})$.*

Proof. Noting that $c_p(E^*) = (-1)^p c_p(E)$ (see, e.g., [18, p.411]), we have from (5.4)

$$(1+(-1)^{p+1})\langle c_p(E),[M]\rangle = \langle P(\xi_\theta)^{[2p]},[S(N)]\rangle, \tag{5.7}$$

for $1 \le p \le \ell$, from which the results follow. □

Further enumeration of the righthand side of (5.7), can be carried out under the conditions of §2.2 which we will deal with next.

5.2. Constant rank case

Suppose now that $\mathrm{rank}_{\mathbb{C}}\psi|_{\Omega(\ell-1)} = s < \ell$, is constant (so that $\Omega(s) = \Omega(\ell-1)$), and the bundle map ψ is taken as a clutching map between E and F, as in §2.2. In this case, there is the class of the $2p$-component given by $P(\xi_\theta)^{[2p]} = c_p(E,\psi,F)$. Also, Theorem §2.2 applies to give

$$\langle P(E)-P(F),[X]\rangle = k\langle P(L),\Omega(\ell-1)]\rangle, \tag{5.8}$$

where, as before $L = \psi(E|_{\Omega(\ell-1)}) \subset F|_{\Omega(\ell-1)}$, with $\mathrm{rank}_{\mathbb{C}}L = \ell-1$. In particular, $P(L) \in H^*(\Omega(\ell-1),\mathbb{Z})$; so knowing the cohomology of $\Omega(\ell-1)$ gives us a handle on the class $P(L)$. Applying Theorem 2.2 (cf. Theorem 3.1), we then have

$$\langle c_p(E)-c_p(F),[M]\rangle = k\langle c_{p-1}(L),[N]\rangle. \tag{5.9}$$

Example 5.2. In view of the above remarks, let us return to the context of Example 5.1. Here we have $L = \psi(E|_{\Omega(\ell-1)}) \subset E^*|_{\Omega(\ell-1)}$, and the cohomology class $P(L) = P(c_1(E^*),\ldots,c_{\ell-1}(E^*))$. On applying (5.9), we thus obtain for p odd,

$$2\langle c_p(E),[M]\rangle = k\langle c_{p-1}(E^*),[N]\rangle = \kappa\langle P(c_1(E^*),\ldots,c_{\ell-1}(E^*)),[N]\rangle, \tag{5.10}$$

for some constants k and κ.

Example 5.3. There are analogous results in [19] for the cohomology class $\{\Omega(s)\}$ in the case of skew-symmetric maps (morphisms) $\psi : E \longrightarrow E^*$. The cases $\psi : E \longrightarrow E^* \otimes \mathcal{L}$, for $\mathcal{L}$ a complex line bundle, are also studied in the symmetric and skew-symmetric cases. For instance, when $\psi : E \longrightarrow E^* \otimes \mathcal{L}$ is a general symmetric bundle map, then the cohomology class $\{\Omega(s)\}$ is given by (5.3), but now taking $c_i = c_i(E^* \otimes \sqrt{\mathcal{L}})$ [19, Th. 10]. The (general) skew -symmetric case can likewise be treated.

Example 5.4 (Application to variation of Hodge structure following [19]**).** Consider a family $\varpi : \mathcal{C} \longrightarrow X$ of curves of genus g. Let $\mathsf{H}^{1,0}, \mathsf{H}^{0,1}$ denote the corresponding Hodge bundles. We have then a period map $\Upsilon : X \longrightarrow \mathrm{Gr}(g, 2g)/\Gamma$, where $\mathrm{Gr}(g, 2g)$ denotes a certain isotropic Grassmannian, and $\Gamma \subset \mathrm{Aut}(\mathrm{Gr}(g, 2g))$ is a discrete subgroup. Consequently, there is a bundle morphism $\psi : TX \longrightarrow \mathrm{Hom}(\mathsf{H}^{1,0}, \mathsf{H}^{0,1})$, that can be expressed alternatively as a symmetric bundle map $TX \longrightarrow S^2(\mathsf{H}^{1,0})^*$ [19].

When X is an algebraic curve of genus g_X, and there are no singular fibres of ϖ, then one can enumerate matters as follows. Setting $E = \mathsf{H}^{1,0}$ (so $\mathrm{rank}_{\mathbb{C}} E = g$), we have from [19, p.82] $c_1(\det(S^2 E^* \otimes \mathcal{O}(1)) \geq 0$. Observing that $c_1(S^2 E^*) = (g+1)c_1(E^*)$, then this previous expression simplifies to $g(g_X - 1) \geq c_1(E)$.

It can be argued that if the variation of Hodge structure over X is nontrivial, then by the local Torelli theorem, the period map Υ has maximal rank at some point of X, and by [19, Th. 10], the degeneracy locus $\Omega(g-1)$ in this case, is not all of X. In the context of a general symmetric bundle map, here given by $\psi : E \longrightarrow E^* \otimes \mathcal{O}(-1)$, and from the remarks in Example 5.3 above, it follows that the cohomology class $\{\Omega(g-1)\} = -2c_1(E \otimes \mathcal{O}(-\frac{1}{2})) = -2(c_1(E) - g(g_X - 1))$. This provides us with an enumeration of (5.9) with $M = X$, $N = \Omega(g-1)$, and $F = \mathcal{O}(-\frac{1}{2})$, in the case of $p = 1$ and $s = g - 1$. In this case there is just a single constant $k = k_\nu = -\frac{1}{2}$.

5.3. Flagged bundles

Suppose now we consider, as in [11], the more general situation of §5.1 for a morphism $\psi : E \longrightarrow F$, over X, for which

$$\begin{aligned} E_1 \subset E_2 \subset \cdots \subset E_u = E \\ F = F_v \twoheadrightarrow F_{v-1} \twoheadrightarrow \cdots \twoheadrightarrow F_1 \end{aligned} \tag{5.11}$$

are flags of subbundles and quotient bundles, respectively. Here we will take integers $s(\alpha, \beta) \in \mathbb{N}$ specified across the intervals $1 \leq \alpha \leq u$, and $1 \leq \beta \leq v$, and the degeneracy locus is then defined by

$$\Omega(\mathbf{s}) := \{x \in X : \mathrm{rank}_{\mathbb{C}}(E_\alpha(x) \longrightarrow F_\beta(x)) \leq s(\alpha, \beta), \forall \alpha, \beta\}, \tag{5.12}$$

where $\mathbf{s}$ is regarded as a certain rank function. As shown in [11], conditions on $\mathbf{s}$ determine the irreducibility of $\Omega(s)$ as a projective variety, and further, the cohomology class $\{\Omega(\mathbf{s})\}$ can be determined in terms of the Chern classes of E and F.

In the case $\psi|_{\Omega(\mathbf{s})}$ has constant rank $s(\alpha,\beta)$, and ψ is a clutching map as before, we apply Theorem §2.2 to obtain $\langle P(E) - P(F), [X]\rangle = k\langle P(L), [\Omega(\mathbf{s})]\rangle$.

5.4. Complete flags

Following [11], we will exemplify matters in the case of 'complete flags' for the data $u = v = m$, and E_i, F_i having (complex) rank i. In this case, $\Omega(\mathbf{s})$ is characterized by permutations in the symmetric group S_m. Given $w \in S_m$, let $\ell(w)$ be the length of w (in other words, the number of inversions). Let $\mathbf{s}_w(\beta,\alpha) = \operatorname{card}\{i \leq \beta : w(i) \leq \alpha\}$, and

$$x_i = c_1(\operatorname{Ker}(F_i \twoheadrightarrow F_{i-1})), \text{and } y_i = c_1(E_i/E_{i-1}), \text{ for } 1 \leq i \leq m. \tag{5.13}$$

Then one restricts attention to

$$\Omega(w) = \Omega(\mathbf{s}_w) := \{x \in X : \operatorname{rank}_{\mathbb{C}}(E_\alpha(x) \longrightarrow F_\beta(x)) \leq \mathbf{s}_w(\beta,\alpha), \forall \alpha,\beta\}. \tag{5.14}$$

Here we make several observations from [11]:

(i) The space $\Omega(w)$ has a natural structure of a scheme given by the vanishing of the induced maps from $\bigwedge^{\mathbf{s}_w(\beta,\alpha)+1}(E_\alpha) \longrightarrow \bigwedge^{\mathbf{s}_w(\beta,\alpha)+1}(F_\beta)$.

(ii) The expected (maximum) value of $r = \operatorname{codim}_{\mathbb{C}}\Omega(w)$, is $r = \ell(w)$.

(iii) The cohomology class $\{\Omega(w)\} = \mathfrak{S}_w(x,y)$, where

$$\mathfrak{S}_w(x,y) = \mathfrak{S}(x_1,\ldots,x_m,y_1,\ldots,y_m), \tag{5.15}$$

is the double Schubert polynomial for w, this being a homogeneous polynomial in the $2m$ variables of degree $\ell(w)$ (see [11] for details of the latter).

Theorem 2.1 applies directly to give

$$\langle P(E) - P(F), [X]\rangle = \langle P(\xi_\theta), [S(\Omega(\mathbf{s}))]\rangle, \tag{5.16}$$

In the case $\psi|_{\Omega(w)}$ has constant rank (less than maximal), we deduce from Theorem 2.2, that

$$\langle P(E) - P(F), [X]\rangle = \langle \widehat{\mathfrak{S}}_w(x,y), [\Omega(\mathbf{s})]\rangle, \tag{5.17}$$

where $\widehat{\mathfrak{S}}_w(x,y)$ is a double Schubert polynomial in the class $\{\Omega(w)\}$. Thus, with respect to the cycles $[M]$ and $[N]$ as previously defined, we have

$$\langle c_p(E) - c_p(F), [M]\rangle = \langle (\widehat{\mathfrak{S}}_w(x,y))^{[2p]}, [N]\rangle. \tag{5.18}$$

5.5. Final remark and a further example

We have already mentioned, in the Introduction, the modification of the main result of [12] to the topology of generalized monoidal transformations [14, 15]. In closing, we should add that there are likely to be further situations to which Theorem §2.1 can be applied. As an example of such a situation, in a similar vein to the development of §5.1, consider the following.

Example 5.5. This follows from [27]. Let $\mathcal{L} \longrightarrow X$ be complex line bundle (X here can be a complex manifold), and let $h : \mathcal{L} \longrightarrow \mathbb{C}^{\ell+1}$ be a transversal linear system on X in the sense of [27]. Let $E = Q\mathcal{L}$ be the vector bundle on X whose sections consist of the $\mathbb{C}$-invariant vector fields on $\mathcal{L}$, and let $F = \operatorname{Hom}(Q\mathcal{L}, \mathbb{C}^{\ell+1})$. From h, one can define a complex vector bundle morphism $\psi : E \longrightarrow F$, whose singular set, called the Jacobian set $J(h)$, can be formulated in a similar way to (5.1) (and plays a similar role to $\Omega(s)$). The main results of §2 likewise apply to the adapted pair $(X, J(h))$ (cf. (5.2)), and further enumeration in the constant rank case produces a formula similar to (5.9) for Chern classes of appropriate order (cf. [27]).

Acknowledgment

It is a pleasure for us to dedicate this paper in recognition of Professor Seade's remarkable contributions to research up to his 60-th year, and hopefully beyond as well. Our contribution to the Proceedings benefitted enormously from a substantial and painstakingly detailed report from an anonymous referee who pointed out several corrections, and suggested significant improvements in the general presentation. Thus we express our sincere gratitude to the referee for providing such an excellent report towards revising an earlier version of this paper.

We also thank Professor Jawad Snoussi for managing our contribution, and JFG thanks Professor Roland Roeder for some discussions about this topic.

A. Verjovsky was financed by grant IN103914, PAPIIT, DGAPA, of the Universidad Nacional Autónoma de México.

References

[1] M. Abate, F. Bracci, and F. Tovena, *Index theorems for holomorphic self-maps.* Ann. of Math. **159**(2) (2004), 819–864.

[2] P. Aluffi, *Chern classes of blow-ups.* Math. Proc. Cambridge Philos. Soc. **148** no. 2 (2010), 227-242.

[3] M. F. Atiyah, *K-Theory*, Benjamin, New York, 1967.

[4] A. M. Bonifant and J. E. Fornæss, *Growth or degree for iterates of rational maps in several variables.* Indiana Univ. Math. J. **49**(2) (2000), 751–778.

[5] R. Bott and L. W. Tu, *Differential Forms in Algebraic Topology.* Grad. Texts in Math. **82**. Springer-Verlag, New York Heidelberg Berlin, 1982.

[6] J.-P. Brasselet, *Sur une formule de M. H. Schwartz relative aux revêtements ramifiés.* C. R. Acad. Sci. Paris Sér. A-B **283** (2) (1976), A41–A44.

[7] S. S. Chern, *Complex Manifolds without Potential Theory.* Springer-Verlag, Berlin, New-York, 1979.

[8] J. Diller and C. Favre, *Dynamics of bimeromorphic maps of surfaces.* Amer. J. Math. **123** (2001), 1135–1169.

[9] S. Eilenberg and N. Steenrod, *Foundations of Algebraic Topology.* Princeton Mathematical Series 15, Princeton Univ. Press, Princeton, NJ, 1952.

[10] J. E. Fornæss and N. Sibony, *Complex dynamics in higher dimension II.* Ann. Math. Studies **137**, Princeton Univ. Press, Princeton, NJ, 1995.

[11] W. Fulton, *Flags, Schubert polynomials, degeneracy loci, and determinantal formulas.* Duke Math. J. **65** (1992), no. 3, 381-420.

[12] S. Gitler, J. F. Glazebrook, and A. Verjovsky, *On the generalized Riemann-Hurwitz formula.* Boletin de la Sociedad Matemática Mexicana **30**(1) (1985), 1–11.

[13] S. Gitler, *The cohomology of blow ups.* Papers in honor of José Adem (Spanish). Bol. Soc. Mat. Mexicana(2) **37** (1992), no. 1–2, 167-175.

[14] J. F. Glazebrook and A. Verjovsky, *Residual circuits in generalized monoidal transformations.* Boletin de la Sociedad Matemática Mexicana **33**(1) (1988), 19–25.

[15] J. F. Glazebrook and A. Verjovsky, *Homology theory formulas for generalized Rieman-Hurwitz and generalized monoidal transformations.* In preparartion.

[16] M. Goresky and R. MacPherson, *Intersection homology theory.* Topology **19** (1980), 135–162.

[17] H. Grauert and K. Fritzsche, *Several Complex Variables.* Graduate Texts in Mathematics **38**. Springer-Verlag, New York-Heidelberg, 1976.

[18] P. A. Griffiths and J. Harris, *Principles of Algebraic Geometry.* Wiley, New York, 1978.

[19] J. Harris and L. W. Tu, *On symmetric and skew-symmetric determinantal varieties.* Topology **23** (1984), no. 1, 71–84.

[20] H. Hironaka, *Flattening theorem in complex algebraic geometry.* Amer. J. Math. **97** (1975), 503–547.

[21] F. Hirzebruch, *Topological Methods in Algebraic Geometry.* Grundlehren **131**, 3rd Ed., Springer Verlag 1966.

[22] M. Karoubi, *K-Theory, an introduction.* Grundlehren der matematische Wissenschaften **226**, Springer-Verlag, Berlin-Heidelberg-New York, 1978.

[23] S. Lefschetz, *Algebraic Topology.* American Math. Soc., Providence RI, 1942.

[24] D. Morrison, *The geometry of K3 surfaces.* Lectures delivered at the Scuola Matematica Interuniversitaria Cortona, Italy, July 31-August 27, 1988.

[25] Ngô Van Quê, *Generalisation de la formula de Riemann-Hurwitz.* Canadian J. Math. **24**(5) (1972), 761–767.

[26] I. R. Porteous, *Blowing up Chern classes.* Proc. Camb. Phil. Soc. **56** (1960), 118–124.

[27] I. R. Porteous, *Todd's canonical classes.* Proceedings of Liverpool Singularities Symposium I, (1969/70), pp. 308–312. Lecture Notes in Math. **192**, Springer, Berlin, 1971.

[28] M.-H. Schwartz, *Champs de repères tangents à une variété presque complexe.* Bull. Soc. Math. Belg. **19** (1967), 389–420.

[29] R. Thom, *Les singularités des applications différentiables.* Ann. Inst. Fourier **6**, (1955-56) 43–87.

James F. Glazebrook
Department of Mathematics and Computer Science, Eastern Illinois University
600 Lincoln Avenue, Charleston, IL 61920–3099, USA
and
Adjunct Faculty, Department of Mathematics
University of Illinois at Urbana–Champaign
Urbana, IL 61801, USA
e-mail: `jfglazebrook@eiu.edu`

Alberto Verjovsky
Instituto de Matemáticas
Universidad Autónoma de México
Av. Universidad s/n, Lomas de Chamilpa
Cuernavaca CP 62210, Morelos Mexico
e-mail: alberto@matcuer.unam.mx

On Singular Varieties with Smooth Subvarieties

María del Rosario González-Dorrego

Dedicated to Professor José Seade

Abstract. Let k be an algebraically closed field of characteristic 0. Let C be an irreducible nonsingular curve such that $rC = S \cap F$, $r \in \mathbb{N}$, where S and F are two surfaces and all the singularities of F are of the form $z^3 = x^s - y^s$, s prime, with $\gcd(3, s) = 1$. We prove that C can never pass through such kind of singularities of a surface, unless $r = 3a$, $a \in \mathbb{N}$. The case when the singularities of F are of the form $z^3 = x^{3s} - y^{3s}$, $s \in \mathbb{N}$, were studied in [3]. Next, we study multiplicity-r structures on varieties for any positive integer r. Let Z be a reduced irreducible nonsingular $(n-1)$-dimensional variety such that $rZ = X \cap F$, where X is a normal n-fold with certain type of singularities, F is a $(N-1)$-fold in $\mathbb{P}^N$, such that $Z \cap \text{Sing}(X) \neq \emptyset$. We study the singularities of X through which Z passes.

Mathematics Subject Classification (2000). Primary 14B05, 14E15, 32S25, 14J17, 14J30; Secondary 14J35, 14J40, 14J70.

Keywords. Brieskorn singularities, fundamental cycle, maximal cycle, resolution of singularities.

1. Introduction

The simplest cases of smooth subvarieties passing through the singular locus of varieties are nonsingular curves passing through rational double points of a surface. For example, there are irreducible nonsingular curves of degree 8 and genus 5 passing through the 16 nodes of a quartic Kummer surface in $\mathbb{P}^3$, [1]. If an irreducible nonsingular curve of degree 6 and genus 3 lies on a singular cubic surface F in $\mathbb{P}^3$, F only has rational double points [4]. An irreducible nonsingular curve C of degree 6 and genus 3 on a quartic surface S with t nodes, $10 \leq t \leq 16$, must pass through 10 of these nodes; C satisfies that $2C = S \cap F$, where F is a cubic surface of Cayley, and C passes through its 4 nodes [4].

To define a multiplicity-2 structure $\tilde{Y}$ on a codimension 2 nonsingular variety Y is equivalent to defining a subbundle $L \subset N_{Y|\mathbb{P}^n}$

We believe that our study of nonsingular varieties which are complete intersections with a non reduced structure on them could be used in the construction of vector bundles in $\mathbb{P}^n$. The study of multiplicity-r structures passing through the singular locus of another variety provides a better understanding of the geometry.

We would like to thank the Department of Mathematics at the University of Toronto for their hospitality during the preparation of this manuscript and Mark Spivakovsky for useful discussions. We also thank the referee for his comments.

2. Surface singularities of type $(V_{3,s,s}, O)$, s prime, gcd$(3, s) = 1$

Definition 1. Let X be an n-dimensional normal variety and P a point of X. Let P be an n-fold isolated singularity (that is, the spectrum of an equicharacteristic local noetherian complete ring of Krull dimension n, without zero divisors, whose closed point P is singular). Let $\pi : \tilde{X} \to X$ be the minimal desingularization of X at P. The **genus** of a normal singularity P is defined to be $\dim_k (R^{n-1}\pi_* O_{\tilde{X}})_P$. If the genus is 0, the singularity is said to be **rational**. If the genus is 1, it is **elliptic**.

Notation 2. Let F be a reduced surface and P a point of F. Let (F, P) be a surface singularity (that is, the spectrum of an equicharacteristic complete local ring of Krull dimension 2 whose closed point P is singular). Let $\pi : \tilde{F} \to F$ be the minimal desingularization of F at P.

Definition 3. Let F be a reduced surface with a singular point at the origin O. We shall denote by (V_{a_0,a_1,a_2}, O) the singularity of the form

$$(V_{a_0,a_1,a_2}, O) = \{(x_0, x_1, x_2) \in U \subset k^3 \text{ such that } {x_0}^{a_0} + {x_1}^{a_1} = {x_2}^{a_2}\}$$

with $2 \leq a_0 \leq a_1 \leq a_2$ and U is a small neighborhood of O.

Notation 4. Let $l = gcd(a_0, a_1, a_2)$, $l_i = \frac{gcd(a_j, a_k)}{l}$, $\alpha_i = \frac{a_i}{l_j l_k l}$, i, j, $k \in \{0, 1, 2\}$, $i \neq j$, $i \neq k$, $j \neq k$.

Proposition 5. (V_{a_0,a_1,a_2}, O), $2 \leq a_0 \leq a_1 \leq a_2$, *is a Kodaira singularity if and only if* $\alpha_0\alpha_1 l_2 \leq \alpha_2$.

Proof. See [6, Prop.4.4]. □

Proposition 6. $(V_{3,s,s}, O)$, *s prime, with* $gcd(3, s) = 1$, *is not a Kodaira singularity.*

Proof. According to 4, $l = 1$, $l_0 = s$, $l_1 = l_2 = 1$, $\alpha_0 = 3$, $\alpha_1 = \alpha_2 = 1$. By Proposition 5,$(V_{3,s,s}, O)$, s prime, with gcd$(3, s) = 1$, is not a Kodaira singularity since $\alpha_0\alpha_1 l_2 = 3 > \alpha_2 = 1$. □

Definition 7.
- We call **maximal cycle** $Z_{\tilde{F}}$ the cycle $Z_{\tilde{F}} = \sum m_i E_i$, defined by the divisorial part of $MO_{\tilde{F}}$, where M is the maximal ideal $MaxO_{F,P}$ of $O_{F,P}$; the E_i are the irreducible components of dimension 1 of the exceptional fiber $\pi^{-1}(P)$ and the m_i are nonnegative integers. A component E_j such that $m_j = 1$ is called a **reduced component** of the cycle.

- Consider positive cycles $Z = \sum r_i E_i$, $r_i \geq 0$ such that $(Z.E_i) \leq 0$, for all i. The unique smallest cycle Z satisfying $(Z.E_i) \leq 0$, for all i, is called **the fundamental cycle** of $\tilde{F}$.
- We denote by $p_f(Z)$ the arithmetic genus of the fundamental cycle Z. We call it the **fundamental genus**. Since it is independent of the resolution we can also write $p_f(F, O)$.

Proposition 8. (a) *Let (F, O) be a normal surface singularity. Let $p_a(F, O)$ be the arithmetic genus of F. Let $p_g(F, O)$ be the genus defined in* (1). *We have that*

$$0 \leq p_f(F, O) \leq p_a(F, O) \leq p_g(F, O).$$

(b) *Let $l_2 \leq \alpha_2$. The fundamental genus of (V_{a_0,a_1,a_2}, O), $2 \leq a_0 \leq a_1 \leq a_2$, is*

$$p_f(Z) = \frac{1}{2}\{(a_0 - 1)(a_1 - 1) - 2[\frac{\alpha_0 \alpha_1 l_2}{\alpha_2}] - 1) \gcd(a_0, a_1) + 1\},$$

where $[a] =: \min\{n \in \mathbb{Z} | n \geq a\}$, $a \in \mathbb{R}$.

Proof. (a) See [8].
(b) See [6, Theorem 1.7]. □

Lemma 9. *The fundamental genus of $(V_{3,s,s}, O)$, s prime, with* $\gcd(3, s) = 1$, *is* $s - 3$.

Proof. According to 4, $l = 1$, $l_0 = s$, $l_1 = l_2 = 1$, $\alpha_0 = 3$, $\alpha_1 = \alpha_2 = 1$. By Proposition 8 (b), $(V_{3,s,s}, O)$, s prime, with $\gcd(3, s) = 1$, $p_f(Z) = s - 3$. □

2.1. Study of the fundamental cycle for $(V_{3,s,s}, O)$, s prime, $\gcd(3, s) = 1$

Let us denote $x_0 = z$, $x_1 = y$, $x_2 = x$. We consider the surface singularity (F, O) given by $z^3 = x^s - y^s$. We want to desingularize F at O.

- **Case $s = 5$:** We consider the surface singularity (F, O) given by $z^3 = x^5 - y^5$. We want to desingularize F at O. Applying to $z^3 = x^5 - y^5$ the change $z_{(1)} = \frac{z}{x}$, $y_{(1)} = \frac{y}{x}$, $x_{(1)} = x$, we obtain in $F_{(1)}$ that ${z_{(1)}}^3 = {x_{(1)}}^2(1 - {y_{(1)}}^5)$; $\pi_1 : F_{(1)} - -- > F$, $E = {\pi_1}^{-1}(O)$ exceptional divisor given by $x_{(1)} = 0$. Let us apply to ${z_{(1)}}^3 = {x_{(1)}}^2(1 - {y_{(1)}}^5)$ the change $z_{(2)} = z_{(1)}$, $x_{(2)} = \frac{x_{(1)}}{z_{(1)}}$, $y_{(2)} = y_{(1)}$. We obtain $z_{(2)} = {x_{(2)}}^2(1 - {y_{(2)}}^5)$ in $F_{(2)}$; $\pi_2 : F_{(2)} - -- > F_{(1)}$. The fundamental cycle Z is $\sum_{j=1}^{5} E_j + 3F$, in a star shape with 5 arms, where $F.F = -2$, $E_j.E_j = -3$, $1 \leq j \leq 5$, $E_i.E_j = 0$, $i \neq j$. $F.E_j = 1$. In the picture below, F is denoted by a star and the E_j, $1 \leq j \leq 5$, are denoted by a circle. It has 5 reduced components. Note that $Z^2 = -3$.

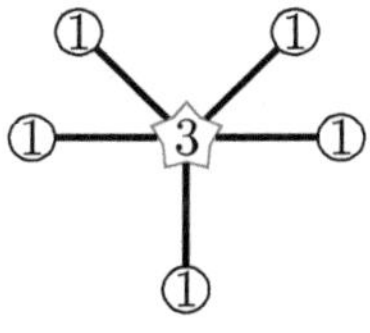

- **Case** $s = 7$**:** We consider the surface singularity (F, O) given by $z^3 = x^7 - y^7$. We want to desingularize F at O. Applying to $z^3 = x^7 - y^7$ the change $z_{(1)} = \frac{z}{x}$, $y_{(1)} = \frac{y}{x}$, $x_{(1)} = x$, we obtain in $F_{(1)}$ that ${z_{(1)}}^3 = {x_{(1)}}^4(1 - {y_{(1)}}^7)$; $\pi_1 : F_{(1)} --- > F$, $E = {\pi_1}^{-1}(O)$ exceptional divisor given by $x_{(1)} = 0$. Let us apply to ${z_{(1)}}^3 = {x_{(1)}}^4(1 - {y_{(1)}}^7)$ the change $x_{(2)} = x_{(1)}$, $z_{(2)} = \frac{z_{(1)}}{x_{(1)}}$, $y_{(2)} = y_{(1)}$. We obtain ${z_{(2)}}^3 = x_{(2)}(1 - {y_{(2)}}^7)$ in $F_{(2)}$; $\pi_2 : F_{(2)} --- > F_{(1)}$. Let us apply to ${z_{(2)}}^3 = x_{(2)}(1 - {y_{(2)}}^7)$ the change $z_{(3)} = z_{(2)}$, $x_{(3)} = \frac{x_{(2)}}{z_{(2)}}$, $y_{(3)} = y_{(2)}$. We obtain ${z_{(3)}}^2 = x_{(3)}(1 - {y_{(3)}}^7)$ in $F_{(3)}$; $\pi_3 : F_{(3)} --- > F_{(2)}$. Let us apply to ${z_{(3)}}^2 = x_{(3)}(1 - {y_{(3)}}^7)$ the change $z_{(4)} = z_{(3)}$, $x_{(4)} = \frac{x_{(3)}}{z_{(3)}}$, $y_{(4)} = y_{(3)}$. We obtain $z_{(4)} = x_{(4)}(1 - {y_{(4)}}^7)$ in $F_{(4)}$; $\pi_4 : F_{(4)} --- > F_{(3)}$. The fundamental cycle Z is $3F + \sum_{j=1}^{7} 2E_{j1} + \sum_{j=1}^{7} E_{j2}$, where $F.F = -5$, $E_{ji}.E_{ji} = -2, 1 \leq j \leq 7$ and $1 \leq i \leq 2$. For each j, $E_{j1}.E_{j2} = 1$. $E_{j1}.E_{k2} = 0$, $1 \leq k \leq 7$, $k \neq j$. For each j, $F.E_{j1} = 1$ and $F.E_{j2} = 0$. Note that $Z^2 = -3$. The fundamental cycle has a star shape with 7 arms; each arm is the weighted dual graph of an A_2 singularity. It has 7 reduced components. In the picture below, F is denoted by a star and the E_{ji}, $1 \leq j \leq 7$, $1 \leq i \leq 2$, are denoted by a circle.

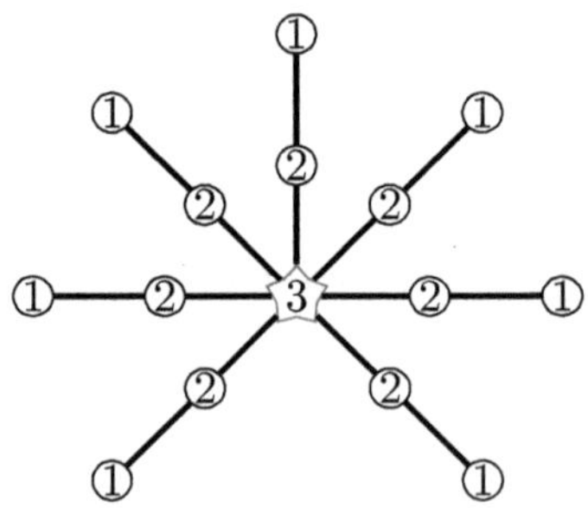

- **Case** $s = 3t + 2$, t **odd:** We consider the surface singularity (F, O) given by $z^3 = x^s - y^s$. We want to desingularize F at O. Applying to $z^3 = x^s - y^s$ the change $z_{(1)} = \frac{z}{x}$, $y_{(1)} = \frac{y}{x}$, $x_{(1)} = x$, we obtain in $F_{(1)}$ that ${z_{(1)}}^3 = {x_{(1)}}^{s-3}(1 - {y_{(1)}}^s)$; $\pi_1 : F_{(1)} --- > F$, $E = {\pi_1}^{-1}(O)$ exceptional divisor given by $x_{(1)} = 0$. Let us apply to ${z_{(1)}}^3 = {x_{(1)}}^{s-3}(1 - {y_{(1)}}^s)$ the change $x_{(2)} = x_{(1)}$, $z_{(2)} = \frac{z_{(1)}}{x_{(1)}}$, $y_{(2)} = y_{(1)}$. We obtain ${z_{(2)}}^3 = x_{(2)}^{s-6}(1 - {y_{(2)}}^s)$ in $F_{(2)}$; $\pi_2 : F_{(2)} --- > F_{(1)}$. We keep on doing similar changes. $x_{(t-1)} = x_{(t-2)}$, $z_{(t-1)} = \frac{z_{(t-2)}}{x_{(t-2)}}$, $y_{(t-1)} = y_{(t-2)}$. We obtain ${z_{(t-1)}}^3 = x_{(t-1)}^5(1 - {y_{(t-1)}}^s)$. We make the change $x_{(t)} = x_{(t-1)}$, $z_{(t)} = \frac{z_{(t-1)}}{x_{(t-1)}}$, $y_{(t)} = y_{(t-1)}$. We obtain ${z_{(t)}}^3 = x_{(t)}^2(1 - {y_{(t)}}^s)$. We make the change $z_{(t+1)} = z_{(t)}$, $x_{(t+1)} = \frac{x_{(t)}}{z_{(t)}}$, $y_{(t+1)} = y_{(t)}$. We obtain $z_{(t+1)} = x_{(t+1)}^2(1 - {y_{(t+1)}}^s)$. The fundamental cycle Z is $3F + \sum_{j=1}^{s} E_j$, where $F.F = -(t+1)$, $E_j.E_j = -3$, $1 \leq j \leq s$. $E_i.E_j = 0$, $i \neq j$. $F.E_j = 1$. It has a star shape with s arms. It has s reduced components. Note that $Z^2 = -3$.

- **Case $s = 3t + 1$, t even:** We consider the surface singularity (F, O) given by $z^3 = x^s - y^s$. We want to desingularize F at O. Applying to $z^3 = x^s - y^s$ the change $z_{(1)} = \frac{z}{x}$, $y_{(1)} = \frac{y}{x}$, $x_{(1)} = x$, we obtain in $F_{(1)}$ that ${z_{(1)}}^3 = {x_{(1)}}^{s-3}(1 - {y_{(1)}}^s)$; $\pi_1 : F_{(1)} --- > F$, $E = {\pi_1}^{-1}(O)$ exceptional divisor given by $x_{(1)} = 0$. Let us apply to ${z_{(1)}}^3 = {x_{(1)}}^{s-3}(1 - {y_{(1)}}^s)$ the change $x_{(2)} = x_{(1)}$, $z_{(2)} = \frac{z_{(1)}}{x_{(1)}}$, $y_{(2)} = y_{(1)}$. We obtain ${z_{(2)}}^3 = x_{(2)}^{s-6}(1 - {y_{(2)}}^s)$ in $F_{(2)}$; $\pi_2 : F_{(2)} --- > F_{(1)}$. Let us apply to ${z_{(2)}}^3 = x_{(2)}^{s-6}(1 - {y_{(2)}}^s)$ the change $z_{(3)} = z_{(2)}$, $x_{(3)} = \frac{x_{(2)}}{z_{(2)}}$, $y_{(3)} = y_{(2)}$. We obtain ${z_{(3)}}^3 = x_{(3)}^{s-9}(1 - {y_{(3)}}^s)$ in $F_{(3)}$; $\pi_3 : F_{(3)} --- > F_{(2)}$.

  We keep on doing similar changes. $x_{(t-1)} = x_{(t-2)}$, $z_{(t-1)} = \frac{z_{(t-2)}}{x_{(t-2)}}$, $y_{(t-1)} = y_{(t-2)}$. We obtain ${z_{(t-1)}}^3 = x_{(t-1)}^4(1 - {y_{(t-1)}}^s)$. We make the change $x_{(t)} = x_{(t-1)}$, $z_{(t)} = \frac{z_{(t-1)}}{x_{(t-1)}}$, $y_{(t)} = y_{(t-1)}$. We obtain ${z_{(t)}}^3 = x_{(t)}(1 - {y_{(t)}}^s)$. We make the change $z_{(t+1)} = z_{(t)}$, $x_{(t+1)} = \frac{x_{(t)}}{z_{(t)}}$, $y_{(t+1)} = y_{(t)}$. We obtain $z_{(t+1)}^2 = x_{(t+1)}(1 - {y_{(t+1)}}^s))$. We make the change $z_{(t+2)} = z_{(t+1)}$, $x_{(t+2)} = \frac{x_{(t+1)}}{z_{(t+1)}}$, $y_{(t+2)} = y_{(t+1)}$. We obtain $z_{(t+2)} = x_{(t+2)}(1 - {y_{(t+2)}}^s)$. The fundamental cycle Z is $3F + \sum_{j=1}^{s} 2E_{j1} + \sum_{j=1}^{s} E_{j2}$, where $F.F = -(2t + 1)$, $E_{ji}.E_{ji} = -2$, $1 \leq j \leq s$ and $1 \leq i \leq 2$. For each j, $E_{j1}.E_{j2} = 1$. $E_{j1}.E_{k2} = 0$, $1 \leq k \leq s$, $k \neq j$. For each j, $F.E_{j1} = 1$ and $F.E_{j2} = 0$. The fundamental cycle has a star shape with s arms; each arm is the weighted dual graph of an A_2 singularity. It has s reduced components. Note that $Z^2 = -3$.

Proposition 10. *The families of smooth curves on a normal surface singularity are in one-to-one correspondence with the reduced components of the maximal cycle of its minimal desingularization* π.

Proof. See [5, 1.14]. □

Proposition 11. *For a surface singularity of the type* (V_{a_0,a_1,a_2}, O), *the maximal cycle of* π *and the fundamental cycle of its weighted dual graph coincide if and only if* $\alpha_2 \geq l_2$.

Proof. See [6, Th. 3.6]. □

Corollary 12. *For a surface singularity of the type* $(V_{3,s,s}, O)$, *s prime, with* $\gcd(3, s) = 1$, *the maximal cycle of* π *coincides with the fundamental cycle of its weighted dual graph.*

Proof. It follows from Proposition 11 since $\alpha_2 = 1 \geq l_2 = 1$. □

Proposition 13. *The self-intersection of the fundamental cycle for a surface singularity of the type* (V_{a_0,a_1,a_2}, O), $\alpha_2 \leq l_2$ *is*

$$-Z^2 = l\alpha_0\alpha_1\alpha_2.$$

Proof. See [6, Prop. 1.6]. □

Corollary 14. *For a surface singularity of the type* $(V_{3,s,s}, O)$, *s prime, with* $\gcd(3,s) = 1$, $Z^2 = -3$, *where* Z *is the fundamental cycle of its weighted dual graph.*

Proof. It follows from Proposition 13 since $\alpha_2 = 1 = l_2 = 1$, $l = \alpha_1 = 1$, $\alpha_0 = 3$. □

Theorem 15. *Let* C *be an irreducible nonsingular curve such that* $rC = S \cap V$, *where* S *and* W *are surfaces and* $r \in \mathbb{N}$. *Then* C *cannot pass through any singular point of the surfaces of the type* $(V_{3,s,s}, O)$, *s prime, with* $\gcd(3,s) = 1$, *unless* $r = 3a$, $a \in \mathbb{N}$.

Proof. Let C be an irreducible nonsingular curve $rC = S \cap W$ where S and W are two surfaces and W has at most points of type $V_{3,s,s}$, s prime, with $\gcd(3,s) = 1$, . Let us suppose that C passes through such a point O of W. Let $\tilde{W}$ be the minimal desingularization of W at O, $\pi : \tilde{W} \to W$. Let F_k, $1 \leq k \leq n$, be the irreducible components of the exceptional divisor. The total transform $\pi^*(rC) = \sum_{j=1}^{n} \beta_k F_k + rE$, where E is the strict transform of C, $\beta_k \in \mathbb{N}$. Consider the fundamental cycle of a surface singularity of the type $(V_{3,s,s}, O)$, s prime, with $\gcd(3,s) = 1$. By Proposition 10 and Corollary 12, if an irreducible nonsingular curve C passes through a singularity of W, then its strict transform must intersect transversally only one exceptional divisor of multiplicity one in the fundamental cycle. Moreover, since our curve C may not be a Cartier divisor, we consider rC, $r \in \mathbb{N}$. Let us consider its total transform; it must have intersection 0 with each exceptional divisor.

- **Case** $s = 5$**:** Recall that $F.F = -2$, $E_j.E_j = -3$, $1 \leq j \leq 5$, $E_i.E_j = 0$, $i \neq j$. $F.E_j = 1$. Let $a \in \mathbb{N}$. If C would pass through a singularity of type $(V_{3,5,5}, O)$, we should be able to find a cycle like the one below:

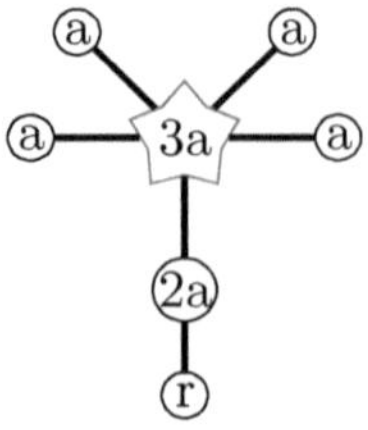

In fact, for $r = 3a$, $a \in \mathbb{N}$, such a cycle exists. For example, for $a = 1$,

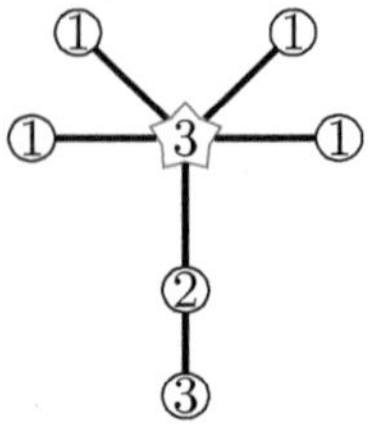

Similarly, if the circle with multiplicity r were attached to another reduced component of the fundamental cycle.

- **Case** $s = 7$:Recall that $F.F = -5$, $E_{ji}.E_{ji} = -2$, $1 \le j \le 7$ and $1 \le i \le 2$. For each j, $E_{j1}.E_{j2} = 1$. $E_{j1}.E_{k2} = 0$, $1 \le k \le 7$, $k \ne j$; for each j, $F.E_{j1} = 1$ and $F.E_{j2} = 0$. Let $a \in \mathbb{N}$. If C would pass through a singularity of type $(V_{3,7,7}, O)$, we should be able to find a cycle like the one below:

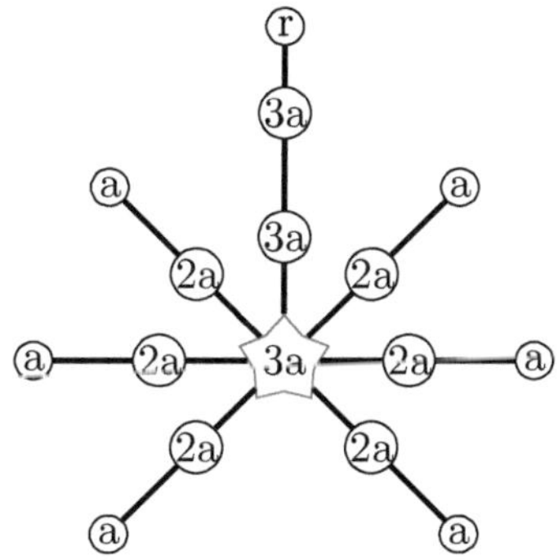

In fact, for $r = 3a$, $a \in \mathbb{N}$, such a cycle exists. For example, for $a = 1$,

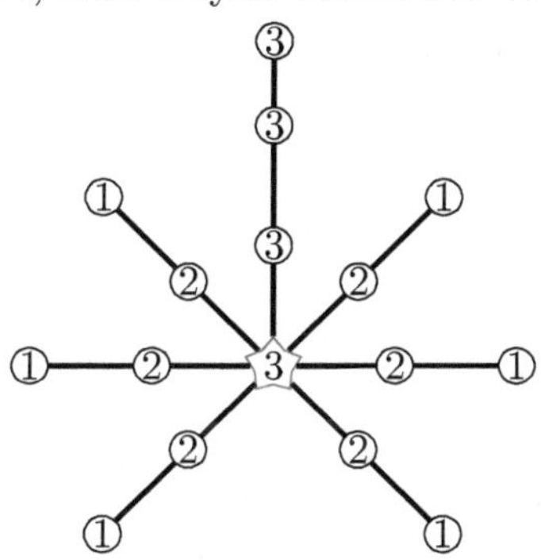

Similarly, if the circle with multiplicity r were attached to another reduced component of the fundamental cycle.
- **Case** $s = 3t+2$, **t odd:** Similar reasoning to Case $s = 5$. Recall that $F.F = -2$, $E_j.E_j = -3$, $1 \le j \le s$, $E_i.E_j = 0$, $i \ne j$. $F.E_j = 1$.
- **Case** $s = 3t+1$, **t even:** Similar reasoning to Case $s = 7$. Recall that $F.F = -5$, $E_{ji}.E_{ji} = -2$, $1 \le j \le s$ and $1 \le i \le 2$. For each j, $E_{j1}.E_{j2} = 1$. $E_{j1}.E_{k2} = 0$, $1 \le k \le s$, $k \ne j$. For each j, $F.E_{j1} = 1$ and $F.E_{j2} = 0$. □

3. Smooth subvarieties through certain type of singularities

Definition 16. (a) Let $(O_{X,P}, M_P)$ be the local ring of a point $P \in X$ of a k-scheme. Let $V \subset M_P$ be a finite dimensional k-vector space which generates M_P as an ideal of $O_{X,P}$. By a **general hyperplane through** P we mean the subscheme H defined in a suitable open neighbourhood U of P by the ideal $(v)O_X$, where $v \in V$ is a k-point of a certain dense Zariski open set in V. [7, (2.5)]. By a **general linear variety of codimension** r **through** P we mean the subscheme $L \subset U$ defined in a suitable open neighbourhood U of P by the ideal $(v_1, \ldots, v_r)O_X$, where $v_1, \ldots, v_r \in V$ are k-points of a certain dense Zariski open set in V.

(b) Let X be a singular n-fold. We say that a point $Q \in \mathrm{Sing}(X)$ is **a general point of Sing**(X) if, for a general hyperplane H such that $Q \in H$ and for some divisorial resolution $f : V \to X$, the preimage $f^{-1}(Q)$ of Q and the strict transform $f_*^{-1}(X \cap H)$ satisfy $f^{-1}(Q) \subset f_*^{-1}(X \cap H)$.
See [2, 2.4].

Definition 17. Let X be a singular n-fold embedded in $\mathbb{P}^N$ such that $\dim \mathrm{Sing}(X) > 0$.

(a) Let H be a hyperplane in $\mathbb{P}^N$ such that $H \cap \mathrm{Sing}(X) \neq \emptyset$. Denote $X \cap H$ by X_0. We say that X_0 is **a general hyperplane section meeting Sing**(X) if it is irreducible and, for some divisorial resolution $f : V \to X$, the total transform $f^*(X_0)$ is equal to the strict transform $f_*^{-1}(X_0)$.

(b) Let L_{r+1} be a linear variety of codimension $r+1$, $0 \leq r \leq n-3$, in $\mathbb{P}^N$ such that $L_{r+1} \cap \mathrm{Sing}(X) \neq \emptyset$. Denote $X \cap L_{r+1}$ by W_r, $0 \leq r \leq n-3$. We say that W_r is **a general linear section meeting Sing**(X) if it is irreducible and, for some divisorial resolution $f : V \to X$, the total transform $f^*(W_r)$ is equal to the strict transform $f_*^{-1}(W_r)$.

Definition 18. Let X be a n-fold in in $\mathbb{P}^N$. A point $P \in X$ is called a **linear compound V_{3s} singularity or a lcV_{3s} point**, s prime, with $\gcd(3, s) = 1$, if, for a general linear variety W of codimension $n-2$ through P, $P \in W$, $X \cap W$ is a singularity of the type $V_{3,s,s}$, s prime, with $\gcd(3, s) = 1$.

Remark 19. $P \in X$ is lcV_{3s} if it is locally analytically isomorphic to the hypersurface singularity of the form $f + g$, where f is a polynomial in $k[x_1, x_2, x_3]$ of the form $c_3(x_3)^3 = c_1(x_1)^s + c_2(x_2)^s + h(x_1, x_2, x_3)$, s prime, with $\gcd(3, s) = 1$, $c_1, c_2, c_3 \in k$, and $g \in k[x_1, ..., x_N]$, $\mathrm{mult}\, g > \deg f \geq 2s$. This is so because any monomial w of h would be divisible either by $(x_3)^2$ or $(x_1)^{s-1}$ or $(x_2)^{s-1}$ which would make possible to absorb w respectively in either $(x_3)^3$ or $(x_1)^s$ or $(x_2)^s$.

Definition 20. Let us consider the d-uple embedding $\rho_d : \mathbb{P}^N \to \mathbb{P}^M$, $M = \binom{n+d}{d} - 1$. Let Y be a complete intersection nonsingular variety in $\mathbb{P}^N$ defined by l equations $\sum_{i=0}^{M} a_{ij} y_i$, $1 \leq j \leq l$, $1 \leq l \leq N-1$, where all the y_i, $0 \leq i \leq M$, are monomials of degree d in $x_0, \ldots, x_N$, for some $l, d \in \mathbb{N}$. We call Y a (d, l) **complete intersection nonsingular variety**. See [3, Def. 7].

Definition 21. Let X be a normal singular variety of dimension n in $\mathbb{P}^N$. Let Y be a (d, l) complete intersection nonsingular variety in $\mathbb{P}^N$ such that $Y \cap \mathrm{Sing}(X) \neq \emptyset$. Let us consider the d-uple embedding $\rho_d : \mathbb{P}^N \to \mathbb{P}^M$, $M = \binom{n+d}{d} - 1$. We have that, as abstract varieties, $X \cap Y = \rho_d(X) \cap \tilde{Y}$, where $\tilde{Y}$ is a linear variety in $\mathbb{P}^M$ obtained from Y using ρ_d. We say that Y is **a** (d, l) **general complete intersection nonsingular variety in $\mathbb{P}^N$ meeting Sing** X if $\tilde{Y}$ is a general linear variety in $\mathbb{P}^M$ meeting $\mathrm{Sing}\, \rho_d(X)$, according to Definition 18. See [3, Def. 9].

Definition 22. Let X be a n-fold. A point $P \in X$ is called a (d, l) **complete intersection compound** V_{3s} **singularity or a dlcic**V_{3s} **point** if, for a general (d, l) complete intersection nonsingular variety Y of codimension $n-2$ through P, $P \in Y$, $X \cap Y$ is a singularity of type $V_{3,s,s}$, s prime, with $\gcd(3, s) = 1$.

Proposition 23. *Let X be a quasiprojective scheme over any field. Let R_{l+1} be a general linear variety of codimension $l+1$ in $\mathbb{P}^N$, $0 \leq l < N-1$. Let $T_l = X \cap R_{l+1}$. Let $r \in \mathbb{N}$. If X satisfies Serre's condition S_r, then so does T_l. Thus, if X is a normal variety, then so is T_l.*

Proof. See [2, 3.4]. □

Proposition 24. *Let Z be a reduced irreducible nonsingular $(n-1)$-dimensional variety such that $rZ = X \cap F$, $r \in \mathbb{N}$, $r \geq 2$, where X is an n-fold and F is a $(N-1)$-fold in $\mathbb{P}^N$, X normal with a lcV_{3s} singularity P and such that $Z \cap \mathrm{Sing}(X) \neq \emptyset$. Let W_{n-3} be a general linear variety as above. Then Z has empty intersection with a lcV_{3s} singularity of X such that $W_{n-3} \cap X = Q$, where Q is a surface singularity of type $V_{3,s,s}$, s prime, with* $\gcd(3, s) = 1$, *unless $r = 3a$, $a \in \mathbb{N}$.*

Proof. Let us define recursively W_t. Let H_{t+1}, $0 \leq t \leq n-4$, be a general hyperplane meeting $\mathrm{Sing}(W_t) \cap F$. Given $rZ = X \cap F$, $r \in \mathbb{N}$, $r \geq 2$, we intersect it with H_t, $0 \leq t \leq n-3$, as follows: $rZ \cap H_0 \cap \cdots \cap H_{n-3} = F \cap X \cap H_0 \cap \cdots \cap H_{n-3}$. We obtain a nonsingular curve C such that $rC = F \cap X \cap H_0 \cap \cdots \cap H_{n-3}$ and that $C \cap \mathrm{Sing}(W_{n-3}) \neq \emptyset$. We apply Theorem 15 to obtain the result. □

Corollary 25. *Let Z be a reduced irreducible nonsingular $(n-1)$-dimensional variety such that $2Z = X \cap F$, where X is an n-fold and F is a $(N-1)$-fold in $\mathbb{P}^N$, X normal with a* dlcicV_{3s} *singularity P, s prime, with* $\gcd(3, s) = 1$, *and such that $Z \cap \mathrm{Sing}(X) \neq \emptyset$. Let Y be a general (d, l) complete intersection nonsingular variety of codimension $n-2$. Then Z has empty intersection with a* dlcicV_{3s} *singularity of X such that $Y \cap X = Q$, where Q is a surface singularity of type $V_{3,s,s}$, s prime, with* $\gcd(3, s) = 1$, *unless $r = 3a$, $a \in \mathbb{N}$.*

Proof. Let us consider the d-uple embedding $\rho_d : \mathbb{P}^N \to \mathbb{P}^M$, $M = \binom{n+d}{d} - 1$. Let Y defined above. We have that, as abstract varieties, $X \cap Y = \rho_d(X) \cap \tilde{Y}$, where $\tilde{Y}$ is a linear variety in $\mathbb{P}^M$. To obtain the result we follow the proof of Proposition 24 substituting W_{n-3} by $\tilde{Y}$. □

References

[1] M.R. González-Dorrego, *Degree 8 and genus 5 curves in $\mathbb{P}^3$ and the Horrocks-Mumford bundle*, in: Algebraic Geometry and Singularities, Birkhäuser, Progress in Mathematics 134 (1996), 303–310.

[2] M.R. González-Dorrego, *Smooth double subvarieties on singular varieties*, RACSAM 108 (2014), 183–192.

[3] M.R. González-Dorrego, *Smooth double subvarieties on singular varieties III*, accepted for publication.

[4] M.R. González-Dorrego, *On the normal bundle of curves on nodal quartic surfaces*, Communications in Algebra, 28(12) (2000), 5837–5855.

[5] G. González-Sprinberg and M. Lejeune-Jalabert, *Families of smooth curves on singularities and wedges*, Ann. Pol. Math. LXVII.2 (1997), 179–190.

[6] K. Konno and D. Nagashima, *Maximal ideal cycles over normal surface singularities of Brieskorn type*, Osaka J. Math. 49 (2012), 225–245.

[7] M. Reid, *Canonical threefolds*, in: Journées de Géometrie Algébrique et Singularités, Angers 1979, A. Beauville ed., Sijthoff and Noordhoff (1980), 273–310.

[8] T. Tomaru, *Complex surface singularities and degenerations of compact complex curves*, Demonstratio Mathematica XLIII (2) (2010).

[9] P. Wagreich, *Elliptic singularities of surfaces*, Amer. J. Math. 92 (1970), 421–454.

María del Rosario González-Dorrego
Departamento de Matemáticas
Facultad de Ciencias
Universidad Autónoma de Madrid
28049 Cantoblanco-Madrid
Spain
e-mail: mrosario.gonzalez@uam.es

On Polars of Plane Branches

A. Hefez, M.E. Hernandes and M.F. Hernández Iglesias

Dedicated to Pepe Seade on occasion of his sixtieth birthday anniversary.

Abstract. It is well known that the equisingularity class, which is the same as the topological type, of the general polar curve of a plane branch is not the same for all branches in a given equisingularity class, but depends upon the analytic type of the branch. It was shown in [1] that, for sufficiently general branches in a given equisingularity class, the topology of the general polar is constant. The aim of this paper is to go beyond generality and show how one could describe the topology of the general polars of all branches in a given equisingularity class, making use of the analytic classification of branches as described in [5]. We will show how this works in some particular equisingularity classes for which one has the complete explicit analytic classification, and in particular for all branches of multiplicity less or equal than four, based on the classification given in [4].

Mathematics Subject Classification (2000). Primary 32S15; Secondary 14H15.
Keywords. Polar curves, analytic classification of curves, equisingularity.

1. Introduction

Let $f \in \mathbb{C}\{x, y\}$ be such that $(0, 0)$ is an isolated singular point of the germ of the curve (f): $f = 0$. Notice that because the singular point is isolated, we may assume that all series under consideration are just formal power series (cf. [13]). A germ of a curve at the origin of $\mathbb{C}^2$ determines its reduced equation up to multiplication by a unit. An irreducible germ of curve will be called a *branch.* Two germs of curves (f) and (g) are said *analytically equivalent* if they are isomorphic as immersed germs, that is, if there exists a germ of analytic diffeomorphism φ at the origin of $\mathbb{C}^2$ that transforms one germ into the other. In terms of equations, this translates into the fact that there also exists a unit u in $\mathbb{C}\{x, y\}$ such that $g = uf \circ \varphi$. In

The first two authors were partially supported by CNPq Grants and the third author by a fellowship from CAPES and ARAUCARIA foundations.

such a case we say that the series f and g are *contact equivalent.* When the above φ is only a homeomorphism, we say that (f) and (g) are topologically equivalent or *equisingular.* The *equisingularity class* of a curve is its class under this last equivalence relation.

The polar curve of f in the direction $(a\colon b) \in \mathbb{P}^1_{\mathbb{C}}$ is the germ of curve defined by the equation $af_x+bf_y = 0$. It is known (cf. [14, page 430] or [2, Theorem 7.2.10]) that, except for a finite set of directions, the polar is reduced and its equisingularity class is constant, although its analytic type depends essentially upon the direction $(a : b)$, as we will see in an example at the end of the paper. From now on, when we refer to the polar of f, we mean its *general polar.* Also, the equisingularity class of the polar of f is constant in the contact class of f (see for example [2, Corollary 8.5.8]), but it is not constant in the equisingularity class of f, as one can easily check by considering for example the curves $y^3 - x^{11}$ and $y^3 - x^{11} + x^8y$ which are equisingular but have no equisingular polars (cf. [12]). So, the topological type of the polar of a given curve is not determined only by the topological type of the curve, but it is determined by its analytical type. In the next section we will see to what extent the analytic type of the curve will influence the topology of its polar.

We refer to [15] for the definitions and basic results we will use in the sequel. It is a classical result that the equisingularity class of a branch (f) is determined by its semigroup of values Γ_f, which is the subset of the integers consisting of all intersection numbers $\mathrm{I}(f,h)$, for $h \in \mathbb{C}\{x,y\}$ and not a multiple of f, where

$$\mathrm{I}(f,h) = \dim_{\mathbb{C}} \frac{\mathbb{C}\{x,y\}}{\langle f,h\rangle},$$

and $\langle f,h\rangle$ denotes the ideal generated by f and h in $\mathbb{C}\{x,y\}$.

The equisingularity class of a reduced curve given by $f = f_1 \cdots f_r$, where the f_i's are irreducible, is determined by the semigroups of the branches (f_i)'s and their mutual intersection numbers $\mathrm{I}(f_i, f_j)$, for $i \neq j$.

The semigroup of values Γ of a branch will be given by its minimal set of generators $\Gamma = \langle v_0, v_1, \ldots, v_g\rangle$ and the integer g will be called the *genus* of the branch. Such semigroup has a conductor μ, that is $\mu-1 \notin \Gamma$ and $\nu \in \Gamma$ for all $\nu \geq \mu$. If $v_0 > 2$, then the equisingularity class that Γ determines may be parametrized by a constructible set $\mathcal{E}$ in $\mathbb{C}^{\mu-v_1-1}$, whose points are the coefficients of the Newton-Puiseux parametrization

$$x(t) = t^{v_0}, \quad y(t) = t^{v_1} + \sum_{i=v_1+1}^{\mu-1} c_i t^i,$$

in the sense that any element in the equisingularity class is analytically equivalent to one with a Newton-Puiseux parametrization as above.

From a Newton-Puiseux parametrization $x = t^n$, $y = y(t)$, where $n = v_0$, we get an implicit equation $f = 0$ of a branch in the following way:

$$f(x,y) = \prod_{i=1}^{n} (y - y_i(x^{\frac{1}{n}})), \tag{1.1}$$

where $y_i(x^{\frac{1}{n}}) = y(\omega^i x^{\frac{1}{n}})$, and ω is a primitive n-th root of unity. So, the coefficients of $f(x,y)$ as a polynomial in y are $a_j(x) = (-1)^j S_j(x)$, where the S_j's are the elementary symmetric functions in n variables evaluated at $(y_1(x^{\frac{1}{n}}), \ldots, y_n(x^{\frac{1}{n}}))$.

Given an equisingularity class of irreducible curves through a semigroup of values Γ, it was proved in [3] and [5] that the parameter space $\mathcal{E}$ may be decomposed into a finite union of disjoint constructible sets $\mathcal{E} = \mathcal{E}_1 \cup \cdots \cup \mathcal{E}_r$, where on each $\mathcal{E}_\ell$, the set Λ_ℓ of values of Kähler differentials on the corresponding curve, which is an analytic invariant of the curve, is fixed.

Since Γ has a conductor and $\Gamma \setminus \{0\} \subset \Lambda_\ell$, the set Λ_ℓ is determined by the finite set $\Lambda_\ell \setminus \Gamma$. If this last set is not empty, the natural number λ associated to a curve represented by a point of $\mathcal{E}_\ell$, defined as

$$\lambda = \min(\Lambda_\ell \setminus \Gamma) - v_0,$$

is an analytic invariant known as the *Zariski invariant* of the curve.

We will now recall a result that shows that the elements of $\mathcal{E}_\ell$ admit a normal form.

Normal Forms Theorem (cf. [5]). *If C is a curve corresponding to a point in $\mathcal{E}_\ell$, then either C is analytically equivalent to a curve with parametrization (t^{v_0}, t^{v_1}), when $\Lambda_\ell \setminus \Gamma = \emptyset$, or to a curve with a parametrization of the form*

$$x = t^{v_0}, \quad y = t^{v_1} + t^{\lambda} + \sum_i c_i t^i,$$

where the summation is over all indices i which are greater than λ and do not belong to the set $\Lambda_\ell - v_0$. Moreover, two curves C, with a parametrization as above, and C' with a similar parametrization but with coefficients (c_i') instead of (c_i), are analytically equivalent if and only if there exists a complex number ζ such that $\zeta^{\lambda - v_1} = 1$ and for all i, one has $c_i = \zeta^{i - v_1} c_i'$.

At this point it is natural to ask if the equisingularity class of the general polar is constant on each stratum $\mathcal{E}_\ell$.

We will see in the next section that the answer may be negative, but positive for a general member of each irreducible component of the stratum $\mathcal{E}_\ell$. This was shown in [1, Theorem 11.3] in the particular case of the whole space $\mathcal{E}$.

For the convenience of the reader, we will state a well known result about Newton non-degenerate plane curve singularities that will be needed in our analysis.

Following [8], we say that a reduced curve (f), where $f = \sum_{ij} a_{ij} x^i y^j$ is *Newton non-degenerate* if x and y do not divide f and for any side L of its Newton polygon $N(f)$, the polynomial $f_L = \sum_{(i,j)\in L} a_{ij} x^i y^j$ has no critical points outside the curve $xy = 0$. Since f_L is a quasi-homogeneous polynomial, we may rephrase the Newton non-degeneration as follows:

Let $P_k = (i_k, j_k)$ and $P_{k+1} = (i_{k+1}, j_{k+1})$, where $j_{k+1} < j_k$ and $i_{k+1} > i_k$, be the extremal points of L and define

$$p_L(z) = z^{-j_{k+1}} f_L(1, z).$$

Then one has that f_L has no critical points outside the curve $xy = 0$ if and only if p_L has no multiple roots. So, (f) is Newton non-degenerate if and only for each side L of its Newton polygon $N(f)$, the polynomial $p_L(z)$ has no multiple roots. We will call $p_L(z)$ the associated polynomial to the side L.

The following result (cf. [11, Proposition 4.7]) will describe the equisingularity class of a Newton non-degenerate curve (h).

Oka's Lemma. *Let (h) be a reduced plane curve germ and (x, y) be a local coordinate system such that the tangent cone to (h) does not contain the line (x). If (h) is non-degenerate, then for each side L_k of $N(h)$ with extremal points (i_k, j_k) and (i'_k, j'_k), with $j'_k > j_k$ and $i_k > i'_k$, there correspond $d_k = GCD(n_k, m_k)$ branches $\xi_{k,l}$, $1 \leq l \leq d_k$, where $n_k = j'_k - j_k$ and $m_k = i_k - i'_k$, with semigroup generated by $\frac{n_k}{d_k}$ and $\frac{m_k}{d_k}$ (allowing any one of them to be 1). If ξ and ξ' are two of these branches with semigroups generated by α_0, α_1 and β_0, β_1, respectively, then the intersection number of these branches is given by* $\mathrm{I}(\xi, \xi') = \min\{\alpha_0\beta_1, \beta_0\alpha_1\}$.

This paper contains parts of the PhD thesis of the third author under the supervision of the other two (cf. [7]).

2. Polars and Normal Forms

We will see in the following example how the Normal Forms Theorem may be used to describe the equisingularity classes of the general polars of all members of a given equisingularity class.

Example 1. Let $\Gamma = \langle 5, 12\rangle$. The Normal Forms Theorem, together with the algorithm to compute normal forms in [6], give us the complete classification of the curves in the equisingularity class determined by Γ which is summarized in Table 2.1 at the end of the paper.

Now, from formula (1.1) one obtains the implicit equation of the family of branches given by the parametrization in each row of Table 2.1. In what follows we exhibit the polars of the members of these families of branches and describe their equisingularity classes. The equations below describe the polars of the curves given by their parametrizations in Table 2.1 and are numbered according to the rows of the table they correspond. The symbols u_1, u_2, u_3 and u_4 will represent units in $\mathbb{C}\{x, y\}$, with $u_i(0, 0) = 1$, for $1 \leq i \leq 4$, not necessarily the same in all cases.

P1. $af_x + bf_y = 5by^4 - 12ax^{11}$.
P2. $af_x + bf_y = 5by^4 - 50ax^9y^3 - 15bx^{10}y^2 + 100ax^{19}y - 12ax^{11} + 5bx^{20} - 38ax^{37}$.
P3. $af_x + bf_y = 5by^4 - 45ax^8y^3 - 15bx^9y^2 + 90ax^{17}y - 12ax^{11} + 5bx^{18} - 33ax^{32}$.

P4. $af_x + bf_y = 5by^4 - 40ax^7y^3 - 15bx^8y^2 + 80ax^{15}y - 12ax^{11} + 5bx^{16} - 28ax^{27}$.
P5. $af_x + bf_y = 5by^4 - 40acx^7y^3 - 15bcu_1x^8y^2 - 10bu_2x^{10}y - 12au_3x^{11}$.
P6. $af_x + bf_y = 5by^4 - 45acx^8y^3 - (15bc + 50a)u_1x^9y^2 - 10bu_2x^{10}y - 12au_3x^{11}$.
P7. $af_x + bf_y = 5by^4 - 35ax^6y^3 - 15bu_1x^7y^2 - 10bcu_2x^{10}y - 12au_3x^{11}$.
P8. $af_x + bf_y = 5by^4 - 35acu_1x^6y^3 - 15bcu_2x^7y^2 - 10bu_3x^9y - 12au_4x^{11}$.
P9. $af_x + bf_y = 5by^4 - 30ax^5y^3 - 15bu_1x^6y^2 - 10bcu_2x^9y - 12au_3x^{11}$.

For general a and b, any one of the polars in the families **P1**–**P9** has Newton polygon with only one side $L = [(0,4);(11,0)]$ that supports only its extremal points associated to non-zero terms of that polar. This implies that all general polars of branches whose paramatrizations are given in rows $1-9$ in Table 2.1 are Newton non-degenerate, so their Newton polygons determine their equisingularity classes, which in this case is given by only one branch with semigroup $\langle 4, 11\rangle$.

P10. $af_x + bf_y = 5by^4 - 30acu_1x^5y^3 - 15bcu_2x^6y^2 - 10bu_3x^8y - 12au_4x^{11}$.

In this case, for general a and b, the Newton polygon of the polar of the curve given parametrically in row 10 of Table 2.1 has two sides:

$L_1 = [(0,4);(8,1)]$, that supports only its extremal points associated to the monomials y^4 and x^8y.

$L_2 = [(8,1);(11,0)]$, that supports only its extremal points associated to the monomials x^8y and x^{11}.

Again, for general a and b, any curve belonging to this family is Newton non-degenerate, so from Oka's Lemma it has two branches: g_1 with semigroup $\langle 3, 8\rangle$, and g_2 smooth, such that $\mathrm{I}(g_1, g_2) = 8$.

P11.

$$\begin{aligned} af_x + bf_y &= 5by^4 - 30a(c+d)u_1x^5y^3 - 15b(c+d)u_2x^6y^2 \\ &\quad -10b(1+c)u_3x^8y - 5bu_4x^{10}. \end{aligned}$$

P12.

$$\begin{aligned} af_x + bf_y &= 5by^4 + 10a(1-3c-4c^2)u_1x^5y^3 + 5b(1-3c-4c^2)u_2x^6y^2 \\ &\quad -10b(1+c)u_3x^8y - 5bu_4x^{10}. \end{aligned}$$

P13.

$$\begin{aligned} af_x + bf_y &= 5by^4 - \tfrac{5}{2}a(13+12c)u_1x^5y^3 - \tfrac{5}{4}b(13+12c)u_2x^6y^2 \\ &\quad -\tfrac{125}{6}bu_3x^8y - 5bu_4x^{10}. \end{aligned}$$

P14. $af_x + bf_y = 5by^4 - \frac{625}{9}au_1x^5y^3 - \frac{625}{18}bu_2x^6y^2 - \frac{125}{6}bu_3x^8y - 5bu_4x^{10}$.
P15. $af_x + bf_y = 5by^4 - \frac{625}{9}au_1x^5y^3 - \frac{625}{18}bu_2x^6y^2 - \frac{125}{6}bu_3x^8y - 5bu_4x^{10}$.
P16. $af_x + bf_y = 5by^4 - \frac{625}{9}au_1x^5y^3 - \frac{625}{18}bu_2x^6y^2 - \frac{125}{6}bu_3x^8y - 5bu_4x^{10}$.

The Newton polygon for general a and b of any member of the families **P11**–**P16** has only one side $L = [(0,4);(10,0)]$, that supports just its extremal points associated to the non-vanishing terms $5by^4$ and $-5bx^{10}$ that do not depend upon the parameters c, d and e. Therefore, these polars are Newton non-degenerate, so they have two branches with semigroup $\langle 2, 5\rangle$ that intersect with multiplicity 10.

P17. $af_x + bf_y = 5by^4 - 25au_1x^4y^3 - 15bu_2x^5y^2 + 5b(\frac{1-4c}{2})u_3x^8y + \frac{15}{2}bu_4x^{10}$.

For a and b general, the Newton polygon of these curves has only one side $L = [(0,4);(10,0)]$, containing also the point $(5,2)$, with associated polynomial $p_L(z) = 5bz^4 - 15bz^2 + \frac{15}{2}b$. Since $p_L(z)$ has only simple roots, the general polar is Newton non-degenerate, so it has two branches with semigroup $\langle 2,5\rangle$ that intersect with multiplicity 10.

P18.

$$\begin{aligned} af_x + bf_y &= 5by^4 - 25au_1x^4y^3 - 15bu_2x^5y^2 \\ &\quad -10[b(c^2+c+d)x^8 + (b(d^2+e) + 5a(c-1))u_3x^9]y \\ &\quad +5b(1-c)x^{10} + (-5b(c^3+dc+d) - 12au_4)x^{11}. \end{aligned}$$

This is the only stratum in which the equisingularity class of the general polars will depend upon the parameters in $\mathcal{E}_\ell$ to which it belongs. We will split the analysis is several cases.

(i) If $c \neq 1$, then the Newton polygon of the polar has the only side $L = [(0,4);(10,0)]$, that supports the points associated to the terms $5by^4$, $-15bx^5y^2$ and $5b(1-c)x^{10}$. In this case, the associated polynomial is $p_L(z) = z^4 - 3z^2 + (1-c)$, whose discriminant is $-16(c-1)(5+4c)^2$. So, the polar is Newton non-degenerate if and only if $c \neq -\frac{5}{4}$. In this case, the polar has two branches with semigroup $\langle 2,5\rangle$ that intersect with multiplicity 10.

If $c = -\frac{5}{4}$, the parametrization of the polar, computed via Newton's algorithm, is given by

$$x = \frac{3^{20}}{2^{21}}(16d+5)^2t^4, \quad y = \frac{3^{38}}{2^{53}}(16d+5)^5t^{10} + \frac{3^{30}}{2^{60}}(16d+5)^6t^{11} + \cdots.$$

Therefore, when $d \neq -\frac{5}{16}$, the members of the family $\mathcal{E}_{18}$ for which $c = -\frac{5}{4}$ have irreducible polars whose semigroup $\langle 4,10,21\rangle$, hence it is of genus 2. When $d = -\frac{5}{16}$, their general polars have two branches with parametrizations (also computed via Newton's algorithm)

$$x_i = \frac{3}{2}t^2, \quad y_i = \frac{27}{8}t^5 + (-1)^i\frac{27}{640b}(256ab - 125b)^{\frac{1}{2}}t^6 + \cdots, \quad i = 1,2,$$

that is, branches with semigroup $\langle 2,5\rangle$ and with intersection multiplicity 10.

(ii) If $c = 1$, then the polar is given by

$$\begin{aligned} af_x + bf_y &= 5by^4 - 25au_1x^4y^3 - 15bu_2x^5y^2 - 10b(d+2)u_3x^8y \\ &\quad -10b(d^2+e)u_3x^9y - (5b + 10bd + 12au_4)x^{11}. \end{aligned}$$

A verification shows that its Newton polygon has two sides:

$L_1 = [(0,4);(5,2)]$, that supports only its extremal points associated to nonzero terms of the polar; and

$L_2 = [(5,2);(11,0)]$, that supports only its extremal points associated to nonzero terms of the polar.

Therefore, any curve in this family with $c = 1$ has, for general values of a and b, a Newton non-degenerate polar with a branch p having the semigroup $\langle 2,5\rangle$ and two non-singular branches g_1 and g_2 such that $\mathrm{I}(p,g_i) = 5$ and $\mathrm{I}(g_1,g_2) = 3$.

Remark 2.1. The stratum $\mathcal{E}_{18}$ gives us an example in which the equisingularity classes of the general polars of its members are not constant. It also gives us a somewhat unexpected example of a family of curves of genus 1 such that its general member has a general polar of genus 2.

What is remarkable is that the analytic classification of the branches in this equisingularity class allowed us to describe the equisingularity classes of all general polars of its members.

Although, as we saw in the above example, the topological type of the polar may not be constant in a given stratum $\mathcal{E}_\ell$, it is constant in an open dense set of each irreducible component of the stratum, as we will show in general in the sequel.

In fact, for the stratum associated to $\Lambda_1 = \Gamma \setminus \{0\}$, the result follows easily from the first assertion in the Normal Forms Theorem, which is essentially due to Zariski (cf. [16]).

Let us consider a normal form in an equisingularity class parametrized by $\mathcal{E}_\ell$ associated to a set of values of differentials $\Lambda_\ell \neq \Gamma \setminus \{0\}$. Putting $v_0 = n$ and $v_1 = m$, from the Normal Forms Theorem, we have

$$x = t^n, \quad y = t^m + t^\lambda + \sum_{\substack{i>\lambda \\ i \notin \Lambda_\ell - n}} c_i t^i. \tag{2.1}$$

The implicit equations of these curves obtained as in formula (1) are given by

$$f = y^n + a_2(x)y^{n-2} + a_3(x)y^{n-3} + \cdots + a_{n-1}(x)y + a_n(x),$$

which are Weierstrass polynomials, since $\text{ord}_x(a_j(x)) > j$ and $\text{ord}_x a_n(x) = m$. Moreover, the coefficients of the power series $a_j(x)$ are polynomials in the c_i's that appear in (2.1).

Therefore, the polars of the curves in $\mathcal{E}_\ell$ are given by the family

$$\begin{aligned} P(f) &= af_x + bf_y \\ &= bny^{n-1} + aa_2'(x)y^{n-2} + \big((n-2)ba_2(x) + aa_3'(x)\big)y^{n-3} + \cdots \\ &\quad + \big(2ba_{n-2}(x) + aa_{n-1}'(x)\big)y + ba_{n-1}(x) + aa_n'(x). \end{aligned}$$

We will now show that in a dense open Zariski set in any irreducible component of $\mathcal{E}_\ell$ the value of the Milnor number of $P(f)$ is constant.

From the equation of $P(f)$ we have that

$$\begin{aligned} P(f)_x &= aa_2''(x)y^{n-2} + \big((n-2)ba_2'(x) + aa_3''(x)\big)y^{n-3} + \cdots \\ &\quad + \big(2ba_{n-2}'(x) + aa_{n-1}''(x)\big)y + +ba_{n-1}'(x) + aa_n''(x), \quad \text{and} \end{aligned}$$

$$\begin{aligned} P(f)_y &= bn(n-1)y^{n-2} + a(n-2)a_2'(x)y^{n-3} + \cdots \\ &\quad + \big(2ba_{n-2}(x) + aa_{n-1}'(x)\big). \end{aligned}$$

Therefore, one has that $P(f)_y$ is the constant $bn(n-1)$ times a Weierstrass polynomial in y and $P(f)_x \in \mathbb{C}\{x\}[y]$, hence their intersection multiplicity, which is the Milnor number of $P(f)$, is the order in x of their resultant R_y in y. Because $R_y \not\equiv 0$

since the generic polar of f is reduced, we have for every irreducible component $\mathcal{E}_{\ell,j}$ of $\mathcal{E}_\ell$ that

$$R_y\big(P(f)_x, P(f)_y\big) = A_j x^{\nu_j} + \text{higher order terms},$$

where each A_j is a non-zero polynomial in a, b and the c_i's (the coefficients in the normal forms (2.1)). Moreover, these polynomials A_j are homogeneous in a and b. So, there exists a Zariski open set in $\mathcal{E}_{\ell,j}$, where at each point this polynomial A_j in a and b is not identically zero, hence the Milnor number of the general polar of the corresponding curve is constant and equal to ν_j in this open set. From the Lê-Ramanujan Theorem [9, Theorem 2.1], we obtain the following result:

Theorem 2.2. *The equisingularity class of the polar of curves in $\mathcal{E}_\ell$ is constant in an open dense Zariski subset of any of its irreducible components.*

3. Polars of branches up to multiplicity four

We will now give a detailed description of the equisingularity classes of the polars of branches of multiplicity less or equal than four. This will be carried out by using the classification done by the first two authors in [5]. Observe that the general polar of a branch of multiplicity 2 is a smooth branch, so we have only to treat the cases of multiplicities three and four.

3.1. Multiplicity three

For curves of multiplicity three, there is only one analytic representative in each stratum which is determined by Zariski's λ invariant, as shown in Table 3.1.

In the case of the monomial curve $x = t^3$, $y = t^\beta$, corresponding to the first row of Table 3.1, we have that the general polar curve has $d = \gcd(2, \beta - 1)$ branches. When $d = 1$, the branch has semigroup $\langle 2, \beta - 1\rangle$ and when $d = 2$, the two branches are smooth and their intersection multiplicity is $\frac{\beta-1}{2}$.

In the case of the second row of Table 3.1, the implicit equation of the curve is $f = y^3 - 3x^{2q+k+\epsilon}y - x^\beta - x^{\beta+\epsilon+3k}$ and the general polar curve is

$$af_x + bf_y = 3by^2 - 3a(2q + k + \epsilon)x^{2q+k+\epsilon-1}y - 3bux^{2q+k+\epsilon},$$

where u is a unit in $\mathbb{C}\{x\}$. After a direct computation and the use of Oka's Lemma, one sees that the equisingularity class of the general polar may be described by Table 3.2.

3.2. Multiplicity four and genus one

A branch of multiplicity 4 may have genus one or two. For the genus one case, we describe in Table 3.3 the normal forms of such branches, where the symbol $[\alpha]$ denotes the integer part of the rational number α.

We describe below the equisingularity classes of the polars in this situation.

First Normal Form in Table 3.3 (monomial curves)

In this case, the equation of the curve is $y^4 - x^m = 0$, so its polar is $4by^3 - amx^{m-1}$, that has $d = \gcd(3, m-1)$ branches. If $d = 1$, the branch has semigroup $\langle 3, m-1 \rangle$ and when $d = 3$, the three branches are smooth with mutual intersection multiplicity equal to $\frac{(m-1)}{3}$.

Second Normal Form in Table 3.3

This is the more complicated case. According to (1.1), the implicit equation of the curve is given by

$$f = y^4 - S_1(x)y^3 + S_2(x)y^2 - S_3(x)y + S_4(x) = 0,$$

where $S_r(x)$ is the r-th symmetric polynomial computed in $y_l(x^{\frac{1}{4}}) = y(\varepsilon^l x^{\frac{1}{4}})$, $l = 0, 1, 2, 3$, with ε a primitive fourth root of 1.

From the definition of $y(t)$, it is clear that $S_1 = 0$. To determine the Newton polygon of the polar, it is sufficient to consider in each of the polynomials $S_r(x)$, $2 \leq r \leq 4$ only the term that determines its multiplicity.

(I) We first consider the case $a_1 = a_2 = \cdots = a_{j-[\frac{m}{4}]-2} = 0$.

For each fixed j, the implicit equation of the curve is

$$f = y^4 - 4x^{m-j}y^2 - x^m + 2x^{2m-2j} - x^{3m-4j} = 0.$$

Therefore,

$$af_x + bf_y = 4by^3 - 4a(m-j)x^{m-j-1}y^2 - 8bx^{m-j}y - amux^{m-1},$$

where $u \in \mathbb{C}\{x\}$ with $u(0) = 1$.

We have the following cases:

i. Case $\frac{2}{m-j} < \frac{1}{j-1}$.

In this case, the Newton polygon of the polar has only one side L containing only its end points $(0, 3)$ and $(m-1, 0)$, associated to monomials of the polar. The associated polynomial to the side L is $p_L(z) = 4bz^3 - am$. Then for a and b general, $p_L(z)$ has three distinct roots $\{z_1, z_2, z_3\}$. Therefore the general polar has:

a) One branch with semigroup $\langle 3, m-1 \rangle$, if $\gcd(3, m-1) = 1$.

b) Three smooth branches with parametrizations: $(t, z_i t^{\frac{m-1}{3}} + \cdots)$, and mutual intersection numbers $\frac{m-1}{3}$, if $\gcd(3, m-1) = 3$.

ii. Case $\frac{2}{m-j} > \frac{1}{j-1}$.

In this case, the Newton polygon of the general polar has two sides $L_1 = [(0, 3); (m-j, 1)]$ and $L_2 = [(m-j, 1); (m-1, 0)]$, since the point $(m-j-1, 2)$ lies above L_1. The associated polynomials are $p_{L_1}(z) = 4bz^2 - 8b$, and $p_{L_2}(z) = -8bz - am$. Then, we have that:

Associated to L_1 there is one branch p_1 with semigroup $\langle 2, m-j \rangle$ and parametrization $x = t^2$ $y = \sqrt{2}t^{m-j} + \cdots$, if $\gcd(2, m-j) = 1$; or two smooths branches g_1, g_2 with parametrizations $x_1 = t,\ y_1 = \sqrt{2}t^{\frac{m-j}{2}} + \cdots$ and $x_2 = t$, $y_2 = -\sqrt{2}t^{\frac{m-j}{2}} + \cdots$, if $\gcd(2, m-j) = 2$.

Associated to L_2, there is one smooth branch p_2 with parametrization $x = t$, $y = -\frac{am}{8b}t^{j-1} + \cdots$.

Finally, we have that $\mathrm{I}(p_1, p_2) = m - j$ and $\mathrm{I}(g_i, p_2) = \mathrm{I}(g_1, g_2) = \frac{m-j}{2}$.

iii. Case $\frac{2}{m-j} = \frac{1}{j-1}$.

Since $j > 2$, because otherwise $m = 4$, which is not allowed, the Newton polygon of the polar has only one side L with tree points with associated polynomial $p_L(z) = 4bz^3 - 8bz - am$. Therefore, for a and b general, the polynomial $p_L(z)$ has three distinct roots $\{z_1, z_2, z_3\}$ and since in this case $\gcd(3, m-1) = 3$, we have three smooth branches associated to L with parametrizations $(t, z_i t^{j-1} + \cdots)$, $i = 1, 2, 3$, and mutual intersection numbers $j - 1$.

(II) Now we consider the case where some of the a_i's is non-zero. Set $k = \min\{i;\ a_i \neq 0\}$.

As before, computing the implicit equation for the branch given parametrically in the second row of Table 3.3, we get

$$f = y^4 - (4x^{m-j} + 2a_k^2 u_1 x^{m-2(j-[\frac{m}{4}]-k)})y^2 - 4a_k u_2 x^{m-j+[\frac{m}{4}]+k)}y - u_3 x^m,$$

where $u_i \in \mathbb{C}\{x\}$ with $u_i(0) = 1$ for $i = 1, 2, 3$. Hence, to determine the Newton polygon of the polar $af_x + bf_y = 0$, it is sufficient to consider the polynomial

$$4by^3 - 4a(m-j)x^{m-j-1}y^2 - 8bx^{m-j}y - 4ba_k x^{m-j+[\frac{m}{4}]+k}. \tag{3.1}$$

We now split the analysis of this case into several sub-cases.

i. Case $\frac{2}{m-j} < \frac{1}{[\frac{m}{4}]+k}$.

The Newton polygon of the polar has just one side L, containing only the points $(0, 3)$ and $(m - j + [\frac{m}{4}] + k, 0)$.

Since the polynomial $p_L(z) = 4bz^3 - 4ba_k$ has three distinct roots $\{z_1, z_2, z_3\}$, it follows that the polar has:

a) Only one branch, if $\gcd(3, m-j+[\frac{m}{4}]+k) = 1$, with semigroup $\langle 3, m-j+[\frac{m}{4}]+k\rangle$.

b) Three smooth branches, if $\gcd(3, m - j + [\frac{m}{4}] + k) = 3$, with parametrizations (computed via Newton's algorithm):

$$x_i = t,\quad y_i = z_i t^{\frac{m-j+[\frac{m}{4}]+k}{3}} + \cdots,\qquad i \in \{1, 2, 3\},$$

and mutual intersection numbers $\frac{m-j+[\frac{m}{4}]+k}{3}$.

ii. Case $\frac{2}{m-j} > \frac{1}{[\frac{m}{4}]+k}$.

In this case, the Newton polygon of the general polar has two sides $L_1 = [(0, 3); (m - j, 1)]$ and $L_2 = [(m - j, 1); (m - j + [\frac{m}{4}] + k, 0)]$, with on each side only the extremal points that correspond to nonzero terms of the polar. This is obtained from polynomial (3.1), observing that the point $(m-j-1,2)$ is above the line supporting L_1.

Considering the associated polynomials to these sides, $p_{L_1}(z) = 4bz^2 - 8b$ and $p_{L_2}(z) = -8bz - 4ba_k$; and defining $d = \gcd(2, m-j)$, we have that:

a) Associated to the side L_1, we have a branch p_1 with semigroup $\langle 2, m-j\rangle$ and parametrization $x = t^2,\ \ y = \sqrt{2}t^{m-j} + \cdots$, if $d = 1$; and two smooth branches g_1, g_2, with parametrizations $x_i = t,\ \ y_i = (-1)^{i-1}\sqrt{2}t^{\frac{m-j}{2}} + \cdots$, $i = 1, 2$, if $d = 2$.

b) Associated to the side L_2, we have a smooth branch p_2, with parametrization $x = t,\ \ y = -\frac{a_k}{2}t^{[\frac{m}{4}]+k} + \cdots$.

Finally, one has $\mathrm{I}(p_1, p_2) = m - j$ and $\mathrm{I}(g_i, p_2) = \mathrm{I}(g_1, g_2) = \frac{m-j}{2}$.

iii. Case $\frac{2}{m-j} = \frac{1}{[\frac{m}{4}]+k}$.

In this case, the Newton polygon of the polar has a unique side L containing the three points $(0,3), (m-j,1)$ and $(m-j+[\frac{m}{4}]+k, 0)$, whose associated polynomial is

$$p_L(z) = 4bz^3 - 8bz - 4a_k.$$

When $a_k \neq \frac{4\sqrt{6}}{9}(-1)^\alpha;\ \alpha = 0, 1$, from the condition $\frac{2}{m-j} = \frac{1}{[\frac{m}{4}]+k}$, it follows that the polar is Newton non-degenerate. In this case, the polynomial $p_L(z)$ has three distinct roots $\{z_1, z_2, z_3\}$, then the general polar has three smooth branches with parametrizations $(t, y_i = z_i t^{[\frac{m}{4}]+k} + \cdots),\ \ i = 1, 2, 3$, with mutual intersection numbers equal to $[\frac{m}{4}] + k = \frac{m-j}{2}$.

Now we suppose that $a_k = \frac{4\sqrt{6}}{9}(-1)^\alpha;\ \alpha = 0, 1$.

In this case, $p_L(z)$ has the double root $\frac{\sqrt{6}}{3}(-1)^{\alpha+1}$ and the simple root $2\frac{\sqrt{6}}{3}(-1)^\alpha$. The polar will have a smooth branch f_1 corresponding to the simple root of $p_L(z)$ and some branches corresponding to the double root, which we describe below.

We may suppose that the roots of $p_L(z)$ are $\frac{\sqrt{6}}{3}, \frac{\sqrt{6}}{3}$ and $-2\frac{\sqrt{6}}{3}$, since the other case is analogous.

a) If for all $l > 0$ one has $a_{k+l} = 0$, then the normal form is $y(t) = t^m + t^{3m-4j} + a_k t^{2m-4(j-[\frac{m}{4}]-k)},\ \ 1 \le j \le [\frac{m}{2}]$. Computing the implicit equation for this branch and then its polar, a simple analysis shows that the general polar has the smooth branch f_1 and a branch with semigroup $\langle 2, 2m-3j\rangle$ with intersection number $m - j$.

b) Suppose that there exists $l > 0$ such that $a_{k+l} \neq 0$. We denote the least such l by s. The parametrization in this case is given by

$$y(t) = t^m + t^{3m-4j} + a_k t^{2m-4(j-[\frac{m}{4}]-k)} + a_{k+s} t^{2m-4(j-[\frac{m}{4}]-k-s)} + \cdots.$$

The implicit equation of this branch reads

$$\begin{aligned} f \; &= y^4 + (-4x^{m-j} - 2a_k^2 x^{m-2(j-[\frac{m}{4}]-k)} - 4a_k a_{k+s} x^{m-2(j-[\frac{m}{4}]-k)+s} \\ &-2a_{k+s}^2 x^{m-2(j-[\frac{m}{4}]-k-s)} + \cdots)y^2 - (4a_k x^{m-j+[\frac{m}{4}]+k} \\ &+4a_{k+s} x^{m-j+[\frac{m}{4}]+k+s} + 4a_k x^{2m-2j-(j-[\frac{m}{4}]-k)} \\ &+4a_{k+s} x^{2m-2j-(j-[\frac{m}{4}]-k-s)} + \cdots)y - ux^m, \end{aligned}$$

where $u \in \mathbb{C}\{x\}$ with $u(0) = 1$.

Now, in order to apply the Newton-Puiseux algorithm to the general polar of f at the double root of $p_L(z)$, we have to split our analysis in several subcases.

b.1) $m - 2j > s$.

b.1.1) s odd. Associated to the double root there is a branch g_1 given by

$$x = t^2, \; y = -\frac{\sqrt{6}}{3} t^{m-j} + \frac{\sqrt{a_{k+s}}}{\sqrt[4]{6}} t^{m-j+s} + \cdots .$$

In this case, the polar has the smooth branch f_1 and a branch g_1 with semigroup $\langle 2, m-j+s \rangle$ such that $\mathrm{I}(f_1, g_1) = m - j$.

b.1.2) s even. The polar splits into three smooth branches f_1, g_1 and g_2, such that $\mathrm{I}(f_1, g_i) = \frac{m-j}{2}$ and $\mathrm{I}(g_1, g_2) = \frac{m-j+s}{2}$.

b.2) $m - 2j < s$. In this case, the polar has the smooth branch f_1 and a branch g_1 associated to the double root with semigroup $\langle 2, 2m - 3j \rangle$, such that $\mathrm{I}(f_1, g_1) = m - j$.

b.3) $m - 2j = s$.

b.3.1) If $a_{k+s} \neq \frac{4\sqrt{6}}{81}(-1)^{\alpha+1}$, we have, associated to the double root, a branch g_1 with semigroup $\langle 2, 2m - 3j \rangle$. So, the polar has the smooth branch f_1 and the branch g_1 such that $\mathrm{I}(f_1, g_1) = m - j$.

b.3.2) If $a_{k+s} = \frac{4\sqrt{6}}{81}(-1)^{\alpha+1}$, we have, associated to the double root, two smooth branches g_1 and g_2. So, the polar has three smooth branches f_1, g_1 and g_2 such that $\mathrm{I}(f_1, g_i) = \frac{m-j}{2}$, $i = 1, 2$, and $\mathrm{I}(g_1, g_2) = \frac{m-j}{2} + s$.

Table 3.4 summarizes the above analysis for the second normal form.

Third to Fifth Normal Forms of Table 3.3

With a similar analysis which we omit because is much simpler, one gets that the Newton polygon of the general polar has one side and the associated polynomial has no multiple roots, then this polar is non-degenerate and its topology is summarized in Table 3.5.

3.3. Multiplicity four and genus two

The branches of multiplicity 4 and genus 2 have semigroups of the form $\langle 4, v_1, v_2\rangle$ where $\gcd(4, v_1, v_2) = 1$ and $v_2 > 2v_1$. In this situation one has $v_1 \equiv 2 \mod 4$, and $v_2 \equiv 1 \mod 4$ or $v_2 \equiv 3 \mod 4$. It then follows that

$$\begin{array}{ll} v_1+v_2,\ 3v_1,\ 2v_1+v_2,\ 3(v_2-v_1) & \not\equiv 0 \mod 4, \\ 2v_1,\ 2v_2-v_1,\ v_1+2v_2 & \equiv 0 \mod 4. \end{array}$$

From [4] one has that such equisingularity classes have only one normal form given by

$$x(t)=t^4,\ \ y(t)=t^{v_1}+t^{v_2-v_1}+a_1t^{v_2-4[\frac{v_1}{4}]}+a_2t^{v_2-4([\frac{v_1}{4}]-1)}+\cdots+a_{[\frac{v_1}{4}]-1}t^{v_2-8}.$$

Now, we use (1.1) to determine the implicit equation $f=0$ for this family of branches. A direct computation with $y_i = y(\varepsilon^i x^{\frac{1}{4}})$, $i = 0, 1, 2, 3$, where ε is a fourth primitive root of unity, shows that

$$\begin{array}{l} S_2(x) = \sum y_iy_j = -2x^{\frac{2v_1}{4}} + \cdots, \\ S_3(x) = \sum y_iy_jy_k = 4x^{\frac{2v_2-v_1}{4}} + \cdots. \end{array}$$

Considering k_1 and k_2 such that $v_1 = 2k_1$ and $2v_2 - v_1 = 4k_2$, we get

$$f = y^4 + (-2x^{k_1}+\cdots)y^2 + (-4x^{k_2}+\cdots)y + ux^{2k_1},$$

where u is a unit in $\mathbb{C}\{x\}$ with $u(0) = 1$. It then follows that

$$\begin{array}{ll} af_x & = a[(-2k_1x^{k_1-1}+\cdots)y^2+(-4k_2x^{k_2-1}+\cdots)y+2k_1ux^{2k_1-1}+u'x^{2k_1}], \\ bf_y & = b[4y^3+(-4x^{k_1}+\cdots)y+(-4x^{k_2}+\cdots)]. \end{array} \tag{3.2}$$

Since $2k_1 = v_1 > 4$ and $v_2 > 2v_1$, it follows that $k_2 > \frac{3}{2}k_1$ and $k_2 - 1 > k_1$. From this last inequality and from (3.2), we get that the Newton polygon of $af_x + bf_y = 0$ is determined by the polynomial

$$4by^3 - 2k_1ax^{k_1-1}y^2 - 4bx^{k_1}y - 4bx^{k_2} + 2k_1ax^{2k_1-1}. \tag{3.3}$$

Since $k_1 > 2$, we have that $\frac{2}{k_1} > \frac{1}{k_1-1}$. In this way, if we put $m = \min\{2k_1 - 1, k_2\}$, then, the Newton polygon of the general polar has two sides $L_1 = [(0, 3); (k_1, 1)]$ and $L_2 = [(k_1, 1); (m, 0)]$, since the point $(k_1 - 1, 2)$ lies above L_1.

Associated to the side L_1 we have the polynomial $p_{L_1}(z) = 4bz^2 - 4b$. Therefore, because $gcd(2, k_1) = 1$, there is only one branch g_1 associated to this side with semigroup $\langle 2, k_1\rangle$. This branch admits a parametrization, calculated via Newton's algorithm, given by

$$x = t^2, \quad y = t^{k_1} + \cdots.$$

Associated to the side L_2 we have the polynomial $p_{L_2} = -4bz + c_m$ where c_m is the coefficient of x^m in (3.3). In this way, since we are taking a and b general, associated to L_2 there is a smooth branch g_2 with parametrization $x = t$ and

$$\begin{aligned} y &= \tfrac{k_1 a}{2b} t^{k_1-1} + \cdots, && \text{if } m = 2k_1 - 1 < k_2; \\ y &= \tfrac{k_1 a - 2b}{2b} t^{k_1-1} + \cdots, && \text{if } m = 2k_1 - 1 = k_2; \\ y &= t^{k_1-1} + \cdots, && \text{if } m = k_2 < 2k_1 - 1. \end{aligned}$$

From the condition $\frac{2}{k_1} > \frac{1}{k_1-1}$, it follows that $I(g_1, g_2) = k_1$.

Summarizing the above analysis, we have the following:

The general polar of a branch of multiplicity 4 *and genus* 2 *is always non-degenerate and has two branches: one branch* g_1 *with semigroup* $\langle 2, k_1\rangle$ *and one smooth branch* g_2. *Moreover,* $\mathrm{I}(g_1, g_2) = k_1$.

In the next example we present a curve (f) for which the analytic type of its polar curve $(af_x + bf_y)$ depends essentially on the direction $(a : b)$.

Example 2. Consider the curve (f) given parametrically by $(t^5, t^{12} + t^{21})$, that belongs to the eighth family in Example 1. We know that, in this case, $af_x + bf_y = 5by^4 - 10bux^9y - 12avx^{11}$, where u and v are units in $\mathbb{C}\{x\}$. This polar is irreducible and is analytically equivalent to a branch with parametrization

$$\left(t^4, t^{11} + t^{14} - \frac{1}{2}t^{17} + \frac{15\sqrt[3]{2}}{2}\left(\frac{12a}{5b}\right)^3 t^{21}\right).$$

This is a branch of multiplicity four belonging to the fourth Normal Form in Table 3.1. So from the Normal Forms theorem, two such branches corresponding to directions $(a : b)$ and $(a_1 : b_1)$ with $bb_1 \neq 0$ are analytically equivalent if and only if one has $\frac{a^3}{b^3} = \frac{a_1^3}{b_1^3}$.

As a final remark, we refer to [10] for a rough description of the polars of the members of the equisingularity class determined by the semigroup $\langle 5, 11\rangle$, which could be completely described by the methods we exhibited in the present paper.

Acknowledgment

The authors would like to thank the Referee for a careful reading of the manuscript and for suggesting several improvements for the redaction.

4. Tables

Table 2.1. The normal forms of the equisingularity class of $\langle 5, 12\rangle$:

Normal Form $x(t) = t^5$	$\Lambda_\ell \setminus \Gamma$
1. $y(t) = t^{12}$	$\emptyset$
2. $y(t) = t^{12} + t^{38}$	$\{43\}$
3. $y(t) = t^{12} + t^{33}$	$\{38, 43\}$
4. $y(t) = t^{12} + t^{28}$	$\{33, 38, 43\}$
5. $y(t) = t^{12} + t^{26} + ct^{28}, \quad c \neq 0$	$\{31, 38, 43\}$
6. $y(t) = t^{12} + t^{26} + ct^{33}$	$\{31, 43\}$
7. $y(t) = t^{12} + t^{23} + ct^{26}$	$\{28, 33, 38, 43\}$
8. $y(t) = t^{12} + t^{21} + ct^{23} + dt^{28}$	$\{26, 31, 38, 43\}$
9. $y(t) = t^{12} + t^{18} + ct^{21} + dt^{26}$	$\{23, 28, 33, 38, 43\}$
10. $y(t) = t^{12} + t^{16} + ct^{18} + dt^{23}$	$\{21, 26, 31, 33, 38, 43\}$
11. $y(t) = t^{12} + t^{14} + ct^{16} + dt^{18} + et^{23}$, $c \neq \frac{13}{12}$, $d \neq \frac{4c^2-1}{3}$	$\{19, 26, 31, 33, 38, 43\}$
12. $y(t) = t^{12} + t^{14} + ct^{16} + (\frac{4c^2-1}{3})t^{18} + dt^{23} + et^{28}$, $c \neq \frac{13}{12}$	$\{19, 26, 31, 38, 43\}$
13. $y(t) = t^{12} + t^{14} + \frac{13}{12}t^{16} + ct^{18} + dt^{21}, \quad c \neq \frac{133}{108}$	$\{19, 28, 31, 33, 38, 43\}$
14. $y(t) = t^{12} + t^{14} + \frac{13}{12}t^{16} + \frac{133}{108}t^{18} + ct^{21} + dt^{23}$, $d \neq \frac{34c}{11}$	$\{19, 31, 33, 38, 43\}$
15. $y(t) = t^{12} + t^{14} + \frac{13}{12}t^{16} + \frac{133}{108}t^{18} + ct^{21}$ $+\frac{34}{11}ct^{23} + dt^{28}, \qquad d \neq \frac{81c^2}{32} + \frac{5225}{559872}$	$\{19, 31, 38, 43\}$
16. $y(t) = t^{12} + t^{14} + \frac{13}{12}t^{16} + \frac{133}{108}t^{18} + ct^{21} + \frac{34}{11}ct^{23}$ $+(\frac{81c^2}{32} + \frac{5225}{559872})t^{28} + dt^{33}$	$\{19, 31, 43\}$
17. $y(t) = t^{12} + t^{13} - \frac{1}{2}t^{14} + ct^{16} + dt^{21} + et^{26}$	$\{18, 23, 28, 33, 38, 43\}$
18. $y(t) = t^{12} + t^{13} + ct^{14} + dt^{16} + et^{21}, \ c \neq -\frac{1}{2}$	$\{18, 23, 28, , 31, 33, 38, 43\}$

Table 3.1. Normal forms for multiplicity three:

$\Gamma = \langle 3, \beta \rangle; \quad \beta = 3q + \varepsilon, \quad \varepsilon = 1,\ 2$
$x = t^3, \ y = t^\beta$
$x = t^3, \ y = t^\beta + t^{\beta+\varepsilon+3k}, \ 0 \le k \le q-2$

Table 3.2. The topology of the polars of curves of multiplicity three:

$2q + k + \epsilon = 2I + 1$	One branch with semigroup $\langle 2, 2q + k + \epsilon \rangle$.
$2q + k + \epsilon = 2I$	Two smooth branches with intersection multiplicity I.

Table 3.3. Normal forms for multiplicity four and genus one:

Normal form $x(t) = t^4$	$\Lambda \setminus \langle 4, m \rangle$
1. $y(t) = t^m$	$\emptyset$
2. $y(t) = t^m + t^{3m-4j} + a_1 t^{2m-4(j-[\frac{m}{4}]-1)} + \cdots + a_{j-[\frac{m}{4}]-2} t^{2m-8}, \quad 2 \le j \le [\frac{m}{2}]$	$\{3m - 4s; 1 \le s \le j-1\}$
3. $y(t) = t^m + t^{2m-4j} + a_k t^{3m-(4[\frac{m}{4}]+j+1-k)} + \cdots + a_{j-[\frac{m}{4}]-2} t^{3m-4([\frac{m}{4}]+3-k)}$, $a_k \ne 0, \quad 2 \le j \le [\frac{m}{4}], \ 1 \le k \le [\frac{m}{4}] - j$	$\{2m - 4s; 1 \le s \le j-1\} \cup \{3m - 4s; 1 \le s \le [\frac{m}{4}] + 1 - k\}$
4. $y(t) = t^m + t^{2m-4j} + a_{[\frac{m}{4}]-j+1} t^{3m-8j} + a_{[\frac{m}{4}]-j+2} t^{3m-4(2j-1)} + \cdots + a_{[\frac{m}{4}]-1} t^{3m-4(j+2)}$, $a_{[\frac{m}{4}]-j+1} \ne \frac{3m-4j}{2m}, \quad 2 \le j \le [\frac{m}{4}]$	$\{2m - 4s; 1 \le s \le j-1\} \cup \{3m - 4s; 1 \le s \le j\}$
5. $y(t) = t^m + t^{2m-4j} + \frac{3m-4j}{2m} t^{3m-8j} + a_{[\frac{m}{4}]-j+2} t^{3m-4(2j-1)} + \cdots + a_{[\frac{m}{4}]} t^{3m-(j+1)}, \quad 2 \le j \le [\frac{m}{4}]$	$\{2m - 4s; 1 \le s \le j-1\} \cup \{3m - 4s; 1 \le s \le j-1\}$

Table 3.4. The polars for curves in the second Normal Form of Table 3.3:

$y = t^m + t^{3m-4j} + a_1 t^{2m-4(j-[\frac{m}{4}]-1)} + \cdots + a_{j-[\frac{m}{4}]-2} t^{2m-8};\ 2 \leq j \leq [\frac{m}{2}]$	
$a_1 = a_2 = \cdots = a_{j-[\frac{m}{4}]-2} = 0$	
$\frac{2}{m-j} < \frac{1}{j-1}$	The polar has one branch with semigroup $\langle 3, m-1\rangle$, if $\gcd(3, m-1) = 1$; otherwise it has three smooth branches p_1, p_2, p_3, with $\mathrm{I}(p_i, p_r) = \frac{m-1}{3}$.
$\frac{2}{m-j} > \frac{1}{j-1}$	The polar has one branch p_1 with semigroup $\langle 2, m-j\rangle$ and one smooth branch p_2, with $\mathrm{I}(p_1, p_2) = m - j$, if $\gcd(2, m-j) = 1$; otherwise it has three smooth branches p_1, p_2, p_3, with $\mathrm{I}(p_i, p_r) = \frac{m-j}{2}$.
$\frac{2}{m-j} = \frac{1}{j-1}$	The polar has three smooth branches p_1, p_2, p_3, with $\mathrm{I}(p_i, p_r) = j - 1$.
There exists i; $a_i \neq 0$. Put $k = \min\{i;\ a_i \neq 0\}$	
$\frac{2}{m-j} < \frac{1}{[\frac{m}{4}]+k}$	The polar has one branch with semigroup $\langle 3, m-j+[\frac{m}{4}]+k\rangle$, if $\gcd(3, m-j+[\frac{m}{4}]+k) = 1$; otherwise it has three smooth branches p_1, p_2, p_3 with $\mathrm{I}(p_i, p_r) = \frac{m-j+[\frac{m}{4}]+k}{3}$.
$\frac{2}{m-j} > \frac{1}{[\frac{m}{4}]+k}$	The polar has a branch p_1, with semigroup $\langle 2, m-j\rangle$ and a smooth branch p_2, with $\mathrm{I}(p_1, p_2) = m - j$, if $\gcd(2, m-j) = 1$; otherwise it has three smooth branches p_1, p_2, p_3, with $\mathrm{I}(p_i, p_r) = \frac{m-j}{2}$.
$\frac{2}{m-j} = \frac{1}{[\frac{m}{4}]+k}$	For $a_k \neq \frac{4\sqrt{6}}{9}(-1)^\alpha$; $\alpha = 0, 1$, the polar has three smooth branches p_1, p_2, p_3, with $\mathrm{I}(p_i, p_r) = \frac{m-j}{2}$. For $a_k = \frac{4\sqrt{6}}{9}(-1)^\alpha$; $\alpha = 0, 1$: a) If $a_{k+l} = 0$, $\forall l > 0$, then the polar has a smooth branch f_1 and a branch g_1 with semigroup $\langle 2, 2m-3j\rangle$, with $\mathrm{I}(f_1, g_1) = m - j$. b) There exists $s > 0$ such that $a_{k+s} \neq 0$ (let s be minimum). b.1) $m - 2j > s$. b.1.1) s odd. The polar has a smooth branch f_1 and a branch g_1 with semigroup $\langle 2, m-j+s\rangle$ with $\mathrm{I}(f_1, g_1) = m - j$. b.1.2) s even. The polar has three smooth branches f_1, g_1, g_2 with $\mathrm{I}(f_1, g_i) = \frac{m-j}{2}$ and $\mathrm{I}(g_1, g_2) = \frac{m-j+s}{2}$. b.2) $m - 2j < s$. The polar has a smooth branch f_1 and a branch g_1 with semigroup $\langle 2, 2m-3j\rangle$, with $\mathrm{I}(f_1, g_1) = m - j$. b.3) $m - 2j = s$. b.3.1) $a_{k+s} \neq \frac{4\sqrt{6}}{81}(-1)^{\alpha+1}$. The polar has a smooth branch f_1 and a branch g_1 with semigroup $\langle 2, 2m-3j\rangle$, with $\mathrm{I}(f_1, g_1) = m - j$. b.3.2) $a_{k+s} = \frac{4\sqrt{6}}{81}(-1)^{\alpha+1}$. The polar has three smooth branches f_1, g_1, g_2 such that $\mathrm{I}(f_1, g_i) = \frac{m-j}{2}$ and $\mathrm{I}(g_1, g_2) = \frac{m-j}{2} + s$.

Table 3.5. The polars for curves in the third to fifth Normal Forms of Table 3.3:

$\gcd(3, m-j) = 1$	One branch with semigroup $\langle 3, m-j \rangle$.
$\gcd(3, m-j) = 3$	Three smooth branches with mutual intersection numbers $\frac{m-j}{3}$.

References

[1] Casas-Alvero, E., *Infinitely near imposed singularities and singularities of polar curves.* Math. Ann. **287** (1990), 429–454.

[2] Casas-Alvero, E., *Singularities of Plane Curves.* London Math. Soc. Lecture Notes Series **276** (2000).

[3] Hefez, A.; Hernandes, M.E., *Standard bases for local rings of branches and their module of differentials.* J. Symb. Comput. **42** (2007), 178–191.

[4] Hefez, A.; Hernandes, M.E., *Analytic classification of plane branches up to multiplicity 4.* J. Symb. Comput. **44** (2009), 626–634.

[5] Hefez, A.; Hernandes, M.E., *The analytic classification of plane branches.* Bull. London Math. Soc. **43** (2011), 289–298.

[6] Hefez, A.; Hernandes, M.E., *Algorithms for the implementation of the analytic classification of plane branches.* J. Symb. Comput. **50** (2013), 308–313.

[7] Hernández Iglesias, M.F., *Polar de um germe de curva irredutível plana.* Tese de Doutorado, UFF (2012).

[8] Kouchnirenko, A.G., *Polyèdres de Newton et nombres de Milnor.* Inv. Math. **32** (1976), 1–31.

[9] Lê D. T.; Ramanujan, C. P., *The invariance of Milnor's numbers implies the invariance of the topological type.* American Journal of Mathematics, Vol **98** (1976), 67–78.

[10] Martin, B.; Pfister, G., *The moduli of irreducible curve singularities with the semigroup* $\langle 5, 11 \rangle$. Rev. Roumaine XXXIII **4** (1988), 359–368.

[11] Oka, M., *Non-degenerate complete intersection singularity.* Actualités Mathématiques, Hermann (1997).

[12] Pham F., *Deformations equisinguliéres des idéaux jacobiens des courbes planes.* In Proc of Liverpool Symposium on Singularities II, Volume **209** of Lect Notes in Math, Springer Verlag, Berlin, London, new york (1971), 218–233.

[13] Samuel, P., *Algébricité de certains points singuliers algebroídes* - J. Math. Pures et Appl. **35** (1956), 1–6.

[14] Teissier, B., *Variétés polaires. II. Multiplicités polaires, sections planes, et conditions de Whitney*; Algebraic Geometry (La Rábida, 1981), SLNM, Volume **961** (1982), 314–491.

[15] Zariski, O., *The moduli problem for plane branches.* University lecture series AMS, Volume **39** (2006).

[16] Zariski, O., *Characterization of plane algebroid curves whose module of differentials has maximum torsion.* Proc. of NAS, Volume **56**, N. 3 (1966), 781–786.

A. Hefez
Instituto de Matemática e Estatística
Universidade Federal Fluminense
R. Mario Santos Braga, s/n
Niterói, RJ 24020-140
Brazil
e-mail: hefez@mat.uff.br

M.E. Hernandes and M.F. Hernández Iglesias
Departamento de Matemática
Universidade Estadual de Maringá
Av. Colombo 5790
Maringá-PR 87020-900
Brazil
e-mail: mehernandes@uem.br
mfhiglesias.pma@uem.br

Singular Intersections of Quadrics I

Santiago López de Medrano

Dedicated to Pepe Seade on his 60th anniversary

Abstract. Let $Z \subset \mathbb{R}^n$ be given by $k+1$ equations of the form

$$\sum_{i=1}^{n} A_i x_i^2 = 0, \qquad \sum_{i=1}^{n} x_i^2 = 1,$$

where $A_i \in \mathbb{R}^k$. It is well known that the condition for Z to be a smooth variety (known as *weak hyperbolicity*) is that the origin in $\mathbb{R}^k$ is not a convex combination of any collection of k of the vectors A_i. We interpret this condition as a transversality property in order to approach the case when it is singular and we extend some results known for the smooth case, in particular the computation of the homology groups of Z in terms of the combinatorics of the natural quotient polytope. We show that Z cannot be an exotic homotopy sphere nor a non-simply connected homology sphere and use this to show that, except for some clearly characterized degenerate cases, when Z is not smooth it cannot be a topological or even a homological manifold.

Mathematics Subject Classification (2000). Primary 14P25, 57R17; Secondary 14E15.

Keywords. Intersections of quadrics, moment-angle manifolds, topology of real algebraic varieties, singular varieties.

Introduction

The topology of the intersection of several homogeneous quadrics in $\mathbb{R}^n$ has been studied for many years in the smooth case ([17], [11], [14],[15]) with an important boom in recent years ([4], [7], [12], [8], [9], [3],...), especially after the discovery of the connection with the theory of moment-angle complexes and their generalizations (see [6], [5], [2] for a part of this story).

The author was partially supported by project IN111415-PAPIIT and CEI Campus Iberus, Spain. He is also grateful to the referee for several suggestions that helped to improve the text.

A first element in this study was the characterization of the smooth varieties in terms of the coefficients of the quadrics: the property we call now (after Marc Chaperon) *weak hyperbolicity*. During the first talks I gave on the subject around 1985, in more than one occasion Pepe Seade asked if it could be possible that the varieties were manifolds, even if this condition failed. At the moment I had no answer for that.

In the autumn of 2014 I gave a course on the topology of intersections of quadrics at the University of Zaragoza to an enthusiastic and critical audience. This forced me to rethink many of the aspects of the theory, in particular, the concept of weak hyperbolicity. I realized that it could be expressed as a transversality condition of the associated polytope to the strata of the first orthant.

With these fresh ideas in mind, and with the workshop celebrating Pepe's 60-th anniversary approaching, I recalled his old question and felt that now I had the correct approach and the tools for answering it. At the moment of my talk this goal was not achieved, but by the end of the workshop I was able to offer an answer as a birthday present.

This article begins with a brief introduction to the subject, including the concept of weak hyperbolicity and a new concept named *non-degeneracy* which is unnecessary in the smooth case, but useful in the singular case to delimit some artificial examples. It continues with some crucial singular examples and the proof that an adequate formulation of the homology computation works in the general case. Then it passes to the new results: first a Sphere Theorem showing that the only homotopy or homology spheres that may appear as an intersection of quadrics of the type we study is the standard round sphere. This result is interesting in itself, but also as the main element in the proof of the Singularity Theorem that says that, besides the perfectly delimited degenerate cases, when weak hyperbolicity fails the intersection of quadrics cannot be a topological or even a homological manifold.

It is not only an interesting question that I have to thank Pepe for. Although we have never worked together, in all these 30 years my work in this and other subjects has been encouraged and supported by Pepe in many, many ways.

1. Intersections of Quadrics

We recall here the main definitions, notations, concepts and simple properties regarding the varieties we study. Most of them are essentially well known but are formulated here in a slightly different form. The only new concept is that of *non-degeneracy* that will be useful in section 5 to isolate some artificial cases.

1.1. Basic concepts

$[n]$ will denote the set $\{1, 2, \ldots, n\}$ and $\mathbb{R}_+ = \{x \in \mathbb{R} \,|\, x \geq 0\}$.

We will think of $\mathbb{R}^n_+$ as a stratified space, the strata being the subsets where a certain sub-collection of coordinates are positive and the rest are 0.

Let A be a $k \times n$ matrix with entries in $\mathbb{R}$ and $A_i \in \mathbb{R}^k$ its columns.

Let $V = V(A)$ be the intersection of homogeneous diagonal quadrics in $\mathbb{R}^n$ given by the equations

$$\sum_{i=1}^{n} A_i x_i^2 = 0$$

and $Z = Z(A)$ the intersection of V with the unit sphere

$$\sum_{i=1}^{n} x_i^2 = 1 \, .$$

Let further $\Pi = \Pi(A)$ be the affine subspace of $\mathbb{R}^n$ given by

$$\sum_{i=1}^{n} A_i x_i = 0,$$

$$\sum_{i=1}^{n} x_i = 1,$$

and $P = P(A)$ be the convex polytope $\Pi(A) \cap (\mathbb{R}_+)^n$.

P is homeomorphic to $Z \cap (\mathbb{R}_+)^n$ and, topologically, Z can be recovered from any of them by reflection in all the coordinate hyperplanes.

An interesting type of the above intersections of quadrics are those of the form

$$\sum_{i=1}^{n} A_i |z_i|^2 = 0, \qquad \sum_{i=1}^{n} |z_i|^2 = 1.$$

They have an action of the n-torus with quotient the same polytope $P(A)$, and, when smooth, are known as *moment-angle manifolds.*

It is well known that any convex polytope of dimension d with m facets can be realized as $P(A) \subset \mathbb{R}^n$ for any $n \geq m$ and some $k \times n$ matrix A of rank $k = n - d - 1$. If $n = m$, then we can suppose that the facets of P coincide with its intersections with the coordinate hyperplanes. In that case we say that we have a *geometric embedding* of P.

Observe that a facet of P must have at least one coordinate equal to zero and that if F_1 is a proper face of F_2, the number of zero coordinates is bigger in F_1 than it is in F_2. Inductively, we have then

Proposition 1.1. *A face of P of codimension c has at least c zero coordinates.*

1.2. Smoothness

The fact that Z is a smooth variety can be expressed in terms of the matrix A, the polytope P, or the subspace Π:

Theorem 1.2. *The following are equivalent:*

(WH) *Weak Hyperbolicity: No collection of k columns of A has the origin in its convex hull.*

(T1) *Every point in P has at least $k+1$ positive coordinates.*

(T2) Π *is transversal to all the strata of* $\mathbb{R}^n_+$.

(S1) *V is a smooth variety outside the origin.*
(S2) *Z is a smooth variety.*

Any of them implies that P is a simple polytope[1]*, but not conversely.*

Proof. **(WH)** $\iff$ **(T1):** This is immediate, since points of P are the same as convex combinations of the A_i adding up to zero.

(T1) $\iff$ **(T2):** Let d be the dimension of Π. Then, since $d \geq n-(k+1)$ (because it is given by $k+1$ equations in $\mathbb{R}^n$), Π is transversal to a k-dimensional stratum of $\mathbb{R}^n_+$ only if it does not intersect it, so (T2) implies (T1). Assuming (T1), consider a vertex with c zero coordinates. Then we have more inequalities

$$n-(k+1) \leq d \leq c \leq n-(k+1)$$

where the second one follows from the Proposition in the previous section and the last one is (T1). So they are all equalities and $c = d = n-(k+1)$. This means, by the same Proposition, that Π is transversal to all $(k+1)$-dimensional strata of $\mathbb{R}^n_+$ at vertices of P and that the intersections of Π with those of dimension $k+1+s$ must have dimension s for $s = 1, \ldots, d$, which means that we have (T2). If P is not empty, this also implies that it cannot be contained in an hyperplane of $\mathbb{R}^n_+$ and the matrix A must have rank k.

(WH) $\iff$ **(S1):** Assuming no (WH), we have 0 as a convex combination of k columns of A and, in particular those columns are linearly dependent. The square roots of the coefficients give a non-zero point in V where the Jacobian matrix of the equations of V is singular.

Reciprocally, if at a non-zero point x of V the Jacobian matrix is singular, let $I = \{i \in [n] \mid x_i \neq 0\}$. Then the set of A_i with $i \in I$ lives in a $(k-1)$-dimensional subspace of $\mathbb{R}^k$ and has the origin in its convex hull. Caratheodory's Theorem[2] implies that the origin is in the convex hull of k of the A_i and (WH) fails.

(S1) $\iff$ **(S2)** is immediate since V is transversal to the unit sphere.

To prove the last statement observe that under (WH) any vertex v lies in the positive orthant of a $(k+1)$-dimensional coordinate space L. Since the d-dimensional space Π intersects L in a single point, it projects bijectively onto de complementary space $L^{\perp}$. This means that a neighborhood of P projects bijectively onto the complementary coordinate subspace. Thus a neighborhood of v in P looks exactly like the neighborhood of the origin in $\mathbb{R}^d_+$. Since this origin lies on d of the faces of $\mathbb{R}^d_+$ also v lies in d faces of P which is therefore simple[3]. □

There are many counterexamples of the reciprocal (see the examples below). In fact, any simple polytope can be embedded as $P(A)$ for some matrix A satisfying (WH), but it can also be embedded non-transversally: add a new variable x_{n+1} and a new equation $x_{n+1} = 0$, so P is contained in a coordinate hyperplane. One

[1]Recall that a d-dimensional convex polytope is *simple* if every vertex lies in exactly d facets.
[2]*If a set of points in $\mathbb{R}^m$ has the origin in its convex hull, so does a subset with no more than $m+1$ points.* A simple proof can be found in Wikipedia, *Carathéodory's theorem (convex hull).*
[3]And any other local combinatorial property of P can be deduced from this local model.

can then rotate P into the $x_{n+1} > 0$ region (with a result as in Figure 1b below) so that it becomes non-degenerate in the sense we are going to define next.

1.3. Non-degeneracy

We are going to consider cases when Z is not smooth, but in order to isolate extremely artificial situations (see examples at the end of 2a) and 2b)) in the singular case we make the following:

Definition 1.3. We say that Z is *non-degenerate*[4] if

a) the matrix A has rank k and
b) Z is not contained in a coordinate hyperplane of $\mathbb{R}^n$.

From the theorem in section 1.2 it follows easily that a non-empty smooth Z is always non-degenerate. But an empty Z is always degenerate because of condition b).

The topology of Z can be studied even in the degenerate case, because one can always eliminate redundant equations or variables without changing Z. Some results are valid in general (like the ones in sections 3 and 4), but many others have a simpler formulation in the non-degenerate case. For example:

If Z is non-degenerate, Z, Π and P have dimension $d = n - k - 1$.

On the other hand, when one takes $Z_0 = Z \cap \{x_1 = 0\}$ (see, e.g., [7]) it may happen that Z_0 is degenerate even if Z is not (it can be empty, to begin with) so care must be taken in certain situations.

2. Some singular examples

Before continuing with the theory it is important to have some more examples in mind. The topology of many smooth examples has been described in [11], [15], [4], [7] and [8], so we will concentrate on some singular ones, contrasting them with related smooth ones.

(a) The interval

Figure 1a shows a transverse interval with equations $x_1 - x_2 + x_3 = 0$, $x_1 + x_2 + x_3 = 1$. Reflecting the interval on the $x_1 = 0$ and $x_3 = 0$ planes we get a (piecewise linear) S^1. Reflecting this figure on the $x_2 = 0$ plane gives a second copy. Z itself is diffeomorphic to $S^1 \times S^0$.

Figure 1b shows a non-transverse interval with equations $x_1 - x_2 = 0$, $x_1 + x_2 + x_3 = 1$. P is the interval joining $\mathbf{a} = (1/2, 1/2, 0)$ and $\mathbf{b} = (0, 0, 1)$ and is simple. Reflecting the interval on the $x_1 = 0$ and $x_2 = 0$ planes we get four segments stemming from the point $\mathbf{b}$. Reflecting this figure on the $x_3 = 0$ plane we see that Z is the suspension of its four points on that plane ($\mathbf{a}$ and its reflections) and is not a manifold (not even a homological one).

[4] Equivalently, *Z is degenerate if all of its points are singular.* Observe that in Algebraic Geometry an algebraic set with no regular points is not considered to be a variety.

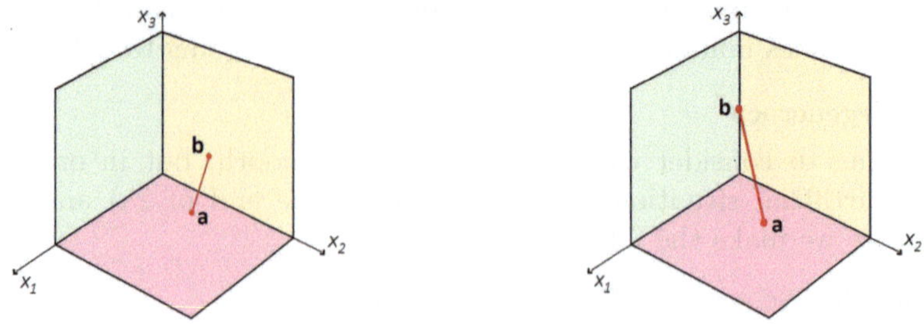

(a) A transverse interval (b) A non-transverse interval

FIGURE 1.

This example generalizes as follows: for $n = p + q$ let P be an interval in $\mathbb{R}^n$ where one vertex has p coordinates equal to zero and the other one has the other q coordinates equal to zero. It is easy to show that Z is the *complete bipartite graph* $K_{2^p,2^q}$, in other words the join $[2^p] * [2^q] = (S^0)^p * (S^0)^q$. Only when $p = q = 1$ do we have that Z is a manifold.

P can also be embedded in $\mathbb{R}^{n+m}$ by reducing it to half its size and making all the new m coordinates equal $1/2m$. In this case $Z = ((S^0)^p) * (S^0)^q) \times (S^0)^m$.

If one further embeds P in $\mathbb{R}^{n+m+r}$ by making all the new r coordinates equal to zero one obtains the same space, a degenerate example.

With these cases we have covered all the possibilities for P one-dimensional.

(b) Triangles in $\mathbb{R}^4$

While in $\mathbb{R}^3$ there is only one way to have P be a triangle, in $\mathbb{R}^4$ there are various possibilities.

b0. The transversal triangle, three transverse vertices.

$$\text{Equations: } x_1 + x_2 + x_3 - x_4 = 0, \;\; x_1 + x_2 + x_3 + x_4 = 1.$$

$$\text{Vertices: } (1/2, 0, 0, 1/2),\ (0, 1/2, 0, 1/2),\ (0, 0, 1/2, 1/2)).$$

$$Z = S^2 \times S^0.$$

This follows from the equations: they imply

$$x_4 = 1/2, \qquad x_1 + x_2 + x_3 = 1/2,$$

so P is a triangle on the space $x_4 = 1/2$. Reflecting P on the $x_1 = 0$, $x_2 = 0$ and $x_3 = 0$ hyperplanes we get a sphere (as for the unit simplex in $\mathbb{R}^3$) and reflecting this sphere on the $x_4 = 0$ hyperplane we get two separate copies.

The topology of Z in the remaining cases follows by the arguments used in examples a and b0.

b1. Two transverse vertices.

$$\text{Equations: } x_2 + x_3 - x_4 = 0 \qquad x_1 + x_2 + x_3 + x_4 = 1.$$

$$\text{Vertices: } (1, 0, 0, 0),\ (0, 1/2, 0, 1/2),\ (0, 0, 1/2, 1/2)).$$

$$Z = \sum (S^1 \times S^0).$$

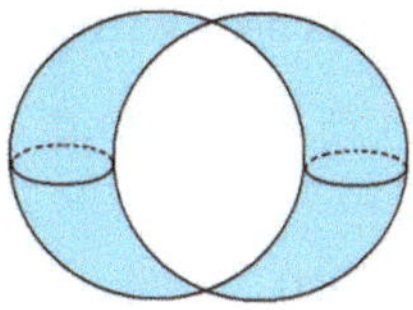

FIGURE 2. Z is the suspension of $S^1 \times S^0$.

b2. One transverse vertex.

$$\text{Equations: } x_3 - x_4 = 0, \qquad x_1 + x_2 + x_3 + x_4 = 1.$$

$$\text{Vertices: } (1,0,0,0),\ (0,1,0,0),\ (0,0,1/2,1/2)).$$

Z is the union of two maximal 2-spheres in S^3 that intersect in a maximal S^1. A projection in $\mathbb{R}^3$ could look like the union of two spheres intersecting transversely in a small circle:

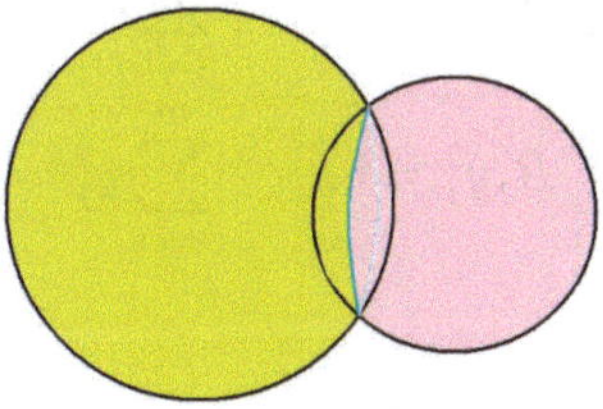

FIGURE 3. Z is the double suspension of $S^0 \times S^0$.

b3. No transverse vertices.

$$\text{Equations: } \quad -x_4 = 0, \qquad x_1 + x_2 + x_3 + x_4 = 1.$$

$$\text{Vertices: } (1,0,0,0),\ (0,1,0,0),\ (0,0,1,0)).$$

$$Z = S^2.$$

One could continue the list of examples by considering the case with equations

$$0 = 0, \qquad x_1 + x_2 + x_3 + x_4 = 1.$$

These two last examples are degenerate: The first is not really a triangle in $\mathbb{R}^4$ but in $\mathbb{R}^3$. The last one is not a triangle, but a tetrahedron and $Z = S^3$, wrong dimension! In both cases Z is not a smooth variety, but is nevertheless a smooth manifold.

If we add to the above list the cases with equations

$$x_1 + x_2 + x_3 + x_4 = 0, \qquad x_1 + x_2 + x_3 + x_4 = 1$$

(where P and Z are empty), and

$$x_1 + x_2 - x_3 - x_4 = 0, \qquad x_1 + x_2 + x_3 + x_4 = 1$$

(where P is a square and Z is a torus), we obtain all cases with $n = 4$ and $k = 1$.

One can construct in the same way many examples where P is the d-simplex in $\mathbb{R}^n$.

(c) The suspension and the join

If to the equations of Z one adds one more variable x_{n+1} to obtain the system

$$\sum_{i=1}^{n} A_i x_i^2 = 0, \qquad \sum_{i=1}^{n+1} x_i^2 = 1,$$

the new intersection of quadrics is the suspension of Z and the new polytope is the pyramid on P.

More generally, if we have two varieties $Z(A)$ and $Z(B)$ given by

$$\sum_{i=1}^{n} A_i x_i^2 = 0, \qquad \sum_{i=1}^{n} x_i^2 = 1,$$
$$\sum_{j=1}^{m} B_j y_j^2 = 0, \qquad \sum_{j=1}^{m} y_j^2 = 1,$$

one can build up the system

$$\sum_{i=1}^{n} A_i x_i^2 = 0,$$
$$\sum_{j=1}^{m} B_j y_j^2 = 0,$$
$$\sum_{i=1}^{n} x_i^2 + \sum_{j=1}^{m} y_j^2 = 1,$$

which clearly represents the join $P(A) * P(B)$ of the polytopes and the join of the varieties $Z(A) * Z(B)$.

Compare with the equations of the product $Z(A) \times Z(B)$ obtained by adding to the above the equation

$$\sum_{i=1}^{n} x_i^2 - \sum_{j=1}^{m} y_j^2 = 0.$$

(d) The non-simple polytopes geometrically embedded

Recall this means that each intersection with a coordinate hyperplane is a facet. Many singular examples of this type can be constructed by taking cones or joins of smooth examples. For example, if P is the pyramid with base a square, then Z is the suspension of a torus which is singular but non-degenerate.

The algebraic topology of the corresponding singular moment-angle manifolds has been studied by A. Ayzenberg and V. Buchstaber in [1].

Other interesting examples to consider are those with only one non-transverse vertex which has k positive coordinates. The case where P is geometrically embedded appears in the study of *wall-crossing* in [4, p. 70–82], where the combinatorial relation between the two different nearby transverse polytopes is described (*flip*) as well as the topological relation between the corresponding moment-angle manifods (*equivariant surgery*). Observe that the singular moment-angle manifolds have non-isolated singularities (due to the torus symmetry), so if seen as intersections of quadrics in $\mathbb{R}^{2n}$ they are not of the simplest singular kind. The information about the polytopes should be helpful for the study of the corresponding intersections of quadrics. The analogous results, as well as the description of the singular varieties themselves, have not yet been considered.

3. The Homology Splitting

Here we compute the homology of Z in terms of the combinatorics of P as embedded in $\mathbb{R}^n_+$. We reproduce essentially the proof in [11], which works equally well for any $k > 2$ in the smooth case ([3]) and, with some adaptations, in the general case:

Let us call g_i the reflection on the subspace $\{x_i = 0\}$ and for $I \subset [n]$ let us define the composition

$$g_I = \prod_{i \in I} g_i.$$

For $i \in [n]$ let P_i be the face[5] of P:

$$P_i = P \cap \{x_i = 0\}$$

(which is the fixed point set of g_i) and P_I be the union of faces:

$$P_I = \bigcup_{i \in I} P_i.$$

[5] Every facet of P is one of the P_i, and two different facets cannot be the same P_i. However, two different P_i can be the same facet and a P_i can also be empty, all of P or a lower-dimensional face. The reader may review section 2 to avoid having the wrong intuition (especially if he is familiar with the smooth case where the non-empty P_i are in one-to-one correspondence with the facets of P), but may also skip this since in all that follows these possibilities need not be considered separately.

We now consider the cell decomposition of Z consisting of P and all its reflections $g_J(P)$. The lower-dimensional cells are all of the form $g_I(F)$ where F is a face so these elements generate the groups of chains of Z.

But they do not form a basis since there are repetitions: for example, if a face F is in P_i, then $g_iF = F$ for $i \in I$, since in that case g_i leaves F fixed. But if we define for each face F

$$J(F) = \{j \in [n] \mid x_j = 0 \text{ in } F\} = \{j \in [n] \mid F \subset P_j\},$$

we have a basis for the chains $C_*(Z)$ of Z consisting of the elements

$$g_I(F),\ I \cap J(F) = \emptyset\,.$$

Still, the boundary operation remains complicated: the boundary of a face F consists of intersections of F with other coordinate subspaces, that is, faces with a *bigger* $J(F)$, so the boundary of $g_I(F)$ may contain summands of the form g_IF' with $J(F') \cap I \neq \emptyset$, which are not in our basis and the boundary formula becomes complicated. In particular, the set of $g_I(F)$ with a given I do not generate a chain subcomplex of $C_*(Z)$.

This problem disappears if we construct a different basis as follows: g_i induces a chain map on $C_*(Z)$ which we will denote again by g_i, and we can consider the chain maps h_i and their compositions:

$$h_i = (1 - g_i),$$

$$h_I = \prod_{i \in I} h_i\,.$$

Then we have a new basis[6] of $C_*(Z)$ formed by the elements

$$h_I(F),\ I \cap J(F) = \emptyset.$$

Now $h_iF = 0$ when $i \in J(F)$, so the terms of the boundary of $h_I(F)$ of the form h_IF' with $J(F') \cap I \neq \emptyset$ are 0, so only the terms h_IF' with $J(F') \cap I = \emptyset$ remain. So the basis elements $h_I(F)$ with a fixed I generate a chain subcomplex of $C_*(Z)$ and we have a direct sum of subcomplexes:

$$C_*(Z) = \bigoplus_{I \subset [n]} h_I C_*(P)\,.$$

Furthermore, $h_IC_*(Z)$ can be seen as the chains of P which have null coefficients in faces P_i with $i \in I$, so

$$h_IC_*(Z) = C_*(P, P_I)$$

and

$$C_*(Z) = \bigoplus_{I \subset [n]} C_*(P, P_I)\,.$$

[6] Any member of one collection can be expressed as a combination of members of the other one. But observe that g_I is induced by an actual map, while h_I is just a chain map.

This splitting at the chain level implies the homology splitting

$$H_*(Z) = \bigoplus_{I\subset[n]} H_*(P, P_I),$$

but also splittings for homology or cohomology with any ring of coefficients.

From the homology exact sequence of (P, P_I) we get (since P is contractible)

$$H_i(P, P_I) = \tilde{H}_{i-1}(P_I)$$

and we can also read directly the homology from the collection of subcomplexes P_I of ∂P:

Theorem 3.1. $H_*(Z) = \bigoplus_{I\subset[n]} H_*(P, P_I) = \bigoplus_{I\subset[n]} \tilde{H}_{*-1}(P_I)$.

Observe that there is always a *top class* $h_{[n]}P$ associated with the top class $[P] \in H_d(P, \partial P)$ which g_i multiplies by -1 and h_i multiplies by 2, so these numbers can be thought of as their *degrees*.

In the smooth case Z is homeomorphic to a generalized moment-angle complex so behind the above homology splitting lies the very general stable homotopy splitting of [2]. The more general homology splitting proved here suggests that this could also be true for a singular Z. However, Ayzenberg and Buchstaber have shown in [1] that when P is geometrically embedded in $\mathbb{R}^n_+$, but not simple, Z is *homotopically equivalent* (but not always homeomorphic) to a generalized moment-angle complex. According to Anton Ayzenberg (private conversation) their argument is valid in our case, so the stable homotopy splitting is valid also for the suspension of any Z and the homology splitting is valid for any generalized homology theory.

3.1. Examples of homology computations

a) The non transversal interval in $\mathbb{R}^3$

In this case (see Figure 1b) the subcomplexes P_I of P are:

$$P_\emptyset = \emptyset, \quad P_1 = \{b\}, \quad P_2 = \{b\}, \quad P_3 = \{a\},$$

$$P_{12} = \{b\}, \quad P_{13} = \{a, b\}, \quad P_{23} = \{a, b\}, \quad P_{123} = \{a, b\}.$$

$P_\emptyset$ gives a basis of $H_0(Z) = \mathbb{Z}$ and the three last ones give a basis of $H_1(Z) = \mathbb{Z}^3$, while the other four are contractible and give nothing.

Compare with the case where P is a transversal interval (Figure 1a.):

$$P_\emptyset = \emptyset,\ P_1 = \{b\},\ P_2 = \emptyset,\ P_3 = \{a\},$$

$$P_{12} = \{b\},\ P_{13} = \{a, b\},\ P_{23} = \{a\},\ P_{123} = \{a, b\}.$$

Here $P_\emptyset$ and P_2 provide the basis of $H_0(Z) = \mathbb{Z}^2$ and the P_{13} and P_{123} a basis of $H_1(Z) = \mathbb{Z}^2$, while the other four complexes give nothing (recall that Z is $S^1 \times S^0$).

The reader may enjoy verifying the homology computations in the other examples where P is an interval (section 2).

b) The triangles in $\mathbb{R}^4$

We detail two of the examples in section 2 b):

b1. Two transverse vertices.

Vertices of P: $A = (1,0,0,0)$, $B = (0,1/2,0,1/2)$, $C = (0,0,1/2,1/2))$.

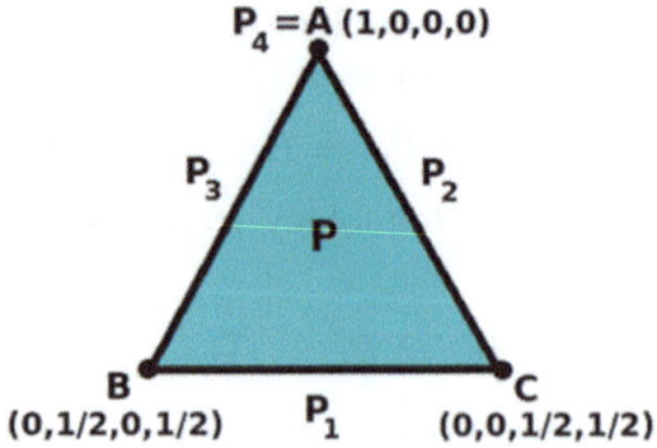

Two transverse vertices.

In the figure, P_1, P_2 and P_3 are the sides of the triangle while P_4 is just the vertex A. The reader can verify that there are only three non-contractible, non-empty unions P_I: the one with $I = \{1,4\}$ (the union of the segment BC and the vertex A) and those with $I = \{1,2,3\}$ and $I = \{1,2,3,4\}$ (which are both the boundary of P). They give the generators of $H_1(Z) = \mathbb{Z}$ and $H_2(Z) = \mathbb{Z} \oplus \mathbb{Z}$, respectively. (Compare with Figure 2.)

b2. One transverse vertex.

Vertices of P: $(1,0,0,0)$, $(0,1,0,0)$, $(0,0,1/2,1/2))$.

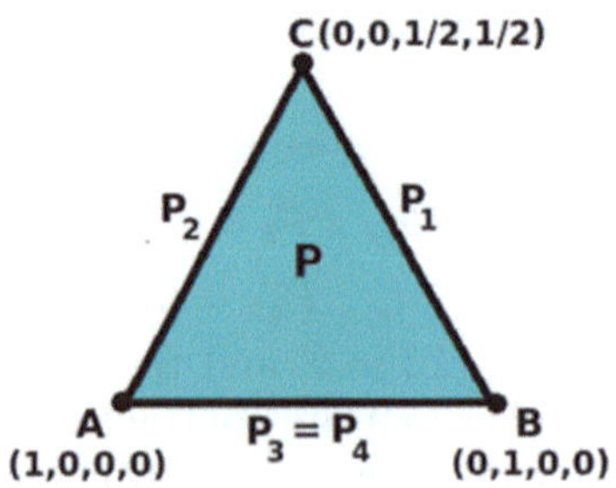

One transverse vertex.

In the figure P_1, P_2 and P_3 are the sides of the triangle but P_4 is again the side AB. There are only three non-contractible, non-empty unions P_I: those with $I = \{1,2,3\}$, $I = \{1,2,4\}$ and $I = \{1,2,3,4\}$ (which are all the boundary of P). They give the three generators of $H_2(Z) = \mathbb{Z} \oplus \mathbb{Z} \oplus \mathbb{Z}$. (Compare with Figure 3.)

Examples b1 and b2 are typical of what happens in the homology of a non-transverse polytope: in section 4 we will see that there are always at least two independent elements in the top dimension, showing that the corresponding Z cannot be a homology sphere.

3.2. A connectivity theorem

Theorem 3.2. *For any Z the following are equivalent:*

(i) *Any c of the faces P_i has a non-empty intersection.*
(ii) $H_i(Z) = 0$ *for* $i = 0, \dots, c-1$.

Proof. Let s_1, s_2 be the largest integers such that $c = s_1$ satisfies (i) and $c = s_2$ satisfies (ii). Then the following lemma implies that $s_1 = s_2$: □

Lemma 3.3. *Let $X_1, \dots, X_q$ be $q \geq 2$ compact convex sets in $\mathbb{R}^n$ $X = \bigcup_{i=1}^q X_i$ and $Y = \bigcap_{i=1}^q X_i$. Then:*

(a) *If Y is not empty, then X is contractible.*
(b) *If any $q-1$ of the X_i have a non-empty intersection but Y is empty, then $\tilde{H}_i(X)$ is trivial for $i < q-2$ and $\tilde{H}_{q-2}(X) = \mathbb{Z}$.*

(a) is immediate since X is star-shaped from any point in Y and (b) can be proved by induction using the Mayer-Vietoris sequence. They also follow from the *Nerve Theorem* (see [16] for a proof).

Remark 3.4. One can also add to the Theorem (iii) that *Z is $(c-1)$-connected* as an equivalent statement. This is well known in the smooth case and not difficult to prove, but we will actually use only the following:

Corollary 3.5. *If $H_i(Z) = 0$ for $i = 0, \dots, c-1$, then any c of the* **facets** *P_i has a non-empty intersection.*

This is true, because the facets of P are included in the set of P_i.

In the smooth and geometrically embedded cases, the converse of the corollary is true, because all P_i are facets (or are empty). But, for Z singular, that converse is not necessarily true (in example b1 every pair of facets has a non-empty intersection, but $H_1(Z)$ is not 0).

4. The Sphere Theorem

Let Z be any intersection of quadrics in $\mathbb{R}^n$ with polytope P, both of dimension d. We start by a general

Proposition 4.1. *Assume that*

1) *for some $i \in [n]$ we have that P_i is not a facet, or*
2) *for some $i, j \in [n]$, $i \neq j$ we have $P_i = P_j$.*

Then the Betti number β_d is larger than 1.

Proof. In case 1), for $J = [n] \setminus \{i\}$ we have that P_J is the whole boundary of P (because every facet is a P_j which must happen for $j \neq i$) and $h_J(P)$ defines a class in $H_d(Z)$ which is independent of the top class $h_{[n]}(Z)$ so $\beta_d \geq 2$.

In case 2), for $J = [n] \setminus \{i\}$ and $J = [n] \setminus \{j\}$ we have that P_J is the whole boundary of P which, together with the top class, make three independent elements of $H_{[n]}(Z)$, so $\beta_d \geq 3$. □

Remarks 4.2. a) Observe that case 1) includes the case where P_i is empty, and that if in case 2) P_i is empty, then $\beta_d \geq 4$.

b) Cf. cases 1) and 2) with the examples in section 3.1: in b.1, P_4 is a vertex and $\beta_2 = 2$; in b.2, $P_3 = P_4$ and $\beta_2 = 3$.

Now we can prove the

Sphere Theorem. *The following are equivalent:*

i) *Z is the unit sphere S^d in a $(d+1)$-dimensional coordinate subspace.*
ii) *Z is a homotopy sphere.*
iii) *Z is a homology sphere.*

Proof. Clearly i) implies ii) implies iii). Now, assuming iii), the corollary in section 3.2 implies that any d facets of P have non-empty intersection. This implies easily that P is a simplex (see also statement 4 in [10, p. 123]).

Now, assume that $n > d + 1$. Since there are many ways of embedding a d-simplex in $\mathbb{R}^n$ for $n > d+1$ (see section 2, examples a) and b)) we have to check that none of them give homology spheres. Now P falls in one of the two cases of the previous proposition, because if all of the n P_i are facets, they cannot all be different. But then the proposition contradicts the fact that Z is a homology sphere. So the only possibility left is $n = d + 1$, so $k = 0$ and P is the unit simplex. □

This result shows that Z is never an exotic sphere or a non-simply connected homology sphere. This result is new even in the smooth case (but compare with [12, p. 243]. and [4, p. 68, Proposition 2]).

5. The Singularity Theorem

We will now prove that singular non-degenerate Z cannot be a manifold.

Singularity Theorem. *For any non-degenerate Z the following are equivalent:*

(i) *A is weakly hyperbolic.*
(ii) *Z is a smooth manifold.*
(iii) *Z is a topological manifold.*
(iv) *Z is a homology manifold.*

Proof. When $d = 0$ every Z is a manifold, but a singular one must have at least one zero coordinate so it is degenerate, so the Theorem is true in this case. So we can assume from now on that $d > 0$.

Assume now that Z is non-degenerate and let v be a vertex of P having p positive coordinates. We have seen in sections 1.1 and 1.2 that $p \leq k + 1$ where equality holds if, and only if, v is a transverse vertex.

The plane Π touches in one point the coordinate space S of the p positive coordinates of v. This means that it projects injectively[7] onto a subspace $\bar{\Pi}$ on

[7] Compare with the last part of the proof in section 1.2.

the complementary coordinate space $S^\perp$. The image of P under the projection is a polytope $\bar{P}$ equivalent to P. The neighborhood of v in P is combinatorially equivalent to the neighborhood of 0 in $\bar{P}$. This means that the link of v in P is equivalent to the intersection of $\bar{\Pi}$ with the unit simplex in $S^\perp$. This is the standard polytope of an intersection of quadrics obtained by reflecting it in all coordinate subspaces of $S^\perp$. This intersection of quadrics is clearly diffeomorphic to the link of v in Z. (We could now write explicit equations for this link but for our present purposes this is not relevant.)

Now assume that the link of v in Z is a homology sphere. Then by the Sphere Theorem this link is a round sphere in some coordinate subspace of $S^\perp$ of dimension d. Since P is not contained in a coordinate hyperplane of $\mathbb{R}^n$, then that d-dimensional coordinate subspace must be all of $S^\perp$, so $n-p=d$, $p=k+1$ and v is a transverse vertex. Applying this to every vertex of P, and using the equivalences of 1.2, we have proved that under the assumption that Z is non-degenerate part (iv) implies (i), the only implication that is not evident in the Theorem. □

Conclusion

So it is possible to construct examples where weak hyperbolicity fails and yet Z is a smooth manifold, but only in two specific and very artificial ways: starting from any smooth non-empty Z_1, one can do one of the following operations (or both):

a) add to the equations of Z_1 new homogeneous equations with all coefficients equal to zero or
b) add to the equations of Z_1 new variables y_i and equations $y_i^2 = 0$.

Then Z_1 is Z (in case b) embedded in a larger space), so it is a smooth manifold, but not a smooth variety.

In a colloquial way we can say that a singular Z can be a manifold, but only in the artificial cases where k *is not really* k or n *is not really* n.

This is my answer to Pepe's question.

References

[1] Ayzenberg, A. and Buchstaber, V. *Nerve complexes and moment-angle spaces of convex polytopes*, Proceedings of the Steklov Institute of Mathematics, **275** (2011), 15–46.

[2] Bahri, A., Bendersky, M., Cohen, F.R., and Gitler, S. *The polyhedral product functor: a method of computation for moment-angle complexes, arrangements and related spaces*, Adv. in Math. **22** (2010), 1634–1668.

[3] Barreto, Y., López de Medrano, S. and Verjovsky, A., *Some Open Book and Contact Structures on Moment-Angle Manifolds*, arXiv:1510.07729v2, to appear in Boletín de la Sociedad Matemática Mexicana.

[4] Bosio, F. and Meersseman, L., *Real quadrics in* $\mathbb{C}^n$, *complex manifolds and convex polytopes*, Acta Mathematica **197** (2006) 53–127.

[5] Buchstaber, V. and Panov, T., *Torus actions and their applications in topology and combinatorics*, University Lecture Seriers, AMS (2002).

[6] Davis, M. and Januszkiewicz, T., *Convex polytopes, Coxeter orbifolds and torus actions*, Duke Math. Journal **62** (1991) 417–451.

[7] Gitler, S., López de Medrano, S. *Intersections of Quadrics, Moment-Angle Manifolds and Connected Sums*, Geometry and Topology **17** (2013) 1497–1534.

[8] Gómez Gutiérrez, V. and López de Medrano, S., *Topology of the Intersection of Quadrics II*, Bol. Soc. Mat. Mex. **20** (2014) 237–255.

[9] Gómez Gutiérrez, V., López de Medrano, S. *Surfaces as Complete intersections*, in *Riemann and Klein Surfaces, Automorphisms, Symmetries and Moduli Spaces*, Contemporary Mathematics Volume 629 (2014) 171–180.

[10] Grünbaum, B., *Convex polytopes*, Graduate Texts in Mathematics (221), 2003.

[11] López de Medrano, S. *The Topology of the Intersection of Quadrics in* $\mathbb{R}^n$, en *Algebraic Topology* (Arcata Ca,1986), Springer-Verlag Lecture Notes in Mathematics **1370** (1989), 280–292.

[12] López de Medrano, S., The space of Siegel leaves of a holomorphic vector field, in Holomorphic Dynamics (Mexico, 1986), Springer-Verlag Lecture Notes in Math., 1345 (1988), 233–245.

[13] López de Medrano, S., *Singularities of real homogeneous quadratic mappings*, Revista de la Real Academia de Ciencias Exactas, Fisicas y Naturales. Serie A. Matematicas. 108 (2014) 95–112. Springer-Verlag.

[14] S. López de Medrano, A. Verjovsky, *A new family of complex, compact, nonsymplectic manifolds*. Bol. Soc. Brasil. Mat. **28** (1997), 253–269.

[15] Meersseman, L. *A new geometric construction of compact complex manifolds in any dimension*, Math. Ann. **317** (2000), 79–115.

[16] McCord, M.C., *Homotopy type comparison of a space with complexes associated with its open covers,* Proc. Amer. Math. Soc. **18** (1967), 705–708.

[17] Wall, C.T.C., *Stability, pencils and polytopes*, Bull. London Math. Soc. **12** (1980), 401–421.

Santiago López de Medrano
Instituto de Matemáticas, UNAM
Área de la Investigación Cientíŋfica
Circuito Exterior, Ciudad Universitaria
04510 México, D.F.
México
e-mail: santiago@im.unam.mx

A New Conjecture, a New Invariant, and a New Non-splitting Result

David B. Massey

Abstract. We prove a new non-splitting result for the cohomology of the Milnor fiber, reminiscent of the classical result proved independently by Lazzeri, Gabrielov, and Lê in 1973–74.

We do this while exploring a conjecture of Fernández de Bobadilla about a stronger version of our non-splitting result. To explore this conjecture, we define a new numerical invariant for hypersurfaces with 1-dimensional critical loci: the beta invariant. The beta invariant is an invariant of the ambient topological-type of the hypersurface, is non-negative, and is algebraically calculable. Results about the beta invariant remove the topology from Bobadilla's conjecture and turn it into a purely algebraic question.

Mathematics Subject Classification (2000). 32B15, 32C35, 32C18, 32B10.

Keywords. Milnor fiber, non-splitting, 1-dimensional critical locus, hypersurface, invariant.

1. Introduction

Throughout this paper, we suppose that $\mathcal{U}$ is an open neighborhood of the origin in $\mathbb{C}^{n+1}$, and that $f : (\mathcal{U}, \mathbf{0}) \to (\mathbb{C}, 0)$ is a complex analytic function with a 1-dimensional critical locus at the origin, i.e., $\dim_{\mathbf{0}} \Sigma f = 1$. We use coordinates $(z_0, \dots, z_n)$ on $\mathcal{U}$ and, to omit the non-reduced curve case, we assume that $n \geq 2$, which implies that f is reduced.

We assume that L is a linear form which is generic enough so that

$$\dim_{\mathbf{0}} \Sigma\big(f_{|_{V(L)}}\big) = 0.$$

For convenience, possibly after a linear change of coordinates, we may assume that L is the first coordinate z_0, so that we have $\dim_{\mathbf{0}} \Sigma\big(f_{|_{V(z_0)}}\big) = 0$.

We assume that $\mathcal{U}$ is chosen (e.g., as a small enough open ball) so that, for each irreducible component C of Σf:

- C contains the origin;
- C is contained in the vanishing locus $V(f)$ of f; and
- $C - \{\mathbf{0}\}$ is homeomorphic to a punctured disk.

Furthermore, we assume that $\mathcal{U}$ is so small that, for each irreducible component C of Σf:

- the isomorphism-type of the reduced integral cohomology groups $\widetilde{H}^*(F_{f,\mathbf{p}}; \mathbb{Z})$ is independent of the choice of $\mathbf{p} \in C - \{\mathbf{0}\}$. This is the same isomorphism-type as the reduced cohomology at $\mathbf{p}$ of the Milnor fiber of the hyperplane slice $f_{|_{V(z_0 - z_0(\mathbf{p}))}}$, for $\mathbf{p} \in C - \{\mathbf{0}\}$ close enough to $\mathbf{0}$. Such a slice yields an isolated critical point at $\mathbf{p}$, and so this cohomology is non-zero in a single degree, namely degree $(n-1)$, and
$$\widetilde{H}^{n-1}(F_{f,\mathbf{p}}; \mathbb{Z}) \;\cong\; \mathbb{Z}^{\mathring{\mu}_C},$$
where $\mathring{\mu}_C$ is the Milnor number at $\mathbf{p}$ of $f_{|_{V(z_0 - z_0(\mathbf{p}))}}$.

We can now state the classic non-splitting result, proved independently by Lê in [8], Lazzeri in [5], and Gabrielov in [4] in 1973–1974.

Theorem 1.1 (Lê-Lazzeri-Gabrielov). *Suppose that the Milnor number of $f_{|_{V(z_0)}}$ at the origin is equal to*
$$\sum_C (C \cdot V(z_0))_{\mathbf{0}}\, \mathring{\mu}_C,$$
where the sum is over the irreducible components C of Σf and $(C \cdot V(z_0))_{\mathbf{0}}$ is the intersection number (which would be the multiplicity of C at $\mathbf{0}$ if z_0 were generic enough). That is, suppose that the Milnor number in the $z_0 = 0$ slice "splits" over the critical points of f in the slice where $z_0 = t$ for a small value of $t \neq 0$.

Then, in fact, Σf has a single irreducible component which is smooth and is transversely intersected by $V(z_0)$ at $\mathbf{0}$.

Remark 1.2. We have stated the above theorem in a slightly more general form than the original statements, but the proofs remain the same.

We should also comment that there is a pleasant priority "dispute" as to which of Lê, Lazzeri, or Gabrielov first proved the above result. Many years ago, we contacted all three authors, and each one claimed that one of the other two proved the result first.

Now, roughly 40 years after Theorem 1.1 was proved, Javier Fernández de Bobadilla has made a conjecture, which looks like it should be related to Theorem 1.1:

Conjecture 1.3 (Fernández de Bobadilla). *Suppose that the critical locus of f has a single irreducible component C and that the isomorphism-type of the cohomology groups $\widetilde{H}^*(F_{f,\mathbf{p}}; \mathbb{Z})$ is independent of the choice of $\mathbf{p} \in C$, i.e., suppose that*

$\widetilde{H}^*(F_{f,\mathbf{0}};\mathbb{Z})$ is non-zero only in degree $(n-1)$ and

$$\widetilde{H}^{n-1}(F_{f,\mathbf{0}};\mathbb{Z}) \cong \mathbb{Z}^{\overset{\circ}{\mu}_C}.$$

Then, C is smooth.

However, this conjecture does **not** follow from Theorem 1.1 (in any way that has yet been tried) and the conjecture remains a conjecture.

Before we continue, we give the obvious **generalized Bobadilla conjecture** for the case where Σf may have more than a single irreducible component:

Conjecture 1.4. *Suppose that $\widetilde{H}^*(F_{f,\mathbf{0}};\mathbb{Z})$ is non-zero only in degree $(n-1)$ and*

$$\widetilde{H}^{n-1}(F_{f,\mathbf{0}};\mathbb{Z}) \cong \bigoplus_C \mathbb{Z}^{\overset{\circ}{\mu}_C},$$

where the sum is over all irreducible components C of Σf. Then, Σf has a single irreducible component, which is smooth.

This paper represents our initial attack on the problem. Here, after recalling earlier definitions and results, we obtain the following new results:

1. In Definition 3.1, we define a new invariant $\beta_f = \beta_{f,z_0}$, which is algebraically calculable.
2. In Theorem 4.1, we prove that β_f is an invariant of the the ambient topological-type of the hypersuface and, in particular, is independent of the linear form z_0.
3. In Theorem 5.2, we show that $\beta_f \geq 0$, and that $\beta_f = 0$ implies that $\widetilde{H}^n(F_{f,\mathbf{0}};\mathbb{Z}) = 0$.
4. We prove in Theorem 5.4 that, in fact, $\beta_f = 0$ is precisely equivalent to the hypotheses of our generalized Bobadilla conjecture, Conjecture 1.4. We thus remove the topology from the hypotheses of the conjecture.

   Furthermore, we prove in this theorem that $\beta_f = 0$ implies that Σf has a single irreducible component, i.e., the cohomology does not "split" over various components. Hence, we are back in the setting of Bobadilla's original conjecture.
5. In Theorem 5.6, we discuss the case where $\beta_f = 1$ and show that, in this case, the critical locus must have precisely two irreducible components.

We thank Javier Fernández de Bobadilla for discussing his conjecture with us, and Lê Dũng Tráng for valuable conversations on this topic.

2. Notation and Known Results

Our assumption that

$$\dim_{\mathbf{0}} \Sigma\left(f_{|_{V(z_0)}}\right) = \dim_{\mathbf{0}} V\left(z_0, \frac{\partial f}{\partial z_1}, \ldots, \frac{\partial f}{\partial z_n}\right) = 0$$

is precisely equivalent to saying that $V\left(\frac{\partial f}{\partial z_1}, \dots, \frac{\partial f}{\partial z_n}\right)$ is purely 1-dimensional (and not empty) at the origin and is properly intersected at the origin by $V(z_0)$.

In terms of analytic cycles,

$$V\left(\frac{\partial f}{\partial z_1}, \dots, \frac{\partial f}{\partial z_n}\right) = \Gamma^1_{f,z_0} + \Lambda^1_{f,z_0},$$

where Γ^1_{f,z_0} is the relative polar curve of f, which consists of components not contained in Σf, and Λ^1_{f,z_0} is the 1-dimensional Lê cycle, which consists of components which are contained in Σf. See Definition 1.11 of [10].

Note that $V\left(\frac{\partial f}{\partial z_0}\right)$ necessarily intersects Γ^1_{f,z_0} properly at $\mathbf{0}$, and that $V(z_0)$ intersects Λ^1_{f,z_0} properly at $\mathbf{0}$ by our assumption.

Letting C's denote the underlying reduced components of Σf at $\mathbf{0}$, at the origin, we have

$$\Lambda^1_{f,z_0} = \sum_C \overset{\circ}{\mu}_C [C],$$

where we use the square brackets to indicate that we are considering C as a cycle, and $\overset{\circ}{\mu}_C$ is the Milnor number of f, restricted to a generic hyperplane slice, at a point $\mathbf{p}$ on $C - \{\mathbf{0}\}$ close to $\mathbf{0}$. See Remark 1.19 of [10].

The intersection numbers $\left(\Gamma^1_{f,z_0} \cdot V\left(\frac{\partial f}{\partial z_0}\right)\right)_{\mathbf{0}}$ and $\left(\Lambda^1_{f,z_0} \cdot V(z_0)\right)_{\mathbf{0}}$ are the Lê numbers λ^0_{f,z_0} and λ^1_{f,z_0} (at the origin). See Definition 1.11 of [10]. Note that

$$\lambda^1_{f,z_0} = \sum_C \left(C \cdot V(z_0)\right)_{\mathbf{0}} \overset{\circ}{\mu}_C.$$

We give here a list of the numbers, other than the beta invariant, which will be used throughout this paper:

1. As we have used several times already, $\overset{\circ}{\mu}_C$ is the Milnor number of a generic hyperplane slice at a point $\mathbf{p} \neq \mathbf{0}$ on the irreducible component C of Σf.
2. We use the Lê numbers λ^0_{f,z_0} and λ^1_{f,z_0}.
3. Throughout, we will use the Betti numbers $\tilde{b}_{n-1}$ and $\tilde{b}_n$ of the reduced integral cohomology of the Milnor fiber $F_{f,\mathbf{0}}$ of f at the origin in degrees $(n-1)$ and n. (We do not need to write "reduced" here and, yet, we do so because we are thinking of the vanishing cycles, not the nearby cycles.)
4. We let $\sigma_f := \sum_C \overset{\circ}{\mu}_C$.
   Note the lack of the intersection multiplicities in this summation. Thus, $\lambda^1_{f,z_0} = \sigma_f$ if and only if each irreducible component C of Σf is smooth and transversely intersected at the origin by $V(z_0)$.
5. Since we are using cohomology, not homology, $\widetilde{H}^{n-1}(F_{f,\mathbf{0}};\mathbb{Z})$ is free Abelian, but $\widetilde{H}^n(F_{f,\mathbf{0}};\mathbb{Z})$ may contain torsion. For each prime p, we let τ_p denote the number of p-torsion direct summands of $\widetilde{H}^n(F_{f,\mathbf{0}};\mathbb{Z})$. With this notation, and

our notation for the Betti numbers, the Universal Coefficient Theorem tells us that

$$\dim \widetilde{H}^{n-1}(F_{f,\mathbf{0}};\mathbb{Z}/p\mathbb{Z}) = \tilde{b}_{n-1}+\tau_p \qquad \text{and} \qquad \dim \widetilde{H}^{n}(F_{f,\mathbf{0}};\mathbb{Z}/p\mathbb{Z}) = \tilde{b}_{n}+\tau_p.$$

6. Finally, we let c_f denote the number of irreducible components of Σf.

In Corollary 10.10 of [10] (though there is an indexing typographical error), we proved a fundamental result linking the Lê numbers and the Betti numbers of the Milnor fiber, which continues to hold with coefficients in $\mathbb{Z}/p\mathbb{Z}$.

Theorem 2.1. *There is an exact sequence*

$$0 \to \widetilde{H}^{n-1}(F_{f,\mathbf{0}};\mathbb{Z}) \to \mathbb{Z}^{\lambda^1_{f,z_0}} \xrightarrow{\delta} \mathbb{Z}^{\lambda^0_{f,z_0}} \to \widetilde{H}^{n}(F_{f,\mathbf{0}};\mathbb{Z}) \to 0, \tag{$\dagger$}$$

and so,

$$\tilde{b}_{n-1} \leq \lambda^1_{f,z_0}, \quad \tilde{b}_{n} \leq \lambda^0_{f,z_0}, \quad \text{and} \quad \tilde{b}_{n}-\tilde{b}_{n-1} = \lambda^0_{f,z_0}-\lambda^1_{f,z_0}.$$

In addition, for each prime number p, there is an exact sequence

$$0 \to \widetilde{H}^{n-1}(F_{f,\mathbf{0}};\mathbb{Z}/p\mathbb{Z}) \to (\mathbb{Z}/p\mathbb{Z})^{\lambda^1_{f,z_0}} \xrightarrow{\delta_p} (\mathbb{Z}/p\mathbb{Z})^{\lambda^0_{f,z_0}} \to \widetilde{H}^{n}(F_{f,\mathbf{0}};\mathbb{Z}/p\mathbb{Z}) \to 0, \tag{$\ddagger$}$$

and so,

$$\tilde{b}_{n-1}+\tau_p \leq \lambda^1_{f,z_0} \quad \text{and} \quad \tilde{b}_{n}+\tau_p \leq \lambda^0_{f,z_0}.$$

3. Definition of the beta invariant and examples

Definition 3.1. *We define the* **beta invariant***:*

$$\begin{aligned}\beta_f = \beta_{f,z_0} &:= \left(\Gamma^1_{f,z_0}\cdot V\left(\frac{\partial f}{\partial z_0}\right)\right)_{\mathbf{0}} - \sum_C \overset{\circ}{\mu}_C\left[(C\cdot V(z_0))_{\mathbf{0}}-1\right] \\ &= \lambda^0_{f,z_0}-\lambda^1_{f,z_0}+\sigma_f \;=\; \tilde{b}_n-\tilde{b}_{n-1}+\sigma_f.\end{aligned}$$

Remark 3.2. Note that the final expression above does not depend on z_0. Thus, the value of β_{f,z_0} is independent of the linear form z_0 (provided that f, restricted to where the linear form is zero, has an isolated critical point). Consequently, we may drop the z_0 from the notation, but sometimes include it to indicate what linear form will actually be used in the calculation of β_f.

In the notation of Siersma in [11], β_f would be the rank of $H^n(F,F')$, and some of our later conclusions can be extracted from the second exact sequence in 9.4 of the Concluding Remarks of that paper.

Example 3.3. Suppose that all of the components C of Σf are smooth and transversely intersected by $V(z_0)$ at $\mathbf{0}$. Then,

$$\beta_{f,z_0} = \left(\Gamma^1_{f,z_0}\cdot V\left(\frac{\partial f}{\partial z_0}\right)\right)_{\mathbf{0}} = \lambda^0_{f,z_0}.$$

Thus, the only time that β_{f,z_0} is really a "new" invariant is when the critical locus itself has a singular component.

Example 3.4. Suppose $f = z^2 + (y^2 - x^3)^d$, where $d \geq 2$. Both $f_{|_{V(x)}}$ and $f_{|_{V(y)}}$ have isolated critical points at the origin. We will calculate both $\beta_{f,x}$ and $\beta_{f,y}$, and see that they are the same.

First, we find that, as sets,

$$\Sigma f = V\left(\frac{\partial f}{\partial x}, \frac{\partial f}{\partial y}, \frac{\partial f}{\partial z}\right)$$
$$= V\big(d(y^2 - x^3)^{d-1}(-3x^2),\ d(y^2 - x^3)^{d-1}2y,\ 2z\big) = V(y^2 - x^3, z).$$

Now, as cycles, we calculate

$$V\left(\frac{\partial f}{\partial y}, \frac{\partial f}{\partial z}\right) = V\big(d(y^2 - x^3)^{d-1}2y,\ 2z\big)$$
$$= V(y, z) + (d-1)V(y^2 - x^3, z) = \Gamma^1_{f,x} + \Lambda^1_{f,x}.$$

Thus, we have $\Gamma^1_{f,x} = V(y, z)$, and that Σf consists of the single component $C = V(y^2 - x^3, z)$, with $\mathring{\mu}_C = d - 1$. Therefore,

$$\beta_{f,x} = \left(\Gamma^1_{f,x} \cdot V\left(\frac{\partial f}{\partial x}\right)\right)_{\mathbf{0}} - \sum_C \mathring{\mu}_C\left[(C \cdot V(x))_{\mathbf{0}} - 1\right]$$
$$= \big(V(y, z) \cdot V(d(y^2 - x^3)^{d-1}(-3x^2))\big)_{\mathbf{0}}$$
$$- (d-1)(V(y^2 - x^3, z) \cdot V(x))_{\mathbf{0}} + (d-1)$$
$$= \big(V(y, z) \cdot V((y^2 - x^3)^{d-1})\big)_{\mathbf{0}} + \big(V(y, z) \cdot V(x^2)\big)_{\mathbf{0}} - 2(d-1) + (d-1)$$
$$= 3(d-1) + 2 - (d-1) = 2d.$$

To calculate $\beta_{f,y}$, we proceed similarly.

As cycles, we calculate

$$V\left(\frac{\partial f}{\partial x}, \frac{\partial f}{\partial z}\right) = V\big(d(y^2 - x^3)^{d-1}(-3x^2),\ 2z\big)$$
$$= 2V(x, z) + (d-1)V(y^2 - x^3, z) = \Gamma^1_{f,y} + \Lambda^1_{f,y}.$$

Thus, we have $\Gamma^1_{f,y} = 2V(x, z)$, and, of course, that Σf consists of the single component $C = V(y^2 - x^3, z)$, with $\mathring{\mu}_C = d - 1$. Therefore,

$$\beta_{f,y} = \left(\Gamma^1_{f,y} \cdot V\left(\frac{\partial f}{\partial y}\right)\right)_{\mathbf{0}} - \sum_C \mathring{\mu}_C\left[(C \cdot V(y))_{\mathbf{0}} - 1\right]$$
$$= \big(2V(x, z) \cdot V(d(y^2 - x^3)^{d-1}(2y))\big)_{\mathbf{0}}$$
$$- (d-1)(V(y^2 - x^3, z) \cdot V(y))_{\mathbf{0}} + (d-1)$$
$$= \big(2V(x, z) \cdot V((y^2 - x^3)^{d-1})\big)_{\mathbf{0}} + (2V(x, z) \cdot V(y)))_{\mathbf{0}} - 3(d-1) + (d-1)$$
$$= 4(d-1) + 2 - 2(d-1) = 2d.$$

As promised, we see that $\beta_{f,x} = \beta_{f,y}$, even though the separate terms in the calculation are different.

4. Invariance

In this short section, we prove the topological invariance of β_f and σ_f.

Theorem 4.1. *If $f : (\mathcal{U}, \mathbf{0}) \to (\mathbb{C}, 0)$ and $g : (\mathcal{U}, \mathbf{0}) \to (\mathbb{C}, 0)$ are reduced with 1-dimensional critical loci at the origin, and $V(f)$ and $V(g)$ have the same local ambient topological-type at $\mathbf{0}$, then $\sigma_f = \sigma_g$ and $\beta_f = \beta_g$.*

Proof. As Lê proved in [6] and [7], the homotopy-type of the Milnor fiber is an invariant of the ambient topological-type for reduced functions f; thus, the topological invariance of β_f would follow from the topological invariance of $\sum_C \mathring{\mu}_C$. However, this latter topological invariance is easy to establish.

The singular set of $V(f)$ must map to the singular set under an ambient homeomorphism and, as we require the origin to map to the origin, the punctured singular set $\Sigma V(f) - \{\mathbf{0}\}$ must map to the punctured singular set, and so the components of Σf at the origin must map bijectively to the components of the singular set at the origin. Now the homotopy-type of the Milnor fiber of f at a point $\mathbf{p} \in \Sigma V(f)$ near $\mathbf{0}$ is invariant under an ambient homeomorphism, and this homotopy-type is that of a bouquet of $\mathring{\mu}_C$ $(n-1)$-spheres, where C is the component of $\Sigma V(f)$ containing $\mathbf{p}$. □

5. Non-negativity and Milnor fiber consequences

In this section, we first need to review a number of known results, and establish some notation.

Recall that our choice of $\mathcal{U}$ implies that, for each irreducible component C of Σf, $C - \{\mathbf{0}\}$ is topologically a punctured disk and so, is homotopy-equivalent to a circle. There is an "internal" (also known as "vertical") monodromy action, h_C, on $\mathbb{Z}^{\mathring{\mu}_C}$ given by traveling once around this circle. This is the monodromy of the local system obtained by considering the complex of sheaves of vanishing cycles along f, and restricting to $C - \{\mathbf{0}\}$.

Now, a result of Siersma in [11], or an easy exercise using perverse sheaves (see the remark at the end of [11]) tells us that

Theorem 5.1 (Siersma). *There is an inclusion*

$$\widetilde{H}^{n-1}(F_{f,\mathbf{0}}; \mathbb{Z}) \hookrightarrow \bigoplus_C \ker\{\mathrm{id} - h_C\},$$

which commutes with the monodromy action on the vanishing cycles along f.

In particular,

$$\operatorname{rank} \widetilde{H}^{n-1}(F_{f,\mathbf{0}}; \mathbb{Z}) \le \sigma_f,$$

and equality implies that each h_C is the identity.

Furthermore, this result holds with $\mathbb{Z}/p\mathbb{Z}$ coefficients, where the proof remains identical. Thus,

$$\dim \widetilde{H}^{n-1}(F_{f,\mathbf{0}};\mathbb{Z}/p\mathbb{Z}) \leq \sigma_f.$$

Thus, we conclude immediately that:

Theorem 5.2. *For all primes p,*

$$\tilde{b}_{n-1} + \tau_p \ \leq \ \sigma_f \ \text{ and } \ \tilde{b}_n + \tau_p \ \leq \ \beta_f.$$

In particular, $0 \leq \tilde{b}_n \leq \beta_f$.

Recall that c_f denotes the number of irreducible components of Σf.

Proposition 5.3. *If* $\operatorname{rank} \widetilde{H}^{n-1}(F_{f,\mathbf{0}};\mathbb{Z}) = \sigma_f$*, then the trace of the Milnor monodromy of f on $\widetilde{H}^{n-1}(F_{f,\mathbf{0}};\mathbb{Z})$ is $(-1)^n c_f$.*

Proof. Under the assumption, Theorem 5.1 tells us that the trace of the Milnor monodromy on $\widetilde{H}^{n-1}(F_{f,\mathbf{0}};\mathbb{Z})$ is the sum of the traces of the Milnor monodromy on each $\ker\{\mathrm{id} - h_C\}$. As h_C is the identity, this is simply the Milnor monodromy of f at a point $\mathbf{p}$ on $C - \{\mathbf{0}\}$ near the origin. By A'Campo's result in [1], this is $(-1)^n$. □

The case where $\beta_f = 0$ is extremely restrictive.

Theorem 5.4. *The following are equivalent:*

1. *$\beta_f = 0$; and*
2. *$\widetilde{H}^n(F_{f,\mathbf{0}};\mathbb{Z}) = 0$, and $\widetilde{H}^{n-1}(F_{f,\mathbf{0}};\mathbb{Z}) \cong \mathbb{Z}^{\sigma_f}$.*

In addition, when these equivalent conditions hold, Σf has a single irreducible component C and the internal monodromy automorphism h_C is the identity.

Proof. That (2) implies (1) follows immediately from Theorem 2.1. That (1) implies (2) follows immediately from Theorem 5.2. That $c_f = 1$ follows from Proposition 5.3 and A'Campo's Theorem applied to the cohomology of $F_{f,\mathbf{0}}$. That the internal monodromy automorphism h_C is the identity is part of Theorem 5.1. □

Remark 5.5. The statement in Theorem 5.4 that Σf must have a single irreducible component at the origin is a "non-splitting result", of the flavor of the result proved independently by Lê in [8], Lazzeri in [5], and Gabrielov in [4]; however, those three works use the cohomology of Milnor fibers of f restricted to hyperplane slices, rather than looking at the cohomology of the Milnor fiber of f itself.

The case where $\beta_f = 1$ is also interesting to consider:

Theorem 5.6. *Suppose that $\beta_f = 1$. Then, $c_f = 2$, and either*

1. *$\tilde{b}_n = 0$, $\tilde{b}_{n-1} = \sigma_f - 1$ and, for all primes p, $\widetilde{H}^n(F_{f,\mathbf{0}};\mathbb{Z})$ has at most one direct summand with p-torsion; or*

2. $\tilde{b}_n = 1$, $\tilde{b}_{n-1} = \sigma_f$, *and* $\widetilde{H}^n(F_{f,\mathbf{0}};\mathbb{Z})$ *has no torsion (and so is isomorphic to* $\mathbb{Z}$*).*

Proof. By Theorem 5.2, $\tilde{b}_n \leq \beta_f$, and we know that $\tilde{b}_n - \tilde{b}_{n-1} = \beta_f - \sigma_f = 1 - \sigma_f$. So we obtain the two cases to consider: (1) where $\tilde{b}_n = 0$ and $\tilde{b}_{n-1} = \sigma_f - 1$ and (2) where $\tilde{b}_n = 1$ and $\tilde{b}_{n-1} = \sigma_f$. The conclusions about torsion in both cases follow from the p-torsion statement in Theorem 5.2. All that remains for us to show is the claim about c_f .

In case (2), by A'Campo's result, the Lefschetz number of the Milnor monodromy on $H^*(F_{f,\mathbf{0}};\mathbb{Z})$ is zero. By Proposition 5.3, the trace of the monodromy of f on $\widetilde{H}^{n-1}(F_{f,\mathbf{0}};\mathbb{Z})$ is $(-1)^n c_f$. Since $\widetilde{H}^n(F_{f,\mathbf{0}};\mathbb{Z}) \cong \mathbb{Z}$, the trace of the monodromy of f in degree n is ± 1. Thus, we obtain that $1 - c_f \pm 1 = 0$. Hence, $c_f = 0$ or 2, but we are assuming that f has a 1-dimensional critical locus, so $c_f \neq 0$.

Case (1) is very similar. Since $\tilde{b}_n = 0$, A'Campo's result tells us that the trace of the monodromy of f on $\widetilde{H}^{n-1}(F_{f,\mathbf{0}};\mathbb{Z})$ is $(-1)^n$. On the other hand, restriction induces the inclusion

$$\widetilde{H}^{n-1}(F_{f,\mathbf{0}};\mathbb{Z}) \hookrightarrow \bigoplus_C \widetilde{H}^{n-1}(F_{f,\mathbf{p}_C};\mathbb{Z}) \cong \mathbb{Z}^{\sigma_f},$$

where $\mathbf{p}_C$ denotes a point of $C - \{\mathbf{0}\}$ close to the origin. This inclusion is compatible with the f monodromy action and, since $\tilde{b}_{n-1} = \sigma_f - 1$, the cokernel is isomorphic to $\mathbb{Z} \oplus T$, where T is pure torsion. The trace of the map induced by the f monodromy on this cokernel is ± 1. Thus, from additivity of the traces, we obtain $(-1)^n = (-1)^n c_f \pm 1$, and conclude once again that $c_f = 2$. □

As we saw in Example 3.3, if all of the components of Σf are smooth at $\mathbf{0}$, then, for generic z_0, $\beta_f = \lambda^0_{f,z_0}$; in this case, results on λ^0_{f,z_0} imply even stronger results when $\beta_f = 0$ or 1.

For instance, the non-splitting result of Lê-Lazzeri-Gabrielov immediately implies the first item below, while the main theorem of [9] immediately implies the second item.

Proposition 5.7. *Suppose that all of the components of* Σf *are smooth at* $\mathbf{0}$.

1. *If* $\beta_f = 0$, *then* Σf *has a unique (smooth) component* C *at the origin, along which the Milnor number of a generic hyperplane slice is constant. In particular,* $\widetilde{H}^n(F_{f,\mathbf{0}}) = 0$ *and* $\widetilde{H}^{n-1}(F_{f,\mathbf{0}}) \cong \mathbb{Z}^{\mathring{\mu}_C}$.
2. *If* $\beta_f = 1$, *then* $\widetilde{H}^n(F_{f,\mathbf{0}}) = 0$ *and* $\widetilde{H}^{n-1}(F_{f,\mathbf{0}}) \cong \mathbb{Z}^{\sigma_f - 1}$.

6. Concluding Remarks

As we stated earlier, our interest in the beta invariant is when the critical locus of f is itself singular, since the reason we defined the beta invariant is because it arose naturally while we were considering the conjecture, Conjecture 1.3, of Fernández de Bobadilla. We now refer to this conjecture as the **beta conjecture** or the β_f **conjecture**.

The beta conjecture is related to another conjecture: Lê's conjecture (see, for instance, [2]). The suspicion is that the proof of the beta conjecture will be very difficult, and will require new techniques. Good candidates for counterexamples are also hard to produce.

We believe that viewing the problem in terms of β_f may help regardless of whether the conjecture is true or false. If the beta conjecture is true, describing the question in terms of β_f may lead to an algebraic proof. If the beta conjecture is false, showing that $\beta_f = 0$ may be the easiest way to verify that one has a counterexample.

However, even if a proof of, or counterexample to, the beta conjecture is difficult, there are other questions which are interesting and, perhaps, more approachable.

Question 6.1. Is the beta conjecture true if f is quasi-homogeneous?

Question 6.2. Is the beta conjecture true if Σf is contained in a smooth surface? That is, after an analytic change of coordinates, is the beta conjecture true if Σf is contained in a 2-plane?

It seems to be difficult to produce hypersurfaces with a critical locus which is 1-dimensional, singular, irreducible, and with a small β_f. The case where $\beta_f = 0$ is what is conjectured not to be possible. The case where $\beta_f = 1$ is not possible by Theorem 5.6. Our question is:

Question 6.3. Is it possible for $\beta_f = 2$ or 3, if f has a critical locus which is 1-dimensional, singular, and irreducible?

Related to the above question, we ask:

Question 6.4. Is there a relationship between β_f and the Milnor number $\mu_{\mathbf{0}}(\Sigma f)$ of the curve Σf, using the Milnor number of Buchweitz and Greuel in [3]? Is there, perhaps, some simple relationship, like something of the form $\beta_f \geq 2\mu_{\mathbf{0}}(\Sigma f)$?

References

[1] A'Campo, N. Le nombre de Lefschetz d'une monodromie. *Proc. Kon. Ned. Akad. Wet., Series A*, 76:113–118, 1973.

[2] Bobadilla, J. A Reformulation of Lê's Conjecture. *Indag. Math.*, 17, no. 3:345–352, 2006.

[3] Buchweitz, R.-O., Greuel, G.-M. The Milnor Number and Deformations of Complex Curve Singularities. *Invent. math.*, 58:241–281, 1980.

[4] Gabrielov, A. M. Bifurcations, Dynkin Diagrams, and Modality of Isolated Singularities. *Funk. Anal. Pril.*, 8 (2):7–12, 1974.

[5] Lazzeri, F. *Some Remarks on the Picard-Lefschetz Monodromy.* Quelques journées singulières. Centre de Math. de l'Ecole Polytechnique, Paris, 1974.

[6] Lê, D. T. Calcul du Nombre de Cycles Évanouissants d'une Hypersurface Complexe. *Ann. Inst. Fourier, Grenoble*, 23:261–270, 1973.

[7] Lê, D. T. *Topologie des Singularites des Hypersurfaces Complexes*, volume 7-8 of *Astérisque*, pages 171–182. Soc. Math. France, 1973.

[8] Lê, D. T. Une application d'un théorème d'A'Campo a l'equisingularité. *Indag. Math.*, 35:403–409, 1973.

[9] Lê, D. T. and Massey, D. Hypersurface Singularities and Milnor Equisingularity. *Pure and Appl. Math. Quart., special issue in honor of Robert MacPherson's 60th birthday*, 2, no.3:893–914, 2006.

[10] Massey, D. *Lê Cycles and Hypersurface Singularities*, volume 1615 of *Lecture Notes in Math.* Springer-Verlag, 1995.

[11] Siersma, D. Variation mappings on singularities with a 1-dimensional critical locus. *Topology*, 30:445–469, 1991.

David B. Massey
Department of Mathematics
Northeastern University
Boston, MA 02115
USA
e-mail: d.massey@neu.edu

Lipschitz Geometry Does not Determine Embedded Topological Type

Walter D. Neumann and Anne Pichon

Dedicated to José Seade for a great occasion. Happy birthday, Pepe!

Abstract. We investigate the relationships between the Lipschitz outer geometry and the embedded topological type of a hypersurface germ in $(\mathbb{C}^n, 0)$. It is well known that the Lipschitz outer geometry of a complex plane curve germ determines and is determined by its embedded topological type. We prove that this does not remain true in higher dimensions. Namely, we give two normal hypersurface germs $(X_1, 0)$ and $(X_2, 0)$ in $(\mathbb{C}^3, 0)$ having the same outer Lipschitz geometry and different embedded topological types. Our pair consist of two superisolated singularities whose tangent cones form an Alexander-Zariski pair having only cusp-singularities. Our result is based on a description of the Lipschitz outer geometry of a superisolated singularity. We also prove that the Lipschitz inner geometry of a superisolated singularity is completely determined by its (non-embedded) topological type, or equivalently by the combinatorial type of its tangent cone.

Mathematics Subject Classification (2000). 14B05, 32S25, 32S05, 57M99.

Keywords. Complex surface singularity, bilipschitz, Lipschitz geometry, embedded topological type, superisolated.

1. Introduction

A complex germ $(X, 0)$ has two natural metrics up to bilipschitz equivalence, the *outer metric* given by embedding $(X, 0)$ in some $(\mathbb{C}^n, 0)$ and taking distance in $\mathbb{C}^n$ and the *inner metric* given by shortest distance along paths in X.

In this paper we investigate the relationships between the Lipschitz outer geometry and the embedded topological type of a hypersurface germ in $(\mathbb{C}^n, 0)$.

It is well known that the Lipschitz outer geometry of a complex plane curve germ determines and is determined by its embedded topological type ([12], see also [5] and [9, Theorem 1.1.]). We prove that this does not remain true in higher dimensions:

Theorem 1.1. *There exist two hypersurface germs in* $(\mathbb{C}^3, 0)$ *having same Lipschitz outer geometry and distinct embedded topological type.*

It is worth noting that for families of isolated hypersurfaces in $\mathbb{C}^3$, the constancy of Lipschitz outer geometry implies constancy of embedded topological type. Indeed, Varchenko proved in [13] that a Zariski equisingular family of hypersurfaces in any dimension has constant embedded topological type and it is proved in [10] that for a family of hypersurface singularities $(X_t, 0) \subset (\mathbb{C}^3, 0)$, Zariski equisingularity is equivalent to constant Lipschitz outer geometry.

It should also be noted that the converse question, which consists of examining which part of the outer Lipschitz geometry of a hypersurface can be recovered from its embedded topological type seems difficult. In particular the outer geometry of a normal complex surface singularity determines its multiplicity ([10, Theorem 1.2 (2)]) so this question somehow contains the Zariski multiplicity question.

In order to prove Theorem 1.1 we construct two germs of hypersurfaces in $(\mathbb{C}^3, 0)$ having the same Lipschitz outer geometry and different embedded topological types. They consist of a pair of superisolated singularities whose tangent cones form an Alexander-Zariski pair of projective plane curves.

A surface singularity $(X, 0)$ is *superisolated* (SIS for short) if it is given by an equation

$$f_d(x, y, z) + f_{d+1}(x, y, z) + f_{d+2}(x, y, z) + \cdots = 0,$$

where $d \geq 2$, f_k is a homogeneous polynomial of degree k and the projective curve $\{f_{d+1} = 0\} \subset \mathbb{P}^2$ contains no singular point of the projective curve $C = \{[x : y : z] : f_d(x, y, z) = 0\}$. In particular, the projectivized tangent cone C of $(X, 0)$ is reduced. In the sequel we will just consider SISs with equations

$$f_d(x, y, z) + f_{d+1}(x, y, z) = 0\,.$$

Definition 1.2 (Combinatorial type of a projective plane curve). The *combinatorial type* of a reduced projective plane curve $C \subset \mathbb{P}^2$ is the homeomorphism type of a tubular neighborhood of it in $\mathbb{P}^2$ (see, e.g., [3, Remark 3]; a more combinatorial version is also given there, which we describe in Section 3).

It is well known that the combinatorial type of the projectivized tangent cone of a SIS $(X, 0)$ determines the topology of $(X, 0)$. In fact, we will show:

Theorem 1.3. (i) *The Lipschitz inner geometry of a SIS determines and is determined by the combinatorial type of its projectivized tangent cone.*

(ii) *There exist SISs with the same combinatorial types of their projectivized tangent cones but different Lipschitz outer geometry.*

Acknowledgments. We are grateful to Hélène Maugendre for fruitful conversations and for communicating to us the equations of the tangent cones for the examples in the proof of Theorem 1.3 (ii). We are also grateful to the anonymous referee who pointed out a missing argument in the last proof. Walter Neumann was supported by NSF grant DMS-1206760. Anne Pichon was supported by ANR-12-JS01-0002-01 SUSI. We are also grateful for the hospitality and support of the following institutions: Columbia University, Institut de Mathématiques de Marseille, Aix Marseille Université and CIRM Luminy, Marseille.

2. Proof of Theorem 1.1

The proof of Theorem 1.1 will need Lemma 2.2 and Proposition 2.3 below, which will be proved in section 4. First a definition:

Definition 2.1. We say that two germs $(C_1, 0)$ and $(C_2, 0)$ of reduced irreducible plane curves are *weak RL-equivalent* if for $i = 1, 2$ there are holomorphic maps $h_i\colon (\mathbb{C}^2, 0) \to (\mathbb{C}, 0)$ with $(h_i^{-1}(0), 0) = (C_i, 0)$, a homeomorphism $\psi\colon (\mathbb{C}^2, 0) \to (\mathbb{C}^2, 0)$, a constant $K \geq 1$ and a neighborhood $\mathcal{U}$ of the origin in $\mathbb{C}^2$ such that for all $a, a' \in \mathcal{U}$.

$$\frac{1}{K}||h_2(\psi(a))(1,\psi(a)) - h_2(\psi(a'))(1,\psi(a'))||_{\mathbb{C}^3} \leq ||h_1(a)(1,a) - h_1(a')(1,a')||_{\mathbb{C}^3}$$
$$\leq K||h_2(\psi(a))(1,\psi(a)) - h_2(\psi(a'))(1,\psi(a'))||_{\mathbb{C}^3}.$$

Lemma 2.2. *Weak RL-equivalence of reduced irreducible plane curve germs $(C_1, 0)$ and $(C_2, 0)$ does not depend on the choice of their defining functions h_1 and h_2. Moreover, it is implied by analytic equivalence of $(C_1, 0)$ and $(C_2, 0)$ in the sense of Zariski* [14] *(also called RL-equivalence or $\mathcal{A}$-equivalence).*

Proposition 2.3. *Let $(X, 0)$ be a SIS with equation $f_d + f_{d+1} = 0$. The Lipschitz outer geometry of $(X, 0)$ is determined by the combinatorial type of its projectivized tangent cone and by the weak RL-equivalence classes of corresponding singularities of the projectivized tangent cones.*

Proof of Theorem 1.1. Recall that a Zariski pair is a pair of projective curves $C_1, C_2 \subset \mathbb{P}^2$ with the same combinatorial type but such that $(\mathbb{P}^2, C_1)$ is not homeomorphic to $(\mathbb{P}^2, C_2)$. The first example was discovered by Zariski: a pair of sextic curves C_1 and C_2, each with six cusps, distinguished by the fact that C_1 has the cusps lying on a quadric and C_2 does not. He constructed those of type C_1 in [14] and conjectured type C_2, confirming their existence eight years later in [15]. He distinguished their embedded topology by the fundamental groups of their complements, but they can also be distinguished by their Alexander polynomials (Libgober [7]) so they are called *Alexander-Zariski pairs.*

Let $(X_1, 0)$ and $(X_2, 0)$ be two SISs whose tangent cones are sextics of types C_1 and C_2 as above. According to [16], the analytic type of a cusp is uniquely determined, so its weak RL-equivalence class is determined (Lemma 2.2). Then by Proposition 2.3, $(X_1, 0)$ and $(X_2, 0)$ are outer Lipschitz equivalent.

On the other hand, Artal showed that $(X_1, 0)$ and $(X_2, 0)$ do not have the same embedded topological type. In fact, he shows ([1, Theorem 1.6 (ii)]) that a Zariski pair is distinguished by its Alexander polynomials if and only if the corresponding SISs are distinguished by the Jordan block decompositions of their homological monodromies. □

3. The inner geometry of a superisolated singularity

We first recall how the topological type of a SIS is determined by the combinatorial type of its projectivized tangent cone. We refer to [2] for details.

A SIS $(X, 0) \subset (\mathbb{C}^3, 0)$ is resolved by blowing up the origin of $(\mathbb{C}^3, 0)$. The exceptional divisor of this resolution of $(X, 0)$ is the projectivized tangent cone C of $(X, 0)$ and one obtains the minimal good resolution by blowing up the singularities of C which are not ordinary double points until one obtains a normal crossing divisor C'. Let Γ be the dual graph of this resolution. Following [4] we say $\mathcal{L}$-curve for a component of C' which is a component of C and $\mathcal{L}$-node any vertex of Γ representing an $\mathcal{L}$-curve.

One can also resolve the singularities of C as a projective plane curve to obtain the same graph Γ except that the self-intersection numbers of the $\mathcal{L}$-curves are different (in the example below the self-intersection number -9 becomes $+3$). The graph Γ with these data is equivalent to the combinatorial type of C.

Example 3.1. Consider the SIS $(X, 0) \subset (\mathbb{C}^3, 0)$ given by $F(x, y, z) = y^3 + xz^2 - x^4 = 0$. Blowing up the origin of $\mathbb{C}^3$ resolves the singularity: using the chart $(x, v, w) \mapsto (x, y, z) = (x, xv, xw)$, the equation of the resolved X^* is $v^3 + w^2 - x = 0$ and the exceptional curve has a cusp singularity $x = v^3 + w^2 = 0$. Blowing up further leads to the following dual graph Γ, the black vertex being the $\mathcal{L}$-node.

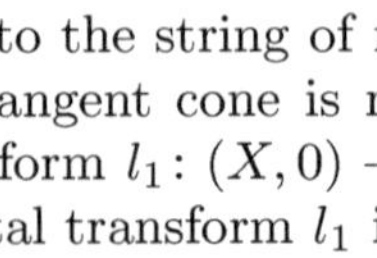

The self-intersection -9 of the $\mathcal{L}$-curve is computed as follows. Let $E_1, \dots, E_4$ be the components of the exceptional divisor indexed so that E_1 is the $\mathcal{L}$-curve and E_2, E_3 and E_4 correspond to the string of non $\mathcal{L}$-nodes indexed from left to right on the graph. Since the tangent cone is reduced with degree 3, the strict transform l_1^* of a generic linear form $l_1 \colon (X, 0) \to (\mathbb{C}, 0)$ consists of three smooth curves transverse to E_1. The total transform l_1 is given by the divisor:

$$(l_1) = E_1 + 3E_2 + 6E_3 + 2E_4 + l_1^* .$$

Since (l_1) is a principal divisor, we have $(l_1).E_1 = 0$, which leads to $E_1.E_1 = -9$.

Proof of Theorem 1.3 (i). Let $(X, 0) \subset (\mathbb{C}^3, 0)$ be a SIS with equation $f_d + f_{d+1} = 0$. We set $f = f_d$ and $g = -f_{d+1}$.

Let $\ell\colon \mathbb{C}^3 \to \mathbb{C}^2$ be a generic linear projection for $(X,0)$, let Π be the polar curve of the restriction $\ell\mid_X$ and $\Delta = \ell(\Pi)$ its discriminant curve.

Let e be the blow-up of the origin of $\mathbb{C}^3$ and let p be a singular point of $e^{-1}(0) \cap X^*$. Without loss of generality, we can assume $\ell = (x,y)$. We can also choose our coordinates so that $p = (1,0,0)$ in the chart (x,v,w) given by $(x,v,w) \mapsto (x,y,z) = (x,xv,xw)$ in the blow-up e (so p corresponds to the x-axis in the tangent cone of X). Then X^* has equation

$$f(1,v,w) - xg(1,v,w) = 0$$

and $g(1,v,w)$ is a unit at p since $\{g=0\} \cap Sing(f=0) = \emptyset$ in $\mathbb{P}^2$.

Let $e_0\colon Y \to \mathbb{C}^2$ be the blow-up of the origin of $\mathbb{C}^2$. We consider e_0 in the chart $(x,v) \mapsto (x,y) = (x,xv)$, we set $q = (1,0) \in Y$ in this chart, and we denote by $\tilde{\ell}\colon (X^*,p) \to (Y,q)$ the projection $(x,v,w) \mapsto (x,v)$. So we have the commutative diagram:

$$\begin{array}{ccc} (X^*,p) & \xrightarrow{\ e\ } & (X,0) \\ \big\downarrow{\scriptstyle \tilde{\ell}} & & \big\downarrow{\scriptstyle \ell} \\ (Y,q) & \xrightarrow{\ e_0\ } & (\mathbb{C}^2,0)\,. \end{array}$$

Now $\Pi = X \cap \{f_z - g_z = 0\}$, so the strict transform Π^* of Π by e has equations:

$$f_w(1,v,w) - xg_w(1,v,w) = 0 \quad \text{and} \quad f(1,v,w) - xg(1,v,w) = 0\,,$$

which are also the equations of the polar curve of the projection $\tilde{\ell}\colon (X^*,p) \to (Y,q)$.

Since $g(1,v,w) \in \mathbb{C}\{v,w\}$ is a unit at p, the quotient $h(v,w) := \frac{f(1,v,w)}{g(1,v,w)}$ defines a holomorphic function germ $h\colon (\mathbb{C}^2_{(v,w)},0) \to (\mathbb{C},0)$. In terms of $h(v,w)$ the above equations for (Π^*,p) can be written:

$$h_w(v,w) = 0 \quad \text{and} \quad h(v,w) - x = 0\,.$$

Consider the isomorphism $proj\colon (X^*,p) \to (\mathbb{C}^2,0)$ which is the restriction of the linear projection $(x,v,w) \mapsto (v,w)$. Then Π^* is the inverse image by $proj$ of the polar curve Π' of the morphism $\ell'\colon (\mathbb{C}^2_{(v,w)},0) \to (\mathbb{C}^2_{(x,v)},0)$ defined by $(v,w) \mapsto (h(v,w),v)$, i.e., the relative polar curve of the map germ $(v,w) \mapsto h(v,w)$ for the generic projection $(v,w) \mapsto v$.

We set $\Delta' = \ell'(\Pi')$ and $q = (1,0)$ in $\mathbb{C}^2_{(x,v)}$. We then have a commutative diagram:

$$\begin{array}{ccccc} (\mathbb{C}^2,\Pi',0) & \xleftarrow{\ proj\ } & (X^*,\Pi^*,p) & \xrightarrow{\ e\ } & (X,\Pi,0) \\ & \searrow^{\ell'} & \big\downarrow{\scriptstyle \tilde{\ell}} & & \big\downarrow{\scriptstyle \ell} \\ & & (Y,\Delta',q) & \xrightarrow{\ e_0\ } & (\mathbb{C}^2,\Delta,0) \end{array}$$

Let $(\Pi_0,0)$ be the part of $(\Pi,0)$ which is tangent to the $x-$axis (i.e., it corresponds to $p \in e^{-1}(0)$ in our chosen coordinates) and let $(\Delta_0,0)$ be its image by ℓ. Let V be a cone around the x-axis in $(\mathbb{C}^3,0)$. As in [4], consider a carrousel decomposition of $(\ell(V), 0)$ with respect to the curve germ $(\Delta_0, 0)$ such that the Δ-wedges around Δ_0 are D-pieces. We then consider the geometric decomposition of $(V, 0)$ into A-, B- and D-pieces obtained by lifting by ℓ this decomposition. Lifting the carrousel decomposition of $\ell(V)$ by e_0 we get a carrousel decomposition of (Y, q) with respect to Δ'. On the other hand the lifting by e of the geometric decomposition of V is a geometric decomposition of (X^*, p) which coincides with the lifting by $\tilde{\ell}$ of the carrousel decomposition of (Y, q) just defined.

By the Lê Swing Lemma [8, Lemma 2.4.7], the union of pieces beyond the first Puiseux exponents of the branches of Δ' at q lift to pieces in X^* which have trivial topology, i.e., their links are solid tori. Therefore these are absorbed by the amalgamation process consisting of amalgamating iteratively any D-piece which is not a conical piece with the neighbor piece using [4, Lemma 13.1].

Moreover, since Δ' is the strict transform of Δ by e_0, the rate of each piece of the obtained decomposition of X^* equals $q+1$, where q is the first Puiseux exponent of a branch of Δ'. Let Γ_p be the minimal resolution graph of the curve $h = 0$ at p. Let us call a *node* of Γ_p any vertex having at least three incident edges including the arrows representing the components of h and the root vertex of Γ_p if $h = 0$ has more than one line in its tangent cone. According to [8, Théorème C], the rate q equals the polar quotient

$$\frac{m_{E_i}(l)}{m_{E_i}(h)}$$

where v_i is the corresponding node in Γ_p and where $l\colon (\mathbb{C}^2_{v,w}, p) \to (\mathbb{C}, 0)$ is a generic linear form at p.

Now, set $\tilde{f}(v, w) = f(1, v, w)$. Since $g(1, v, w)$ is a unit at p, the curves $h = 0$ and $\tilde{f} = 0$ coincide, so $m_{E_i}(h) = m_{E_i}(\tilde{f})$. Since the strict transform of $\tilde{f}$ coincides with the germ of $\mathcal{L}$-curves at p, Γ_p is a connected component of Γ minus its $\mathcal{L}$-nodes with free edges replaced by arrows. Therefore the rates $\frac{m_{E_i}(l)}{m_{E_i}(\tilde{f})}$, and then the inner rate of $(X, 0)$ are computed from Γ. □

Example 3.2. Consider again the SIS $(X, 0)$ of Example 3.1 with equation $xz^2 + y^3 - x^4 = 0$. Its projectivized tangent cone $xz^2 + y^3 = 0$ has a unique singular point, and the corresponding graph Γ_p is the resolution graph of the cusp $w^2 + v^3 = 0$, i.e., the graph Γ of Example 3.1 with the $\mathcal{L}$-node replaced by an arrow. The multiplicity of $\tilde{f}$ along the curve E_3 corresponding to the node of Γ_p equals 6 while that of a generic linear form $(v, w) \mapsto l(v, w)$ equals 2. We then obtain the polar quotient $\frac{m_{E_3}(l)}{m_{E_3}(\tilde{f})} = 1/3$, which gives inner rate $1/3 + 1 = 4/3$.

The Lipschitz inner geometry is then completely described (see [4, Section 15]) by the graph Γ completed by labeling its nodes by the inner rates of the corresponding geometric pieces:

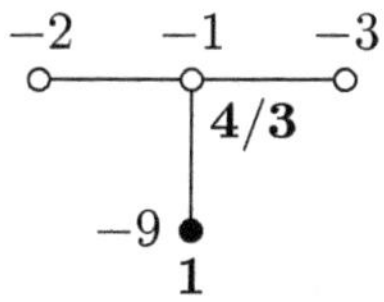

Example 3.3. Consider the SIS $(X, 0)$ with equation $(zx^2+y^3)(x^3+zy^2)+z^7=0$, that we already considered in [4, Example 15.2] and in [10]. The tangent cone consists of two unicuspidal curves C and C' with 6 intersecting points $p_1, \dots p_6$, the germ $(C \cup C', p_1)$ consisting of two transversal cusps, and the remaining 5 points being ordinary double points of $C \cup C'$.

For each $i = 1, \dots, 6$, the tangent cone of $(C \cup C', p_i)$ has two tangent lines and the quotient $m_{E_{v_0}}(l)/m_{E_{v_0}}(\tilde{f})$ at the root vertex v_0 of Γ_{p_i} is then a polar quotient in the sense of [8]. The root vertex v_0 has valency 2 and it corresponds to a special annular piece in the sense of [4], with inner rate $m_{E_{v_0}}(l)/m_{E_{v_0}}(\tilde{f})+1$. For $p_2, \dots, p_6$, we obtain inner rate $1/2+1 = 3/2$ for that special annular piece and for p_1, we obtain $1/4+1 = 5/4$. The inner rates at the two other nodes of Γ_{p_1} both equal $2/10+1 = 6/5$. We have thus recovered the inner geometry:

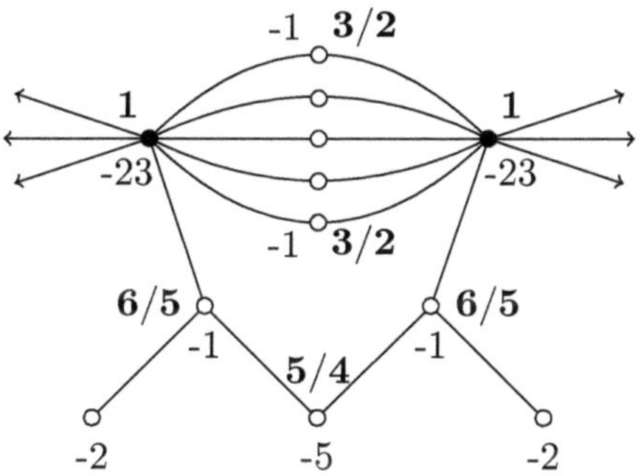

This was also computed in [4] with the help of Maple, in terms of the carrousel decomposition of the discriminant curve of a generic projection of $(X, 0)$.

Proof of Theorem 1.3 (ii). Consider the two SISs $(X_1, 0)$ and $(X_2, 0)$ with equations respectively:

$$X_1: \quad F_1(x,y,z) = (y^3 - z^2x)(y^3 + z^2x) + (x+y+z)^7 = 0,$$
$$X_2: \quad F_2(x,y,z) = (y^3 - z^2x)(y^3 + 2z^2x) + (x+y+z)^7 = 0.$$

We will prove that they have same inner geometry and different outer geometries.

On one hand, the projectivized tangent cones of $(X_1, 0)$ and $(X_2, 0)$ have same combinatorial type, so $(X_1, 0)$ and $(X_2, 0)$ have same Lipschitz inner geometry (Theorem 1.3). The tangent cone consists of two unicuspidal components C and C' with two intersection points: one, p_1, at the cusps, with maximal contact there, and one, p_2, at smooth points of C and C' intersecting with contact 3 there. The inner geometry is given by the following graph. In particular, the inner rates at the two non $\mathcal{L}$-nodes are computed from the corresponding polar rates in the two graphs Γ_{p_1} and Γ_{p_2}. They both equal $1/6 + 1 = 7/6$.

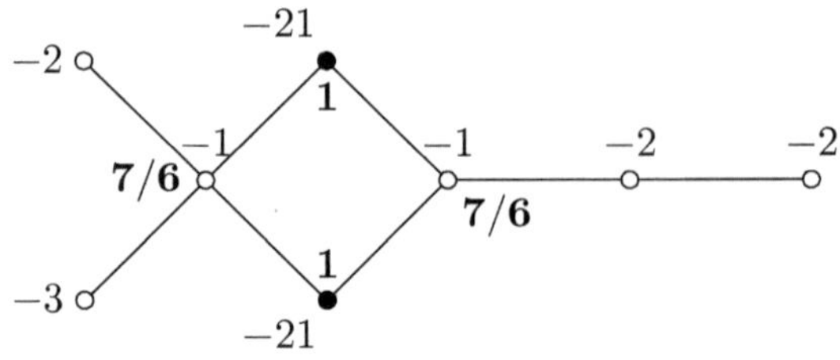

On the other hand, let us compute the multiplicities of the three functions x, y and z at each component of the exceptional locus. We obtain the following triples $(m_{E_j}(x), m_{E_j}(y), m_{E_j}(z))$ for both X_1 and X_2:

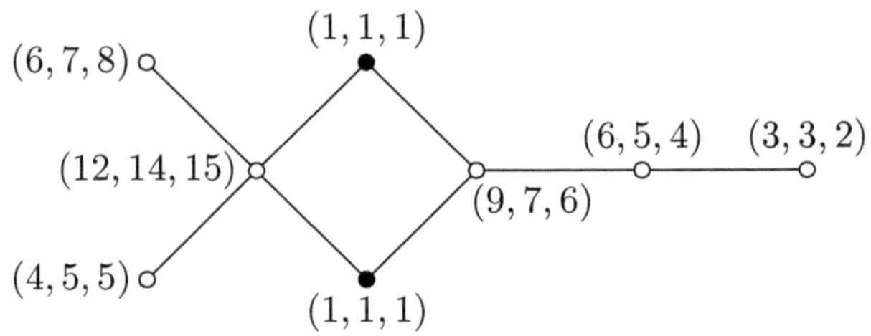

We compute from this the partial derivatives $\frac{\partial F_i}{\partial x}$, $\frac{\partial F_i}{\partial y}$ and $\frac{\partial F_i}{\partial z}$ along the curves of the exceptional divisor. We obtain different values for two multiplicities (in bold) for $(X_1, 0)$ and $(X_2, 0)$, written in that order on the graph:

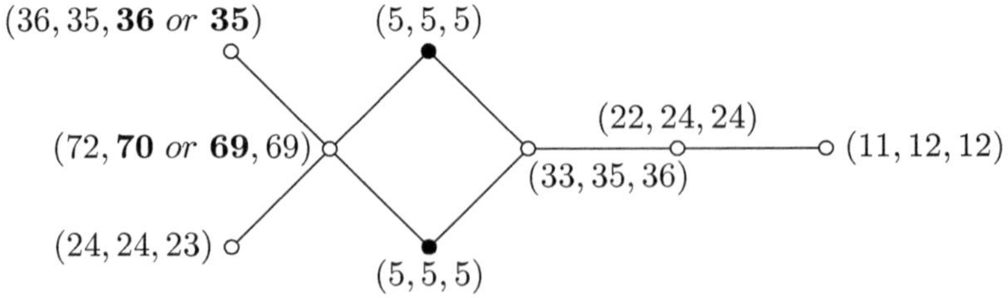

We compute from this the resolution graph of the family of polar curves $a\frac{\partial F_i}{\partial x} + b\frac{\partial F_i}{\partial y} + c\frac{\partial F_i}{\partial z} = 0$. In the X_1 case one has to blow up once more to resolve a basepoint. We then get the resolution graph of the polar curve of a generic plane projection of $(X_1, 0)$ resp. $(X_2, 0)$ (the arrows represent the strict transform, the numbers in parentheses are the multiplicities of the function $a\frac{\partial F_i}{\partial x} + b\frac{\partial F_i}{\partial y} + c\frac{\partial F_i}{\partial z}$ for generic a, b, c and the negative numbers are self-intersections):

$(X_1, 0)$

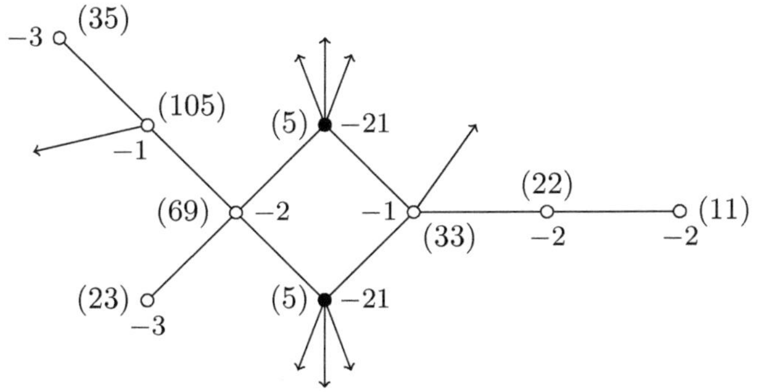

$(X_2, 0)$

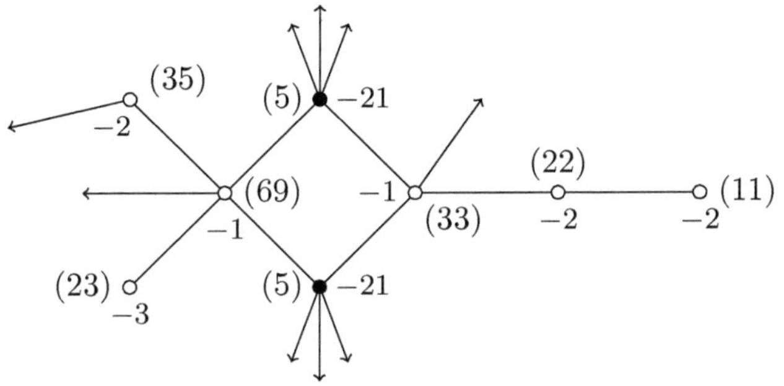

The polar curves of $(X_1, 0)$ and $(X_2, 0)$ have different Lipschitz geometry since they don't even have the same number of components. Therefore, by [10, Theorem 1.2 (6)], $(X_1, 0)$ and $(X_2, 0)$ have different outer Lipschitz geometries. □

4. The outer geometry of a superisolated singularity

Proof of Lemma 2.2. We first re-formulate the definition of weak RL-equivalence. We will use coordinates (v, w) in $\mathbb{C}^2$ and (x, y, z) in $\mathbb{C}^3$. We have functions $h_1(v, w)$ and $h_2(v, w)$ whose zero sets are the curves $(C_1, 0)$ and $(C_2, 0)$, a homeomorphism $\psi\colon (\mathbb{C}^2, 0) \to (\mathbb{C}^2, 0)$ of germs, a constant $K \geq 1$ and a neighborhood $\mathcal{U}$ of the origin in $\mathbb{C}^2$ such that for all $a, a' \in \mathcal{U}$.

$$\frac{1}{K}||h_2(\psi(a))(1,\psi(a)) - h_2(\psi(a'))(1,\psi(a'))||_{\mathbb{C}^3} \leq ||h_1(a)(1,a) - h_1(a')(1,a')||_{\mathbb{C}^3}$$
$$\leq K||h_2(\psi(a))(1,\psi(a)) - h_2(\psi(a'))(1,\psi(a'))||_{\mathbb{C}^3}.$$

For $i = 1, 2$ we define $H_i\colon (\mathbb{C}^2, 0) \to (\mathbb{C}^3, 0)$ by

$$H_i(v, w) = h_i(v, w)(1, v, w)$$

and denote by $(S_i, 0)$ the image of H_i in $(\mathbb{C}^3, 0)$. Note that H_i maps $(C_i, 0)$ to 0 and is otherwise injective. We can thus complete the maps ψ, H_1 and H_2 to a commutative diagram

$$\begin{array}{ccc} (\mathbb{C}^2,0) & \xrightarrow{H_1} & (S_1,0) \\ \Big\downarrow{\scriptstyle\psi} & & \Big\downarrow{\scriptstyle\psi'} \\ (\mathbb{C}^2,0) & \xrightarrow{H_2} & (S_2,0) \end{array} \qquad (\star)$$

and ψ' is bijective. Weak RL-equivalence is now the statement that ψ' is bilipschitz for the outer geometry.

Now write $h_1 = Uh'_1$ and $H_1 = UH'_1$ where $U = U(v,w) \in \mathbb{C}\{v,w\}$ is a unit. Then we obtain a commutative diagram

$$\begin{array}{ccc} (\mathbb{C}^2,0) & \xrightarrow{H'_1} & (S'_1,0) \\ \Big\| & & \Big\downarrow{\scriptstyle\eta} \\ (\mathbb{C}^2,0) & \xrightarrow{H_1} & (S_1,0) \end{array}$$

where η is $(x,y,z) \mapsto U(\frac{y}{x},\frac{z}{x})(x,y,z)$. The factor $U(\frac{y}{x},\frac{z}{x}) = U(v,w)$ has the form $\alpha_0 + \sum_{i,j\geq 0} \alpha_{ih} v^i w^j$ with $\alpha_0 \neq 0$ so if the neighborhood $\mathcal{U}$ is small then the factor is close to α_0, so η is bilipschitz. Thus $\psi' \circ \eta \colon (S'_1,0) \to (S_2,0)$ is bilipschitz, so we have shown that modifying h_1 by a unit does not affect weak RL-equivalence. The same holds for h_2, so weak RL-equivalence does not depend on the choice of defining functions for the curves $(C_1,0)$ and $(C_2,0)$.

It remains to show that analytic equivalence of $(C_1,0)$ and $(C_2,0)$ implies weak RL-equivalence. Analytic equivalence means that there exists a biholomorphic germ $\psi\colon (\mathbb{C}^2,0) \to (\mathbb{C}^2,0)$ and a unit $U \in \mathbb{C}\{v,w\}$ such that $Uh_1 = h_2 \circ \psi$. We have already dealt with multiplication with a unit, so we will assume we have $h_1 = h_2 \circ \psi$. If ψ is a linear change of coordinates, then we get a diagram as in $(\star)$ above, with ψ' given by the corresponding coordinate change in the y,z coordinates of $\mathbb{C}^3$, so ψ' is bilipschitz and we have weak RL-equivalence. For general ψ the same is true up to higher order in v and w, so we still get weak RL-equivalence. □

Proof of Proposition 2.3. Let $(X_1,0)$ and $(X_2,0)$ be two SISs with equations respectively

$$f_1(x,y,z) - g_1(x,y,z) = 0 \text{ and } f_2(x,y,z) - g_2(x,y,z) = 0,$$

where for $i = 1,2$, f_i and g_i are homogeneous polynomials of degrees d and $d+1$ respectively. We can assume that the projective line $x = 0$ does not contain any singular point of the projectivized tangent cones $C_1 = \{f_1 = 0\}$ and $C_2 = \{f_2 = 0\}$. We assume also that C_1 and C_2 have the same combinatorial types and that corresponding singular points of C_1 and C_2 are weak RL-equivalent.

Since the tangent cone of a SIS $(X,0)$ is reduced, the general hyperplane section of $(X,0)$ consists of smooth transversal lines. Therefore, adapting the arguments of [11, Section 4] by taking simply a line as test curve, we obtain that the inner and outer metrics are Lipschitz equivalent inside the conical part of $(X,0)$, i.e., outside cones around its exceptional lines. So we just have to control outer distance inside conical neighborhoods of the exceptional lines of $(X_1, 0)$ and $(X_2, 0)$ whose projective points are corresponding singular points of C_1 and C_2.

Let $p_1 \in Sing(C_1)$ and $p_2 \in Sing(C_2)$ be two singular points in correspondence. After modifying $(X_1, 0)$ and $(X_2, 0)$ by analytic isomorphisms, we can assume that $p_i = (1, 0, 0)$ for $i = 1, 2$. We use again the notations of the proof of Theorem 1.3, and we work in the chart $(x, v, w) = (x, y/x, z/x)$ for the blow-up e.

Set $h_i(v, w) = f_i(1, v, w)/g_i(1, v, w)$. Then the germs (X_i^*, p_i) have equations $h_i(v, w) + x = 0$.

Since C_1 and C_2 are weak RL-equivalent and $h_i = 0$ is an equation of C_i, there exists a local homeomorphism $\psi\colon (\mathbb{C}^2_{(v,w)}, 0) \to (\mathbb{C}^2_{(v,w)}, 0)$, a constant $K \geq 1$ and a neighborhood U of the origin in $\mathbb{C}^2$ such that for all $(v, w), (v', w') \in U$.

$$\begin{aligned}\frac{1}{K}||h_2(\psi(v,w))(1,\psi(v,w)) - h_2(\psi(v',w'))(1,\psi(v',w'))||_{\mathbb{C}^3} \leq \\ ||h_1(v,w)(1,v,w) - h_1(v',w')(1,v',w')||_{\mathbb{C}^3} \leq \qquad (*)\\ K||h_2(\psi(v,w))(1,\psi(v,w)) - h_2(\psi(v',w'))(1,\psi(v',w'))||_{\mathbb{C}^3}\end{aligned}$$

Locally,

$$X_1^* = \{x = h_1(v,w)\} \quad \text{and} \quad X_2^* = \{x = h_2(\psi(v,w))\}\,.$$

As in the proof of Theorem 1.3 we consider the isomorphisms $\mathrm{proj}_i\colon (X_i^*, p_i) \to (\mathbb{C}^2, 0)$ for $i = 1, 2$, the restrictions of the linear projections $(x, v, w) \mapsto (v, w)$. The composition $\mathrm{proj}_2^{-1} \circ \psi \circ \mathrm{proj}_1$ gives a local homeomorphism $\psi'\colon (W_1, p_1) \to (W_2, p_2)$, where W_i is an open neighborhood of p_i in X_i^*. Then, ψ' induces a local homeomorphism $\psi''\colon e(W_1) \to e(W_2)$ such that $\psi'' \circ e = e \circ \psi'$. Notice that each $e(W_i)$ contains the intersection of X_i with a cone in $(\mathbb{C}^3, 0)$ around the exceptional line represented by p_i.

Consider a pair of points $q = (x, xv, xw)$ and $q' = (x', x'v', x'w')$ in $e(W_1)$. By definition of ψ'', we have

$$||q - q'|| = ||h_1(v,w)(1,v,w) - h_1(v',w')(1,v',w')||_{\mathbb{C}^3},$$

$$||\psi''(q) - \psi''(q')|| = ||h_2(\psi(v,w))(1,\psi(v,w)) - h_2(\psi(v',w'))(1,\psi(v',w'))||_{\mathbb{C}^3}.$$

Then $(*)$ implies that the ratio $\frac{||\psi''(q)-\psi''(q')||}{||q-q'||}$ is bounded above and below in a neighborhood of the origin.

Now let $\widetilde{W}_i$ be the union of the W_i's and let $\psi'\colon \widetilde{W}_1 \to \widetilde{W}_2$ be the homeomorphism whose restriction to each W_1 is the local ψ'. Then $\psi''\colon e(\widetilde{W}_1) \to e(\widetilde{W}_2)$ is the outer bilipschitz homeomorphism induced by ψ' and we must extend ψ'' over all of X_1.

Let B be a Milnor ball for X_1 and X_2 around 0. We set

$$\widetilde{Y}_i = \overline{(e^{-1}(B \cap X_i) \setminus \widetilde{W}_i}.$$

For $i = 1, 2$ we can adjust $\widetilde{W}_i$ so that $\widetilde{Y}_i$ is a D^2-bundle over the exceptional divisor C_i minus its intersection with $\widetilde{W}_i$, i.e., over $\widetilde{C}_i := \overline{C_i \setminus \widetilde{W}_i}$, and whose fibers are curvettes of C_i. We want to extend $\psi''\colon e(\widetilde{W}_1) \to e(\widetilde{W}_2)$ to a bilipschitz map over the conical regions $e(\widetilde{Y}_1)$ and $e(\widetilde{Y}_2)$. For this it suffices to extend ψ' by a bundle isomorphism $\widetilde{Y}_1 \to \widetilde{Y}_2$, since the resulting $e(\widetilde{Y}_1) \to e(\widetilde{Y}_2)$ is then bilipschitz.

$(X_1, 0)$ and $(X_2, 0)$ are inner bilipschitz equivalent by Theorem 1.3), so by [4, 1.9 (2)] the image by ψ'' of the foliation of $e(\widetilde{W}_1)$ by Milnor fibers of a generic linear form ℓ_1 has the homotopy class of the corresponding foliation by fibers of ℓ_2 in $e(\widetilde{W}_2)$. Since the projectivized tangent cones C_1 and C_2 are reduced, a fiber of $\ell_i \circ e$ intersects each D^2-fiber over $\partial\widetilde{C}_i$ in one point. This gives a trivialization of the D^2-bundle over each $\partial\widetilde{C}_i$ and therefore determines a relative Chern class for each component of the bundle $\widetilde{Y}_i$ over $\widetilde{C}_i$. The map ψ' restricted to the bundle over $\partial\widetilde{C}_1$ extends to bundle isomorphisms between the components of $\widetilde{Y}_1$ and $\widetilde{Y}_2$ if and only if their relative Chern classes agree. But for $i = 1, 2$ these relative Chern classes are given by the negative of the number of intersection points of ℓ_i^* with each component of C_i (i.e., the degrees of these components of C_i), and these degrees agree since C_1 and C_2 are combinatorially equivalent.

We have now constructed a map $\psi''\colon (X_1, 0) \to (X_2, 0)$ which is outer bilipschitz if we restrict to distance between pairs of points x, y which are either both in a single component of $e(\widetilde{W}_1)$ or both in the conical region $e(\widetilde{Y}_1)$. Let $N\widetilde{Y}_i$ be a larger version of the bundle $\widetilde{Y}_i$, so $e(N\widetilde{Y}_i)$ is a conical neighborhood of $e(\widetilde{Y}_i)$. We still have an outer bilipschitz constant for ψ'' for any x and y which are both in a single component of $e(\widetilde{W}_1)$ or both in the conical region $e(N\widetilde{Y}_1)$. Otherwise, either one of x, y is in $e(\widetilde{W}_1) \setminus e(N\widetilde{Y}_1)$ and the other in $e(\widetilde{Y}_1)$ or x and y are in different components of $e(\widetilde{W}_1)$. The ratio of inner to outer distance is clearly bounded for such point pairs, so since ψ'' is inner bilipschitz, it is outer bilipschitz. $\square$

References

[1] Enrique Artal Bartolo, Forme de Jordan de la monodromie des singularités superisolées de surfaces, Mem. Amer. Math. Soc. **109** (1994), no. 525.

[2] Enrique Artal Bartolo, Ignacio Luengo and Alejandro Melle Hernández, Superisolated surface singularities, Singularities and computer algebra, 1339, London Math. Soc. Lecture Note Ser., 324, Cambridge Univ. Press, Cambridge, 2006.

[3] Enrique Artal Bartolo, José Ignacio Cogolludo and Hiro-o Tokunaga, A survey on Zariski pairs, Algebraic geometry in East Asia—Hanoi 2005, 1–100, Adv. Stud. Pure Math. **50**, Math. Soc. Japan, Tokyo, 2008.

[4] Lev Birbrair, Walter D Neumann and Anne Pichon, The thick-thin decomposition and the bilipschitz classification of normal surface singularities, Acta Math. **212** (2014), 199–256.

[5] Alexandre Fernandes. Topological equivalence of complex curves and bi-Lipschitz maps, Michigan Math. J. **51** (2003), 593–606.

[6] Abramo Hefez and Marcelo Escudeiro Hernandes, The analytic classification of plane branches. Bull. Lond. Math. Soc. **43** (2011), no. 2, 289–298.

[7] A Libgober, Alexander polynomials of plane algebraic curves and cyclic multiple planes, Duke Math. J. **49** (1982), 833–851.

[8] Lê Dũng Tráng, Françoise Michel and Claude Weber, Courbes polaires et topologie des courbes planes, Ann. Sci. École Norm. Sup. **24** (1991), 141–169.

[9] Walter D Neumann and Anne Pichon, Lipschitz geometry of complex curves, Journal of Singularities volume **10** (2014), 225-234.

[10] Walter D Neumann and Anne Pichon, Lipschitz geometry of complex surfaces: analytic invariants and equisingularity (2014), arXiv:1211.4897v2.

[11] Walter D Neumann, Helge Møller Pedersen and Anne Pichon, Minimal surface singularities are Lipschitz normally embedded (2015), 24 pages, arXiv:1503.03301.

[12] Frédéric Pham and Bernard Teissier, Fractions Lipschitziennes d'une algebre analytique complexe et saturation de Zariski. Prépublications Ecole Polytechnique No. M17.0669 (1969). Available at http://hal.archives-ouvertes.fr/hal-00384928/fr/

[13] A. N. Varchenko, Algebro-Geometrical Equisingularity and Local Topological Classification of Smooth Mappings, Proc. Internat. Congress of Mathematicians (Vancouver, B.C., 1974), Vol. 1, pp. 427–431. Canad. Math. Congress, Montreal, Que., 1975.

[14] Oscar Zariski, On the problem of existence of algebraic functions of two variables possessing a given branch curve, Amer. J. Math. **51** (1929), 305–328.

[15] Oscar Zariski, The topological discriminant genus p, Amer. J. Math. **59** (1937), 335–358.

[16] Oscar Zariski, The moduli problem for plane branches, with an appendix by Bernard Teissier, Amer. Math. Soc., University Lecture Series 39 (2006) [translated from original French edition of 1973].

Walter D. Neumann
Department of Mathematics, Barnard College, Columbia University, 2009 Broadway MC4429, New York, NY 10027, USA
e-mail: `neumann@math.columbia.edu`

Anne Pichon
Aix Marseille Université, CNRS, Centrale Marseille, I2M, UMR 7373, 13453 Marseille, France
e-mail: `anne.pichon@univ-amu.fr`

Projective Transverse Structures for Some Foliations

Paulo Sad

A Pepe, com admiração e amizade

Abstract. We construct examples of regular foliations of holomorphic surfaces which are generically transverse to a compact curve and have a projective transverse structure.

Mathematics Subject Classification (2000). Primary 37; Secondary 37F75.

Keywords. Holomorphic foliations, transverse structures, conormal bundles.

1. Introduction

Let us consider a codimension 1 holomorphic foliation $\mathcal{F}$ of a complex manifold M. A (singular) projective transverse structure for $\mathcal{F}$ is defined by the following data:

1. a covering of the complement M^* of a finite set of embedded leaves by open sets $\{U_i\}$; in each U_i there is a trivialization of the foliation.
2. a collection $\{f_i\}$ of holomorphic functions $f_i : U_i \to \bar{\mathbb{C}}$ which are first integrals of $\mathcal{F}$ in each U_i and a collection of Moebius transformations $\{\phi_{ij}\}$ such that $f_i = \phi_{ij} \circ f_j$ whenever $U_i \cap U_j \neq \emptyset$.

In general, the map ϕ_{ij} which relates f_i to f_j is simply a diffeomorphism between open sets of $\bar{\mathbb{C}}$.

The presence of projective transverse structure leads to the existence of a multivalued first integral of $\mathcal{F}$ in M^*; we start with some open set, say U_1, and extend f_1 along paths starting at U_1 just by composing it with a convenient choice of functions ϕ_{ij}.

Let us present some examples.

Example 1.1. A classical example comes from projective structures on Riemann surfaces. Let C be a compact Riemann surface with an atlas of coordinate charts such that all changes of coordinates are given by elements of $\mathrm{PSL}(2,\bar{\mathbb{C}})$ (Moebius transformations of $\bar{\mathbb{C}}$). After applying the process mentioned above we get a developing map $\mathcal{D}: \tilde{C} \to \bar{\mathbb{C}}$ where $\tilde{C}$ is the universal covering of C ($\tilde{C} = \mathbb{D}$ when the genus of C is greater than 1 and $\tilde{C} = \mathbb{C}$ when the genus of C is 1) and a monodromy representation $\rho : \pi_1(C) \to \mathrm{PSL}(2,\bar{\mathbb{C}})$ which are related by the equality $\mathcal{D}(\gamma(p)) = \rho(\gamma)\mathcal{D}(p)$, for $p \in \tilde{C}$ and $\gamma \in \pi_1(C)$ view as a deck transformation.

Then $\pi_1(C)$ acts on $\tilde{C} \times \bar{\mathbb{C}}$: to each $\gamma \in \pi_1(C)$ we associate the map $(p, z) \mapsto (\gamma(p), \rho(\gamma)(z))$. The action preserves the horizontal and vertical fibrations of $\tilde{C} \times \bar{\mathbb{C}}$; therefore, the space of orbits of the action is a compact surface that has a rational fibration over C (coming from the vertical fibration) and a transversely projective foliation $\mathcal{F}$ (coming from the horizontal fibration). The action preserves also the graph $\{(p, \mathcal{D}(p)); p \in \tilde{C}\}$ of $\mathcal{D}$ (because $(p, \mathcal{D}(p)) \mapsto (\gamma(p), \rho(\gamma)\mathcal{D}(p)) = (\gamma(p), \mathcal{D}(\gamma(p)))$, which becames in the quotient a section of the rational fibration, generically transverse to $\mathcal{F}$. The foliation induces in this section the same projective structure of C.

We may adapt this construction to some cases where the projective structure of C has a number of singularities (for example, multivalued maps of the form $z \mapsto z^\alpha$). The local monodromy around the singularity has to be realized as the monodromy of a foliation of $\mathbb{D} \times \bar{\mathbb{C}}$ which is transverse to the fibers $\{z\} \times \bar{\mathbb{C}}$ for $z \neq 0$ and has $\{0\} \times \bar{\mathbb{C}}$ as a leaf; then we glue this foliation with the one constructed as before outside the singularities.

Example 1.2. We mentioned in the last example the difficulties that arise in the presence of singularities of the projective structure. There is a related construction that avoids this problem by using "pre-integration" data. We take a line bundle L over a compact, holomorphic curve C. In some covering $\mathcal{U} = \{U_i\}$ of C we write $\{\lambda_{ij}\}$ for the transition functions of L and take trivializations (x_i, z_i) of L with $z_i = \lambda_{ij} z_j$. Let now $\{\omega_i\}$, $\{\eta_i\}$ and $\{\xi_i\}$ be meromorphic 1-forms defined in the open sets U_i satisfying $\omega_i = \lambda_{ij}\omega_j$, $\eta_i - \eta_j = d\log \lambda_{ij}$ and $\xi_j = \lambda_{ij}\xi_i$. We notice that the equations $dz_i = z_i\eta_i + \frac{z_i^2}{2}\xi_i + \omega_i$ define a foliation $\mathcal{G}$ in L. In fact, we may compactify L as a $\bar{\mathbb{C}}$ over C and extend the foliation to the closure of L. Except by the fibers over some pole (which are leaves), the other fibers are transverse to the leaves. This implies the existence of a transverse projective structure for $\mathcal{G}$ outside the vertical leaves. These are called Ricatti foliations (see [6]).

We may think of course that the pre-integration data produce a (singular) projective structure for the curve C; we explain in Section 2 how this works.

In this paper we combine in Theorem 3.1 features of Examples 1.1 and 1.2. We start with pre-integration data in a curve and obtain a (singular) projective structure. Then we embed the curve into a surface which comes with a regular foliation generically transverse to it, with the aditional property that the structure of the curve can be extended to the surface along the leaves of the foliation. We get many more examples, at the price of not having a compact surface and no fibration over the curve. It is worth noticing that we also describe the singularities of the projective structure.

Example 1.3. Let us consider a codimension 1 foliation defined by some integrable holomorphic 1-form ω. We suppose that this 1-form is completed to a triplet of holomorphic 1-forms with η and ξ such that $d\omega = \eta \wedge \omega$, $d\eta = \omega \wedge \xi$ and $d\xi = \xi \wedge \eta$. According to Darboux (see [5]), we may write locally triplets of holomorphic functions (f, g, h) such that

$$\omega = -gdf, \ \eta = \frac{dg}{g} + h\,\omega, \ \xi = dh + h\,\eta + \frac{h^2}{2}\omega.$$

Furthermore, if $(\bar{f}, \bar{g}, \bar{h})$ is another triplet of functions satisfying the same relations then $f = \phi \circ \bar{f}$ for some Moebius transformation ϕ. This implies that the foliation has a projective transverse structure.

In general, we may work with a triplet of meromorphic 1-forms, and the projective structure is singular. Many authors take the existence of such a triplet as the definition of a transversely projective foliation (see for example [6]). Of course a true transverse projective structure appears for the foliation outside the poles of the 1-forms; but it is not clear if we can produce a triplet out of a transverse projective structure defined outside a divisor. We treat this question for the examples constructed in Theorem 1.3, where we know the nature of the singularities. We give in Section 4 an answer adding a negativity hypothesis for the embedding of the curve. We do not know if this is true for the other cases.

Example 1.4. This is an example of a different nature, but it shows how foliations with transverse projective structures appear quite naturally. Let K be a radial type Kupka component of a codimension 1 foliation in some complex manifold of dimension $n \geq 3$. This means that K is covered by open sets $\{U_i\}$ such that in each U_i the foliation is conjugated by a diffeomorphism Θ_i to the foliation $x_i\, dy_i - y_i d\, x_i = 0$ of $\mathbb{D}^{n-2} \times \mathbb{D}^2$ and $\Theta_i(K \cap U_i) = \mathbb{D}^{n-2} \times \{(0,0)\}$. Now we take the function $f_i = \dfrac{y_i}{x_i} \circ \Theta_i$ defined in $U_i \setminus K$ as a first integral for the foliation in U_i. If $U_i \cap U_j \neq \emptyset$ and $f_i = \phi_{ij} \circ f_j$, we see that ϕ_{ij} is a Moebius transformation because f_i and f_j are surjective onto $\bar{\mathbb{C}}$ (see [4]).

There is a lot of important work done the subject of transversely projective foliations; let us mention [1], [2] and [3] where the structure of these foliations is discussed.

We follow all the time part of the presentation of [4], which relies upon the paper [5]. We are grateful to J.V. Pereira for helping to establish the right setting of our construction. We are also grateful to the referee for valuable comments.

2. Projective Structures for Curves

Let us consider a line bundle L over a compact Riemann surface C; we cover C by a family $\mathcal{U} = \{U_i\}$ of open sets, and write $\{\lambda_{ij}\} \in H^1(\mathcal{U}, \mathcal{O}_C^*)$ for the transition functions of L; the notation $\mathcal{O}_C^*$ stands for the sheaf of non vanishing germs of holomorphic functions of C. We will also write $\mathcal{M}_C$ for the sheaf of germs of meromorphic functions of C and $\mathcal{M}_C^1$ for the sheaf of germs of meromorphic 1-forms of C.

We study now (singular) projective structures of C. We take meromorphic 1-forms $\omega = \{\omega_i\} \in H^0(\mathcal{U}, \mathcal{M}_C^1 \otimes L)$ (that is, $\omega_i = \lambda_{ij}\,\omega_j$ whenever $U_i \cap U_j \neq \emptyset$) and meromorphic 1-forms $\eta = \{\eta_i\} \in C^0(\mathcal{U}, \mathcal{M}_C^1)$ such that $\eta_i - \eta_j = d\log\lambda_{ij}$. We also select meromorphic functions $h = \{h_i\} \in C^0(\mathcal{U}, \mathcal{M})$. Proceeding formally, we define $f = \{f_i\}$ as the solution of the system

$$d\log g_i = \eta_i + h_i\omega_i \tag{2.1}$$

$$\omega_i = -g_i\, df_i \tag{2.2}$$

The elements of f will be taken as coordinate charts for C deprived of a finite set of singularities. We will impose later conditions on ω, η and h in order to assure the existence of f and be able to describe its singularities. For the moment we will compare the elements of f assuming that they are difeomorphisms over open sets of $\bar{\mathbb{C}}$; let $f_i = \phi_{ij}(f_j)$, the functions ϕ_{ij} being difeomorphisms between open sets of $\bar{\mathbb{C}}$. Let us write $\psi_{ij}(t) = d\log\phi'_{ij}(t)$.

Lemma 2.1. *$\psi_{ij}\, df_j = h_i\,\omega_i - h_j\,\omega_j$ in $U_i \cap U_j$.*

Proof. Since $f_i = \phi_{ij}(f_j)$, we have $df_i = \phi'_{ij}(f_j)\, df_j$. From equation (2.2) it follows that $\omega_i/g_i = \phi'_{ij}(f_j)\,\omega_j/g_j$ and therefore $\lambda_{ij}g_j/g_i = \phi'_{ij}(f_j)$. We apply then $d\log$ to both sides and use equation (2.2).

We remark that if $h = 0$, that is, $h_i = 0$ for every i, then the coordinate charts give a (singular) affine structure for C.

Now we will make another choice for h; we consider a collection $\xi = \{\xi_i\} \in C^0(\mathcal{U}, \mathcal{M}_C^1)$ of meromorphic 1-forms and define

$$dh_i + h_i\,\eta_i + \frac{h_i^2}{2}\omega_i = \xi_i. \tag{2.3}$$

Remind that the Schwarz derivative of a holomorphic function l is given by

$$S(l)(t) = \psi'(t) - \frac{\psi(t)^2}{2}$$

where $\psi(t) = d\log l'$.

Lemma 2.2. *$\dfrac{1}{g_j}S(\phi_{ij})(f_j)df_j = \lambda_{ij}\xi_i - \xi_j$ in $U_i \cap U_j$.*

Proof. We have from Lemma 2.1:

$$\phi_{ij}(f_j) = -g_j(h_j - \lambda_{ij}\, h_i).$$

It follows that

$$\frac{1}{g_j}\psi'_{ij}(f_j)df_j = -\xi_j - \frac{h_j^2}{2}\omega_j + \lambda_{ij}(dh_i + h_i\eta_i) + \lambda_{ij}h_ih_j\omega_j$$

and

$$\frac{1}{g_j}\frac{\psi(f_j)^2}{2}\,df_j = -\frac{h_j^2}{2}\omega_j - \lambda_{ij}\frac{h_i^2}{2}\omega_i + \lambda_{ij}h_i\,h_j\omega_j.$$

Consequently the Lemma follows. □

The case which will be of interest for us is when $\xi_j = \lambda_{ij}\xi_i$ whenever $U_i \cap U_j \neq \emptyset$, that is, $\xi \in H^0(\mathcal{U}, \mathcal{M}^1_C \otimes L^*)$. Lemma 2.2 implies that all ϕ_i are Moebius transformations of $\bar{\mathbb{C}}$, so that the collection $f = \{f_i\}$ provides a (singular) projective structure for C.

We proceed now to introduce conditions on $\omega \in H^0(\mathcal{U}, \mathcal{M}^1_C \otimes L)$, $\eta = \{\eta_i\} \in C^0(\mathcal{U}, \mathcal{M}^1_C)$ such that $\eta_i - \eta_j = d\log\lambda_{ij}$ and $\xi \in H^0(\mathcal{U}, \mathcal{M}^1_C \otimes L^*)$ which allow us to analyse more carefully the family f.

We will use $Z(\cdot)$ and $P(\cdot)$ for the set of zeros or poles of a 1-form.

Lemma 2.3. *There exist* 1*-forms as above such that all poles are simple, there are no common poles for any pair of* 1*-forms and*

- $Z(\omega) \neq \emptyset$,
- $Z(\omega) \cap (P(\eta) \cup P(\xi)) = \emptyset$.

Proof. We will use here classical results of Complex Analysis (in the case of curves) such as Riemann-Roch's Theorem and the existence of meromorphic 1-forms with pre-assigned polar parts.

1) Let us start with some $\bar{\omega} \in H^0(\mathcal{U}, \mathcal{M}^1_C \otimes L)$. We denote by $q_1, \ldots, q_s$ the poles of $\bar{\omega}$, $m_1, \ldots, m_s$ being their polar orders. We select a disjoint set of points $p_1, \ldots, p_r, p_{r+1}$ where $\bar{\omega}$ is regular and non vanishing and look for a meromorphic function l of C such that

$$(l) \geq p_{r+1} - \sum p_j + \sum m_i q_i =: -D.$$

The vector space $\mathcal{L}(D)$ of meromorphic functions of C whose divisor is greater or equal to $-D$ has dimension $l(D) \geq \deg(D) - g + 1$, where g is the genus of C. Therefore $l(D) \geq r - \sum m_i - g$ is positive for large r; we take $l \in \mathcal{L}(D)$. Then $\omega =: l\bar{\omega} \in H^0(\mathcal{U}, \mathcal{M}^1_C)$ has certainly a zero at the point p_{r+1} and all possible poles are at the points $p_1, \ldots p_r$ (with order at most 1).

2) A 1-form $\bar{\eta} = \{\bar{\eta}_i\}$ with the property $\bar{\eta}_i - \bar{\eta}_j = d\log\lambda_{ij}$ whenever $U_i \cap U_j \neq \emptyset$ can be obtained as $\bar{\eta}_i = \dfrac{du_i}{u_i}$ where $u = \{u_i\}$ is a meromorphic section of L. We can add to $\bar{\eta}$ a meromorphic 1-form θ of C, which is interesting if we are willing to move poles of $\bar{\eta}$. Notice firstly that $\bar{\eta}$ has simple poles, with residues α_i; of course $\sum \alpha_i = c(L)$, where $c(L)$ is the Chern class of L. We select θ with simple poles at the same points but with residues $-\alpha_i$, besides other poles which can be taken as simple poles with residues μ_k satisfying $-\sum \alpha_i + \sum \mu_k = 0$, or $\sum \mu_k = c(L)$. We ask also $P(\theta) \cap (Z(\omega) \cup P(\omega)) = \emptyset$; finally we take $\eta = \bar{\eta} + \theta$.

3) We apply to some $\bar{\xi} \in H^0(\mathcal{U}, \mathcal{M}^1_C \otimes L^*)$ the same technique as in step 1) of this proof. □

We remark that the only restriction to the residues μ_k in the choice of θ is that $\sum \mu_k = c(L)$. We will demand that $Im\mu_k \neq 0$ for any k.

Lemma 2.4. *Equation* (2.3) *has always meromorphic solutions.*

Proof. First of all we remark that if ω_i, η_i and ξ_i are holomorphic then there exists a (unique) holomorphic solution for a given initial condition $h_i(0)$.We have to deal with the case where some (simple) pole appears.

1. ω_i has a pole. Let us write (2.3) as $y' + b(x)y + a(x)y^2 = c(x)$, with $a(x) = \frac{A(x)}{x}$); $A(x)$ is a holomorphic function and $A(0) \neq 0$. This equation can be written as the system
$$\frac{dx}{dt} = x, \quad \frac{dy}{dt} = xc(x) - (xb(x) + A(x)y)y.$$
Therefore $(0, 0)$ is a saddle-node with a strong separatrix $x \to (x, y_0(x))$ transverse to the vertical line $x = 0$ (which is the weak separatrix). We take then $x \to y_0(x)$ as our solution.
2. η_i has a pole. We write (2.3) as $y' + (\frac{\mu}{x} + B(x))y + a(x)y^2 = c(x)$, where $B(x)$ is a holomorphic function and $\text{Im}(\mu) \neq 0$. The corresponding system is
$$\frac{dx}{dt} = x, \quad \frac{dy}{dt} = xc(x) - (\mu + xB(x))y - xa(x)y^2$$
which has a (unique) non-vertical separatrix $x \to (x, y_1(x))$ at $(0, 0)$. We select $x \to y_1(x)$ as the solution for (2.3).
3. ξ_i has a pole. We write (2.3) as $y' + b(x)y + a(x)y^2 = \frac{C(x)}{x}$; $C(x)$ is a holomorphic function and $C(0) \neq 0$. The corresponding system is
$$\frac{dx}{dt} = x, \quad \frac{dy}{dt} = C(x) - (b(x)x + xa(x))y.$$
We look to this system in a neighborhood of $(0, \infty)$; using $u = y^{-1}$ we get
$$\frac{dx}{dt} = x, \quad \frac{du}{dt} = -uxb(x) + xa(x) + C(x)u^2.$$
This is a saddle-node at $(x, u) = (0, 0)$, with a non-vertical strong separatrix $x \to (x, u_0(x))$ in the coordinates (x, u). We take then as solution $x \to u_0^{-1}(x)$; it is meromorphic with a simple pole at $0 \in \mathbb{C}$.

This ends the proof of the Lemma. □

Lemma 2.4 allows us to describe the singularities of the functions in $f = \{f_i\}$. We have the following cases:

- at a zero of ω_i: let us look to the equation (2.1); since $\eta_i + h_i\omega_i$ is holomorphic, we can say the same for $g_i(x_i) = g_i(0)e^{\int_0^{x_i} \eta_i + h_i\omega_i}$ (x_i is a coordinate in U_i and $g_i(0) \neq 0$). From equation (2.2) it follows that f_i is a holomorphic function of the type $f_i(x_i) = a + x_i^m v(x_i)$ where $v(x_i)$ is a non-vanishing holomorphic function and $m \in \mathbb{N}$ is greater or equal to 2.

- at a pole of ω_i: by our construction $h_i\omega_i$ is holomorhic, and so is $g_i(x_i) = g_i(0)e^{\int_0^{x_i} \eta_i + h_i\omega_i}$. Equation (2.2) gives $f_i(x_i) = a\log x_i + \beta(x_i)$, for $a \in \mathbb{C}$ and $\beta(x_i)$ holomorphic.
- at a pole of η_i: Equation (2.1) gives $log\, g_i = \mu \log x_i + \gamma(x_i)$, so $g_i(x_i) = Cx^{\mu}e^{\int_0^{x_i}}$ and $f_i'(x_i) = x_i^{-\mu}\bar{\gamma}(x_i)$.
- at a pole of ξ_i: now $h_i\omega_i$ has a simple pole. From (2.2) we have $g_i(x_i) = x_i^r\bar{\delta}(x_i)$ for $r \in \mathbb{C}$ and $\bar{\delta}(x_i)$ holomorphic. Consequently $f_i'(x_i) = x_i^{-r}\delta(x_i)$, for $\delta(x_i)$ holomorphic.

We remark that at the singularities of f we should take convenient sectors in order to have well defined branches for each f_i. Another remark that we shall use later is: suppose $c_1(L) \leq g - 1$, where g is the genus of C; then the 1-form ξ can be chosen as a holomorphic 1-form. This is another simple consequence of Riemann-Roch's theorem.

3. Constructing Foliations

We give in Theorem 3.1 below a simple construction of foliations which have a projective transverse structure.

Before doing this, let us remark the following facts. Consider some ω_i in the collection ω with a zero of order k_i, and take the Equation (2.2) $\omega_i = -g_i\, df_i$. The 1-form $-\dfrac{\omega_i}{g_i}$ can be written as $x_i^{k_i}\, a(x_i)dx_i$ where $a(x_i)$ is a nonvanishing holomorphic function. It can be easily shown that there exists a holomorphic local diffeomorphism s_i such that $s_i(0) = 0$, $s_i^{k_i+1} = s_i \circ s_i \circ \cdots \circ s_i = \mathrm{Id}$ (the composition is taken $k_i{+}1$ times) and $f_i \circ s_i = f_i$ (or $s_i^*(df_i) = df_i$). The periodic diffeomorphism s_i is conjugated to the linear rotation $l_i(x_i) = s_i'(0)x_i$ via a diffeomorphism $\phi_i : (\mathbb{D}, 0) \to (\mathbb{D}, 0)$: $\phi_i^*\, s_i = l_i$ $\phi_i'(0) = 1$; furthermore, $(\phi_i^* f_i) \circ l_i = \phi^* f_i$.

Theorem 3.1. *Let C be a smooth, compact, holomorphic curve and $n \in \mathbb{Z}$. Let L be a line bundle over C such that $c_1(L) = n$ and $\omega = \{\omega_i\} \in H^0(C, \mathcal{M}_C^1 \otimes L)$ with $Z(\omega) \neq \emptyset$ and simple poles. There exists a surface S and a embedding of C in S such that:*

1) *C has n as its self-intersection number in S;*
2) *C is generically transverse to a foliation of S which has a (singular) projective transverse structure;*
3) *C is tangent to this foliation only at the points of $Z(\omega)$.*

Proof. The total space of L total space is foliated by the fibers. We will modify this foliation in order to make it tangent to C at the points of $Z(\omega)$. From the analysis we did in Section 1, we have a (singular) projective structure for C given by a collection $f = \{f_i\}$. Let us take one of these elements, say f_1, defined in a neighborhood U_1 of a zero p_1 of order k_1 ($x_1(p_1) = 0$). Consider in a neighborhood of $(0, 0) \in \mathbb{C}^2$ the foliation $\mathcal{I}$ defined by the level curves of $(x, t) \to t - x^{k_1+1}$; we replace near p_1 the foliation by fibers of L by $\mathcal{I}$. This can be done as follows:

- we take an annulus $\mathbb{A} \subset \mathbb{C} \times \{0\}$ with center at $(0,0)$ and a small neighborhood V of this annulus in $\mathbb{C}^2$ saturated by the leaves of $\mathcal{I}$. Let R_V (respectively R) be a holomorphic vector field transverse to A (respect. transverse to $\phi_1(A)$) and tangent to the leaves of $\mathcal{I}$ (respec. tangent to the fibers of L); we denote by $R_V(.,.)$ and $R(.,.)$ their flows. Then V is diffeomorphic to a neighborhood W of $\phi_1(\mathbb{A})$ via a diffeomorphism Φ_1 such that $\Phi_1(R_V((x,0),T) = R((\phi_1(x),0),T)$, where $(x,0) \in \mathbb{A}$ and $T \in \mathbb{C}$ is small; clearly Φ extends ϕ.
- W is a subset of some neighborhood P of p_1 which is saturated by the fibers of L and is diffeomorphic to a polydisc (W itself is diffeomorphic to the product of an annulus by a disc). Similarly, V is a subset of a neighborhood of $(0,0)$ which is diffeomorphic to a polydisc Q to which we may restrict $\mathcal{I}$. Finally we remove all fibers of L which pass through $P \setminus W$ and add Q, using Φ_1 as the gluing map between V and W.

We see that the new surface is foliated with the fibers of L except at some neighborhood of p_1, where now there is a leaf tangent to this point with order k_1. The definition of Φ_1 guarantees that C has $c_1(L)$ as self-intersection number in this surface. Furthermore, f_1 has the same value at each point (close to p_1) of intersection of a leaf with C, which allows us to extend f_1 along the nearby leaves. We can repeat the same construction for any subset of $Z(\omega)$, getting a foliation $\mathcal{F}$. At the other points of C where no modification is made we simply extend the elements of f along the fibers of L. □

4. Projective transverse structures and 1-forms

As we have seen in the Introduction, a projective transverse structure for a codimension 1 foliation $\mathcal{G}$ of a manifold M may be obtained from the folowing data:

1. a covering $\bar{\mathcal{U}} = \{\bar{U}_i\}$ of M;
2. a line bundle $\bar{L} = \{\Lambda_{ij}\} \in H^1(\bar{\mathcal{U}}, \mathcal{O}^*_M)$;
3. a triplet of meromorphic 1-forms $\bar{\omega} = \{\bar{\omega}_i\}$, $\bar{\eta} = \{\bar{\eta}_i\}$ and $\bar{\xi} = \{\bar{\xi}_i\}$ defined in the open sets of $\bar{\mathcal{U}}$ such that $\bar{\omega}_i = 0$ defines $\mathcal{G}$ in each $\bar{U}_i$ and satisfy:

$$d\bar{\omega}_i = \bar{\eta}_i \wedge \bar{\omega}_i \ , d\bar{\eta}_i = \bar{\omega}_i \wedge \bar{\xi}_i \ , d\bar{\xi}_i = \bar{\xi}_i \wedge \bar{\eta}_i$$

and

$$\bar{\omega}_i = \Lambda_{ij}\bar{\omega}_j \ , \bar{\eta}_i - \bar{\eta}_j = d\log\Lambda_{ij} \ , \bar{\xi}_j = \Lambda_{ij}\bar{\xi}_i$$

in each $\bar{U}_i$ and $\bar{U}_i \cap \bar{U}_j$, respectively.

A similar procedure as in Section 1 yields a transverse projective structure for the foliation, at least outside zeroes and poles of the 1-forms. The question we address now is whether the transverse projective structure of the foliations constructed in Theorem 1 may be defined from a triplet of 1-forms as above.

Let us use the same construction and notation of Sections 1 and 2, but with a modification. We will deal here with the case $c_1(L) \leq g-1$, so we can choose $\xi \in H^0(\mathcal{U}, \Omega^1_C \otimes L^*)$ (that is, a holomorphic 1-form); in particular, all the functions of $h = \{h_i\}$ are holomorphic and vanish at the poles of ω and η. We introduce the following notation: if in the open set U_i there is a point of tangency of the foliation we put $\tilde{g}_i = g_i$; otherwise we put $\tilde{g}_i \equiv 1$. We replace then each ω_i by $\frac{\omega_i}{\tilde{g}_i}$; consequently we have to replace ξ_j by $\tilde{g}_j \xi_j$ and η_i by $\eta_i - d\log \tilde{g}_i$. We remark that all $\tilde{g}_i$ are non-vanishing holomorphic functions. The corresponding functions h_i and g_i are affected by these changes: they become respectively $\tilde{g}_i\, h_i$ and $\frac{g_i}{\tilde{g}_i}$, but the functions in the collection f remain the same. The transition functions for the line bundle L become $\frac{\tilde{g}_j}{\tilde{g}_i}\, \lambda_{ij}$. The implication is that in the setting of Theorem 1, we may assume that all 1-forms ω_i and all functions g_i and f_i can be extended to a neighborhood $\bar{U}_i$ of U_i in S along the leaves of $\mathcal{F}$ to $\bar{\omega}_i$, G_i and F_i. Our interest relies in fact in $\bar{\omega}_i$, $d\,G_i$ and $d\,F_i$.

Let us write $\bar{\omega}_i = \Lambda_{ij}\,\bar{\omega}_j$, $\bar{\omega}_i = -G_i\, dF_i$ and $\tilde{\eta}_i = \dfrac{d\,G_i}{G_i}$; it follows that $\tilde{\eta}_i - \tilde{\eta}_j = d\log \dfrac{G_i}{G_j} = d\log \Lambda_{ij} - d\log \dfrac{dF_i}{dF_j}$. this last quotient makes sense since the functions F_i are constant along the leaves of $\mathcal{F}$. As $F_i = \phi_{ij}(F_j)$, it follows that $\dfrac{dF_i}{dF_j} = {\phi_{ij}}'(F_j)$ and $\tilde{\eta}_i - \tilde{\eta}_j = d\log \Lambda_{ij} - \dfrac{\phi''_{ij}(F_j)}{\phi'_{ij}(F_j)}\, dF_j$, so finally

$$\tilde{\eta}_i - \tilde{\eta}_j = d\log \Lambda_{ij} - \psi_{ij}(F_j)dF_j.$$

Theorem 4.1. *Let $g \geq 1$ and suppose that the selfintersection number $C.C$ of C inside S satisfies $C.C < 2-2g$. Then there exists a meromorphic 1-form $\bar{\eta} = \{\bar{\eta}_i\}$ such that $d\bar{\omega}_i = \bar{\eta}_i \wedge \bar{\omega}_i$ for all $\bar{U}_i$ and $\bar{\eta}_i - \bar{\eta}_j = d\log \Lambda_{ij}$ whenever $\bar{U}_i \cap \bar{U}_j \neq \emptyset$.*

Proof. We start noticing that $c_1(L) < g-1$, so that we are in the setting above. We intend to write $\{\psi_{ij} dF_j\}$ as a special coboundary. We notice also that the 1-forms $h_i\omega_i$ defined in the open sets U_i are all holomorphic.

Let us look at the conormal bundle $N^*_{\mathcal{F}}$ of the foliation $\mathcal{F}$ and at the short exact sequence

$$0 \to \mathcal{I}_C.N^*_{\mathcal{F}} \to N^*_{\mathcal{F}} \to N^*_{\mathcal{F}}/\mathcal{I}_C.N^*_{\mathcal{F}} \to 0$$

in some neighborhood $\bar{U} \subset \cup \bar{U}_i$ of C; $\mathcal{I}_C$ is the ideal sheaf of C. We have the associated long exact sequence

$$\cdots \to H^1(\bar{U}, \mathcal{I}_C.N^*_{\mathcal{F}}) \to H^1(\bar{U}, N^*_{\mathcal{F}}) \to H^1(\bar{U}, N^*_{\mathcal{F}}/\mathcal{I}_C.N^*_{\mathcal{F}}) \to \ldots$$

By a theorem of Grauert, $\bar{U}$ may be chosen as a Levi strongly pseudoconvex neighborhood of C (since $C.C < 0$), and $H^1(\bar{U}, \mathcal{I}_C.N^*_{\mathcal{F}}) = 0$ (because we have $C.C < 2-2g$; more generally, $H^1(\bar{U}, \mathcal{I}_C.\mathcal{A}) = 0$ for any coherent sheaf $\mathcal{A}$ defined in $\bar{U}$). Consequently the map

$$H^1(\bar{U}, N^*_{\mathcal{F}}) \to H^1(\bar{U}, N^*_{\mathcal{F}}/\mathcal{I}_C.N^*_{\mathcal{F}})$$

is injective. Let us consider some holomorphic extension $\bar{H}_i\bar{\omega}_i$ of $h_i\omega_i$ to $\bar{U}_i$. Now the cocycle $\{\psi_{ij}(F_j)dF_j\}$ restricted to C coincides with $\{\psi_{ij}(f_j)df_j = h_i\omega_i - h_j\omega_i\}$; by injectivity, $\{\psi_{ij}dF_j\} = \{\bar{H}_i\bar{\omega}_i - \bar{H}_j\bar{\omega}_j\}$ in $H^1(\bar{U}, N^*_{\mathcal{F}})$ since they have the same image in $H^1(\bar{U}, N^*_{\mathcal{F}}/\mathcal{I}_C.N^*_{\mathcal{F}})$. Consequently $\{\psi_{ij}dF_j\}$ is a 1-cobord: $\{\psi_{ij}dF_j\} = \{\tilde{\omega}_i - \tilde{\omega}_j\}$ for a collection $\{\tilde{\omega}_i\} \in C^0(\bar{U}, N^*_{\mathcal{F}})$. We define then $\bar{\eta}_i = \tilde{\eta}_i - \tilde{\omega}_i$. We have $d\bar{\omega}_i = \bar{\eta}_i \wedge \bar{\omega}_i$ since both sides vanish. □

Let us make some remarks:

- in the case we have an affine transverse structure, $\psi_{ij} = 0\ \forall i, j$ such that $U_i \cap U_j \neq \emptyset$. I follows that $\tilde{\eta}_i - \tilde{\eta}_j = d\log\Lambda_{ij}$, and there is no need of the negativity hypothesis on $C.C$.
- take now $g = 0$. Then the same proof applies when $C.C < -1$. If $C.C = -1$, then a neighborhood of C can be blown down to a neighborhood of $(0,0)$ in $\mathbb{C}^2$. We find directly the 1-forms $\bar{\omega}$ and $\bar{\eta}$ such that $\bar{\omega} = 0$ defines the foliation and $d\bar{\omega} = \bar{\eta} \wedge \bar{\omega}$.
- It can be readily seen that the third 1-form of the triplet can be taken as $\bar{\xi}_i = dH_i + H_i\bar{\eta}_i + \frac{H_i^2}{2}\bar{\omega}_i$ where H_i comes from $\tilde{\omega}_i = H_i\bar{\omega}_i\ \forall i$. This illustrates an interesting property of foliations with a projective transverse structure: the existence of the two first 1-forms of the triplet implies the existence of the third 1-form (see [4]).

References

[1] C. Camacho, B. Scardua. *Holomorphic foliations with Liouvillian first integrals*, Ergodic Theory Dynam. Systems 21 (2001) Volume **21**, Issue 03, pp. 717-756.

[2] G. Cousin, J. V. Pereira. *Transversely Affine Foliations on Projective Manifolds*, Mathematical Research Letters (2014) Volume 21 Number 5, pp. 989-1014.

[3] F. Loray, J.V. Pereira and F. Touzet. *Representations of quasiprojective groups, Flat connections and Transversely projective foliations* (arXiv:1402.1382 [math.AG])

[4] M. Brunella. *Sur les feuilletages de l'espace projectif ayant une composante de Kupka*, l'Enseignement Mathematique (2009) Volume **55**, pp. 227-234.

[5] B. Scardua. *Transversely affine and transversely projective holomorphic foliations*, Ann. Sci. de lcole Norm. Sup. (1997) (4) 30, pp. 169-204.

[6] F. Loray and J.V. Pereira. *Transversely Projective Foliations on Surfaces*, International Journal of Mathematics (2007) Vol. **18**, No. 6, pp. 723-747.

Paulo Sad
Instituto de Matemática Pura e Aplicada (IMPA)
Estrada Dona Castorina 110, Jardim Botânico
22460-320 Rio de Janeiro
Brazil
e-mail: `sad@impa.br`

Chern Classes and Transversality for Singular Spaces

Jörg Schürmann

Dedicated to Pepe Seade on his 60th birthday

Abstract. In this paper we compare different notions of transversality for possible singular complex algebraic or analytic subsets of an ambient complex manifold and prove a refined intersection formula for their Chern-Schwartz-MacPherson classes. In case of a transversal intersection of complex Whitney stratified sets, this result is well known. For splayed subsets it was conjectured (and proven in some cases) by Aluffi and Faber. Both notions are stronger than a micro-local "non-characteristic intersection" condition for the characteristic cycles of (associated) constructible functions, which nevertheless is enough to imply the asked refined intersection formula for the Chern-Schwartz-MacPherson classes. The proof is based the multiplicativity of Chern-Schwartz-MacPherson classes with respect to cross products, as well as a new Verdier-Riemann-Roch theorem for "non-characteristic pullbacks".

Mathematics Subject Classification (2000). 14C17, 14C40, 32S60.

Keywords. Chern classes, transversality, Euler obstruction, non-characteristic pullback, characteristic cycle, Verdier-Riemann-Roch.

1. Introduction

In this paper we work in the embedded complex analytic or algebraic context, with X and Y closed (maybe singular) subspaces in the ambient complex manifold M. And we want to show under suitable "transversality assumptions" the following refined intersection formula for their Chern-Schwartz-MacPherson classes:

$$d^! \left(c_*(X) \times c_*(Y)\right) = c(TM) \cap c_*(X \cap Y) \in H_*(X \cap Y)\,. \tag{1.1}$$

Here $H_*(X)$ denotes either the Borel-Moore homology group in even degrees $H_{2*}^{BM}(X, \mathbb{Z})$ or in the algebraic context the Chow group $CH_*(X)$, with

$$d^! : H_*(X \times Y) \to H_*(X \cap Y)$$

the corresponding refined pullback for the (regular) diagonal embedding $d : M \to M \times M$ of the ambient complex manifold M (as recalled in the next section). Note that for X compact, the topological Euler characteristic $\chi(X)$ of X is given by

$$\chi(X) = deg(c_*(X)) = deg(c_0(X)) .$$

So formula (1.1) shows that in general for X and Y compact the Euler characteristic $\chi(X \cap Y)$ of the intersection cannot be given just in terms of $\chi(X)$ and $\chi(Y)$, but that the information of their total Chern-Schwartz-MacPherson classes $c_*(X)$ and $c_*(Y)$ is needed.

For X and Y smooth complex submanifolds, all these different notions of "transversality" for singular subspaces just reduce to the classical notion of transversality, so that $X \cap Y$ also becomes a smooth complex submanifold of M, with normal bundle

$$N_{X\cap Y}M = N_X M|_{X\cap Y} \oplus N_Y M|_{X\cap Y} .$$

And then (1.1) easily follows from the fact, that

$$c_*(Z) = c(TZ) \cap [Z] = c(N_Z M)^{-1} \cap (c(TM) \cap [Z])$$

for Z a closed smooth complex submanifold of M (with c the total Chern class). Also recall that the smooth complex submanifolds X and Y of M intersect transversally, iff the diagonal embedding $d : M \to M \times M$ is transversal to $X \times Y$, with $d^{-1}(X \times Y) = X \cap Y$ and

$$N_{X\cap Y}M = d^*(N_{X\times Y}(M \times M)) = d^*(N_X M \times N_Y M) .$$

And this last viewpoint can be generalized in different ways to singular complex subspaces.

Maybe the best known notion of tranversality for singular X and Y is the transversality as complex Whitney stratified subsets, i.e., both are endowed with complex Whitney b-regular stratifications such that all strata S of X and S' of Y are transversal. Equivalently, the diagonal embedding d is transversal to all strata $S \times S'$ of the induced product Whitney stratification of $X \times Y$. And then the intersection formula (1.1) is well known, see, e.g., [10][Thm. 3.3] or [28][Cor. 0.1 and the discussion afterwards]. Another notion of transversality for singular complex subspaces X and Y in the ambient complex manifold M was studied by Aluffi and Faber [2, 3] (and first introduced and characterized by Faber [11] in the hypersurface case):

Definition 1.1. X and Y are *splayed at a point* $p \in M$, if there is near p a local analytic isomorphism $M = V_1 \times V_2$ of analytic manifolds so that X resp. Y can be defined by an ideal in the coordinates of V_1 resp. V_2. X and Y are *splayed* if they are splayed at all points $p \in X \cap Y$.

Note that also in the complex algebraic context, these local coordinates are only asked for in the local analytic context. And Aluffi and Faber [2, 3] conjectured (and could prove in some cases) the intersection formula (1.1) for X and Y splayed.

The problem is of course, that both notions of transversality cannot be directly compared and are of very different type, with "stratified transversality" more of geometric and "splayedness" more of algebraic nature.

We will gereralize in the next section both notions even to constructible functions $\alpha \in F(X)$ and $\beta \in F(Y)$, showing that both are stronger than the micro-local "non-characteristic intersection" condition, that the diagonal embedding $d : M \to M \times M$ is "non-characteristic" with respect to the support

$$\mathrm{supp}(CC(\alpha \times \beta)) \subset T^*(M \times M)$$

of the characteristic cycle of $\alpha \times \beta$. Nevertheless this micro-local "non-characteristic intersection" condition implies the following generalization of the refined intersection formula (1.1) even for the Chern-Schwartz-MacPherson classes of constructible functions:

Theorem 1.2. *Let X, Y be two closed subspaces of the complex (algebraic) manifold M with given constructible functions $\alpha \in F(X)$ and $\beta \in F(Y)$. Assume that the diagonal embedding $d : M \to M \times M$ is non-characteristic with respect to* $\mathrm{supp}(CC(\alpha \times \beta))$ *(e.g., α and β are splayed or stratified transversal). Then*

$$d^! \left(c_*(\alpha) \times c_*(\beta)\right) = c(TM) \cap c_*(\alpha \cdot \beta) \in H_*(X \cap Y)\,. \tag{1.2}$$

In particular

$$c_*(\alpha) \cdot c_*(\beta) = c(TM) \cap c_*(\alpha \cdot \beta) \in H_*(M)\,. \tag{1.3}$$

By definition of the MacPherson Chern class in [23], $c_*(Z) := c_*(1_Z)$ for $Z = X, Y$ or $X \cap Y$, so that (1.2) implies the formula (1.1) by $1_X \cdot 1_Y = 1_{X\cap Y}$. Let us mention here, that Brasselet and Schwartz [9] (see also [1]) showed that the MacPherson's Chern class $c_*(1_X)$ corresponds to the Schwartz class $c^S(X) \in H^{2*}_X(M)$ (see [34, 35]) by Alexander duality for X embedded in the smooth complex manifold M. That is why the total homology class $c_*(X) = c_*(1_X)$ is called the Chern–Schwartz–MacPherson class of X.

Other natural Chern classes of a singular complex algebraic or analytic set Z are the Aluffi-Chern class $c^A_*(Z) := c_*(\nu_Z)$ defined by the constructible Behrend function ν_X introduced in [4], or for Z pure-dimensional the Mather-Chern class $c^M_*(Z) := c_*(Eu_Z)$ defined by the famous constructible Euler obstruction function Eu_Z of MacPherson [23]. Both functions ν_Z and Eu_Z commute with cross-products, restriction to open subsets and switching from the algebraic to the analytic context, with $Eu_Z = 1_Z = (-1)^{\dim(Z)} \cdot \nu_Z$ for Z smooth.

Corollary 1.3. *Let X, Y be two closed splayed subspaces of the complex manifold M, with $m = dim(M)$. Then also ν_X and ν_Y are splayed, with $\nu_X \cdot \nu_Y = (-1)^m \cdot \nu_{X\cap Y}$ so that*

$$d^! \left(c^A_*(X) \times c^A_*(Y)\right) = (-1)^m \cdot c(TM) \cap c^A_*(X \cap Y) \in H_*(X \cap Y)\,. \tag{1.4}$$

If in addition X and Y are pure-dimensinal, then also Eu_X and Eu_Y are splayed, with $X \cap Y$ pure-dimensional and $Eu_X \cdot Eu_Y = Eu_{X\cap Y}$ so that

$$d^! \left(c_*^M(X) \times c_*^M(Y)\right) = c(TM) \cap c_*^M(X \cap Y) \in H_*(X \cap Y)\,. \tag{1.5}$$

The proof of theorem 1.2 is based on the multiplicativity of Chern-Schwartz-MacPherson classes for cross products (see [21, 22]):

$$c_*(\alpha \times \beta) = c_*(\alpha) \times c_*(\beta)\,,$$

as well as the following new Verdier-Riemann-Roch theorem for "non-characteristic pullbacks" (applied to the diagonal embedding $d : M \to M \times M$):

Theorem 1.4. *Let $f : M \to N$ be a morphism of complex (algebraic) manifolds, with $Y \subset N$ a closed subspace and $X := f^{-1}(Y) \subset M$. Assume that $\gamma \in F(Y)$ is a constructible function such that f is* non-characteristic *with respect to the support* $\mathrm{supp}(CC(\gamma)) \subset T^*N|Y \subset T^*N$ *of the characteristic cycle $CC(\gamma)$ of γ. Then*

$$f^!(c(TN)^{-1} \cap c_*(\gamma)) = c(TM)^{-1} \cap c_*(f^*(\gamma)) \in H_*(X)\,, \tag{1.6}$$

with $f^! : H_*(Y) \to H_*(X)$ the Gysin map induced by the morphism $f : M \to N$ of complex (algebraic) manifolds.

This Verdier-Riemann-Roch theorem for "non-characteristic pullbacks" is the main result of this paper, from which the refined intersection formula (1.2) directly follows in the spirit of Lefschetz' definition of intersection theory via cross-products and (refined) pullbacks for the diagonal map (as explained in the next section). The proof of Theorem 1.4 is based on the micro-local approach to Chern-Schwartz-MacPherson classes via characteristic cycles of constructible functions (as recalled in the last section, and see, e.g., [15, 16, 20, 26, 31]). Note that the micro-local notion of "non-characteristic" has its origin in the theory of (holonomic) $\mathcal{D}$-modules (as in [15, 16]) as well as in the micro-local sheaf theory of Kashiwara-Schapira [19]. Nevertheless, in our context we think of it as a micro-local "transversality condition" fitting nicely with the geometry of Chern-Schwartz-MacPherson classes of constructible functions. As a byproduct of our proof we also get the following "micro-local intersection formula":

Corollary 1.5. *Let M a complex (algebraic) manifold of dimension $m = dim(M)$, with $\alpha, \beta \in F(M)$ given constructible functions. Assume that the diagonal embedding $d : M \to M \times M$ is non-characteristic with respect to* $\mathrm{supp}(CC(\alpha \times \beta))$ *(e.g., $\alpha, \beta \in F(M)$ are splayed or stratified transversal), with* $\mathrm{supp}(\alpha \cdot \beta)$ *compact.*

Then also $\mathrm{supp}(CC(\alpha) \cap CC(\beta)) \subset T^*M$ *is compact, with*

$$\chi(M; \alpha \cdot \beta) = (-1)^m \cdot \deg(CC(\alpha) \cap CC(\beta))\,. \tag{1.7}$$

In the end of the next section we also illustrate some other situations, where such a "non-characteristic condition" follows from suitable "splayedness" assumptions. Characteristic cycles are of "cotangential nature", since they are living in the cotangent bundle T^*M of the ambient manifold M. So for the pullback situation of a holomorphic map $f : M \to N$ of complex manifolds we have to study the associated correspondence

$$T^*M \leftarrow f^*TN \to T^*N$$

of cotangent bundles. And here the "non-characteristic" condition shows automatically up for the definition of the pullback of a characteristic cycle living in T^*N. Let us finally point out, that there are also other *intrinsic* notions of Fulton- and Fulton-Johnson-Chern-classes (see [13][Example 4.2.6])

$$c_*^F(X) = c^*(TM) \cap s_*(C_XM)\,,\; c_*^{FJ}(X) = c^*(TM) \cap s_*(\mathcal{N}_XM) \in H_*(X)$$

of a singular complex variety $X \subset M$ embedded into a complex manifold M, which are more of "tangential nature" defined via Segre-classes of the normal cone C_XM resp. the cone associated to the conormal sheaf $\mathcal{N}_XM$ of X in M. And also for them the refined intersection formula (1.1) holds under the assumption that X and Y are *splayed* in M. For the Fulton-Chern-classes this was shown by Aluffi and Faber [3][Thm. III]. And in another paper we will give a different proof of this result, very close to the ideas of this paper, which will also apply to the Fulton-Johnson-Chern-classes.

Finally let us point out, that results similar to Theorem 1.2 and (1.1) are also true for the *Hirzebruch class transformation* T_{y*} of [7, 32] in the context of complex algebraic mixed Hodge modules, if one asks the "non-characteristic property" for the characteristic variety of the underlying (filtered) $\mathcal{D}$-modules. But in this case the proof is different and follows the lines of [28] (as explained elsewhere).

2. Splayed constructible functions

In this chapter we work in the embedded complex analytic or algebraic context, with X a closed subspace in the complex manifold M. Let $F(X) = F_{\mathrm{alg}}(X)$ or $F(X) = F_{\mathrm{an}}(X)$ be the group of $\mathbb{Z}$-valued constructible function, so that a constructible function $\alpha \in F(X)$ in the complex algebraic (resp. analytic) context is a (locally finite) linear combination of indicator functions 1_Z with $Z \subset X$ a closed irreducible subspace. Viewing an algebraic variety as an analytic variety, one gets a canonical injection $F_{\mathrm{alg}}(X) \hookrightarrow F_{\mathrm{an}}(X)$.

First we extend the definition of *splayedness* form subspaces to constructible functions:

Definition 2.1. Let X, Y be two closed subspaces of M with given constructible functions $\alpha \in F(X)$ and $\beta \in F(Y)$. Then α and β are *splayed at a point* $p \in M$, if there is near p a local analytic isomorphism $M = V_1 \times V_2$ of analytic manifolds so that $\alpha = \pi_1^*(\alpha')$ and $\beta = \pi_2^*(\beta')$ for some $\alpha' \in F(V_1)$ and $\beta' \in F(V_2)$, with $\pi_i : V_1 \times V_2 \to V_i$ the projection ($i = 1, 2$).
$\alpha \in F(X)$ and $\beta \in F(Y)$ are *splayed* if they are splayed at all points $p \in X \cap Y$.

So two algebraically constructible functions α, β are by definition splayed (at a point p), if this is the case for them viewed as analytically constructible functions. Let us give some examples:

Example 2.2. 1. If $\alpha \in F(M)$ is *locally constant*, then α and β are splayed for any $\beta \in F(Y)$. Just take a local isomorphism $U = \{pt\} \times U$ with α constant on U so that $\alpha = \alpha' \cdot 1_U = \pi_1^*(\alpha')$ for $\alpha' \in \mathbb{Z} = F(pt)$, with $\pi_1 : \{pt\} \times U \to \{pt\}$ the projection.

So if one wants to show that two constructible functions $\alpha \in F(X)$ and $\beta \in F(Y)$ are splayed, then one only needs to check this at all points $p \in M$ were α and β are not (locally) constant near p. In particular the choice of the closed subspaces X and Y in Definition 2.1 doesn't matter.

2. Let X, Y be two closed subspaces of M. Then $\alpha = 1_X$ and $\beta = 1_Y$ are splayed as constructible funstions, if and only if X and Y are splayed as closed subspaces. Moreover, in this case also the constructible indicator functions of their complements $1_{M \setminus X}, 1_{M \setminus Y} \in F(M)$ are splayed.

For X a closed subspace in the complex manifold M, let $c_* : F(X) \to H_*(X)$ be the MacPherson Chern class transformation for constructible functions, were $H_*(X)$ denotes either the Borel-Moore homology group in even degrees $H_{2*}^{BM}(X, \mathbb{Z})$ or in the algebraic context the Chow group $CH_*(X)$. In the next chapter we will explain a possible definition of c_* in this embedded context $X \subset M$ via the theory of conic Lagrangian cycles in the cotangent bundle $T^*M|X$. This implies in the algebraic context directly the commutativity of the following diagram, with *cl* the cycle map (as in [13][Chapter 19]):

$$\begin{array}{ccc} F_{\text{alg}}(X) & \longrightarrow & F_{\text{an}}(X) \\ {\scriptstyle c_*}\downarrow & & \downarrow{\scriptstyle c_*} \\ CH_*(X) & \xrightarrow[cl]{} & H_{2*}^{BM}(X, \mathbb{Z}) \,. \end{array} \tag{2.1}$$

One of the main results of this paper is the following

Theorem 2.3. *Let X, Y be two closed subspaces of the complex (algebraic) manifold M with given* splayed *constructible functions $\alpha \in F(X)$ and $\beta \in F(Y)$. Then*

$$d^! \left(c_*(\alpha) \times c_*(\beta)\right) = c(TM) \cap c_*(\alpha \cdot \beta) \in H_*(X \cap Y) \,. \tag{2.2}$$

In particular

$$c_*(\alpha) \cdot c_*(\beta) = c(TM) \cap c_*(\alpha \cdot \beta) \in H_*(M) \,. \tag{2.3}$$

In the algebraic context

$$d^! : CH_*(X \times Y) \to CH_*(X \cap Y)$$

is the *refined Gysin map* ([13][Sec.6.2]) associated to the regular diagonal embedding $d : M \to M \times M$. In the analytic context it is the *refined pullback*

$$d^! : H_{2*}^{BM}(X \times Y, \mathbb{Z}) \to H_{2*}^{BM}(X \cap Y, \mathbb{Z})$$

which under Poincaré duality corresponds to the pullback

$$d^* : H^{2*}_{X\times Y}(M \times M, \mathbb{Z}) \to H^{2*}_{X\cap Y}(M, \mathbb{Z})$$

in cohomology with support.

Corollary 2.4. *Let* X, Y *be two* splayed *closed subspaces of the complex (algebraic) manifold* M. *Then*

$$d^!\,(c_*(X) \times c_*(Y)) = c(TM) \cap c_*(X \cap Y) \in H_*(X \cap Y)\,, \tag{2.4}$$

in particular

$$c_*(X) \cdot c_*(Y) = c(TM) \cap c_*(X \cap Y) \in H_*(M)\,. \tag{2.5}$$

Similarly

$$c_*(M\backslash X) \cdot c_*(M\backslash Y) = c(TM) \cap c_*(M\backslash(X \cup Y)) \in H_*(M)\,. \tag{2.6}$$

The proof of Theorem 2.3 uses first the multiplicativity

$$c_*(\alpha \times \beta) = c_*(\alpha) \times c_*(\beta) \tag{2.7}$$

of the MacPherson Chern class transformation (see [21, 22]), which by induction on the dimension of the support of the constructible functions and resolution of singularties follows from the Chern class formula

$$c(TM \times TM') = c(TM) \times c(TM')$$

for the Chern classes of complex (algebraic) manifolds M, M'.

The second main ingredient (explained in the next section) is a *Verdier-Riemann-Roch Theorem* for the behaviour of MacPherson Chern classes under a *non-characteristic* pullback for a morphism $f : M \to N$ of complex (algebraic) manifolds, based on the Lagrangian approach to MacPherson Chern classes of constructible functions via the *characteristic cycle* map

$$CC : F(Y) \xrightarrow{\sim} L(Y, N)$$

to the group $L(Y,N) = L_{\mathrm{an}}(Y,N)$ (resp. $L(Y,N) = L_{\mathrm{alg}}(Y,N)$) of conic Lagrangian cycles in $T^*N|Y$ for Y a closed subspace of the complex (algebraic) manifold N. The characteristic cycle map CC is characterized by (see [29][(6.35), p. 293 and p. 323–324])

$$CC(Eu_Z) = (-1)^{\dim(Z)} \cdot [T^*_Z N] \tag{2.8}$$

for $Z \subset Y$ a closed irreducible subspace. Here $Eu_Z \in F(Z)$ is the famous *local Euler obstruction* of Z, with $Eu_Z|Z_{\mathrm{reg}}$ constant of value 1, and $T^*_Z N := \overline{T_{Z_{\mathrm{reg}}} N}$ the closure of the conormal space to the regular part of Z. In particular CC is compatible with switching from the complex algebraic to the complex analytic context.

Theorem 2.5. *Let $f : M \to N$ be a morphism of complex (algebraic) manifolds, with $Y \subset N$ a closed subspace and $X := f^{-1}(Y) \subset M$. Assume that $\gamma \in F(Y)$ is a constructible function such that f is* non-characteristic *with respect to the support* $\mathrm{supp}(CC(\gamma)) \subset T^*N|Y \subset T^*N$ *of the characteristic cycle $CC(\gamma)$ of γ. Then*

$$f^!(c(TN)^{-1} \cap c_*(\gamma)) = c(TM)^{-1} \cap c_*(f^*(\gamma)) \in H_*(X)\,, \tag{2.9}$$

with $f^! : H_*(Y) \to H_*(X)$ the Gysin map induced by the morphism $f : M \to N$ of complex (algebraic) manifolds.

Also note that

$$f^!\left(c(TN)^{-1} \cap (-)\right) = f^*(c(TN)^{-1}) \cap f^!(-) = c(f^*TN)^{-1} \cap f^!(-)\,.$$

Before we recall the definition of *non-characteristic*, we need to introduce the following commutative diagram (whose right square is cartesian, see, for example, [19][(4.3.2), p. 199] or [29][(4.15), p. 249]):

$$\begin{array}{ccccc} T^*M|X & \xleftarrow{t=t_f} & f^*(T^*N|Y) & \xrightarrow{f'} & T^*N|Y \\ \downarrow{\scriptstyle \pi_X} & & \downarrow{\scriptstyle \pi} & & \downarrow{\scriptstyle \pi_Y} \\ X & = & X & \xrightarrow{f} & Y\,. \end{array} \tag{2.10}$$

Here f' is the map induced by base change, whereas t is the dual of the differential of f. Then f is by definition *non-characteristic* with respect to a closed conic subset $\Lambda \subset T^*N|Y$ (i.e., a closed complex analytic (or algebraic) subset invariant under the $\mathbb{C}^*$-action given by multiplication on the fibers of the vector bundle $T^*N|Y$), if

$$f'^{-1}(\Lambda) \cap Ker(t) \subset f^*(T^*_N N|Y)\,, \tag{2.11}$$

with $f^*(T^*_N N|Y)$ the zero section of the vector bundle $f^*(T^*N|Y)$ (compare also with [19][Def. 5.4.12] or [29][p. 255]).

If, for example, $f : M \to N$ is a *submersion*, then $t : f^*(T^*N|Y) \hookrightarrow T^*M|X$ is an injection so that $\mathrm{Ker}(t) = f^*(T^*_N N|Y)$ is just the zero section of the vector bundle $f^*(T^*N|Y)$. So in this case f is non-characteristic with respect to any closed conic subset $\Lambda \subset T^*N|Y$.

Corollary 2.6. *Let $f : M \to N$ be a submersion of complex (algebraic) manifolds, with T_f the bundle of tangents to the fibers of f. Let $Y \subset N$ a closed subspace and $X := f^{-1}(Y) \subset M$. Then*

$$c(T_f) \cap f^!(c_*(\gamma)) = c_*(f^*(\gamma)) \in H_*(X) \tag{2.12}$$

for any $\gamma \in F(Y)$, with $f^! : H_*(Y) \to H_*(X)$ the Gysin map induced by the smooth morphism $f : X \to Y$.

Note that $c(TM) = c(T_f) \cup c(f^*TN)$ due to the short exact sequence of vector bundles

$$0 \to T_f \to TM \to f^*TN \to 0\,.$$

This Verdier-Riemann-Roch theorem is true for any smooth morphism $f : X \to Y$ of complex (algebraic) varieties (see [36], [14][p. 111] and [7][Cor. 3.1(3)]). The Verdier-Riemann-Roch theorem for a *smooth* morphism also follows from the existence of a corresponding *bivariant* Chern class transformation as in [5, 8] (as explained in [14][Prop. 6B, p. 67] for a corresponding bivariant Stiefel-Whitney class transformation). But here in this paper we are interested in the case of a closed *embedding* $i : M \to N$ of complex (algebraic) manifolds, which is not covered by the bivariant context (compare [14][Sec. 10.7, p. 112]).

Corollary 2.7. *Let $i : M \to N$ be a closed embedding of complex (algebraic) manifolds, with normal bundle $T_M N$. Let $Y \subset N$ be a closed subspace, with $X := i^{-1}(Y) = M \cap Y \subset M$. Assume that $\gamma \in F(Y)$ is a constructible function such that i is* non-characteristic *with respect to the support* supp$(CC(\gamma))$ *of the characteristic cycle $CC(\gamma)$ of γ. Then*

$$i^!(c_*(\gamma)) = c(T_M N) \cap c_*(i^*(\gamma)) \in H_*(X)\,, \tag{2.13}$$

with $i^! : H_*(Y) \to H_*(X)$ the Gysin map induced by the regular embedding $i : M \to N$.

Note that $c(i^*TN) = c(TM) \cup c(T^*_M N)$ due to the short exact sequence of vector bundles

$$0 \to TM \to i^*TN \to T_M N \to 0\,.$$

Consider, for example, a stratification of Y by locally closed complex (algebraic) submanifolds $S \subset Y$, which is *Whitney a-regular*, i.e., such that

$$\Lambda := \bigcup_S T^*_S N \subset T^*N|Y$$

is *closed*. Then the embedding i is non-characteristic with respect to Λ, if and only if M is *transversal* to all strata S of Y (see [29][p.255]). So the property *non-characteristic* is a micro-local version (or generalization) of a stratified transversality condition. If this is the case, then i is by (2.8) non-characteristic with respect to all $Eu_{\bar{S}}$ and for Y irreducible (or pure dimensional) also to Eu_Y.

If the stratification is also Whitney b-regular (which also implies a-regularity), then

$$\mathrm{supp}(CC(\gamma)) \subset \Lambda = \bigcup_S T^*_S N$$

for any γ which is constructible with respect to this stratification, i.e., such that $\gamma|S$ is locally constant for all S (see [29][sec.5.0.3). So if M is transversal to all strata S of a Whitney b-regular stratification, then i is non-characteristic for all γ which are constructible with respect to this stratification. For example, the ambient manifold $N = P^n(\mathbb{C})$ is a complex projective space and $M = H$ is a *generic* hyperplane.

Remark 2.8. For another approach to Corollary 2.7 based on *Verdier specialization* instead of the theory of "non-characteristic pullback for Lagrangian cycles" see [28][Cor.0.1, p.7].

In this paper we are especially interested in the *diagonal embedding* $d: M \to M \times M$ for M a complex (algebraic) manifold, so that $t: T^*M \times_M T^*M \to T^*M$ is just the addition map in the fibers. Also note that the normal bundle $T_M(M \times M)$ of the diagonal embedding d is isomorphic to TM.

Corollary 2.9. *Let M be a complex (algebraic) manifold, with $d: M \to M \times M$ the diagonal embedding. Let $Y \subset M \times M$ be a closed subspace, with $X := d^{-1}(Y) \subset M$. Assume that $\gamma \in F(Y)$ is a constructible function such that d is* non-characteristic *with respect to the support* $\mathrm{supp}(CC(\gamma))$ *of the characteristic cycle $CC(\gamma)$ of γ. Then*

$$d^!(c_*(\gamma)) = c(TM) \cap c_*(d^*(\gamma)) \in H_*(X)\,, \tag{2.14}$$

with $d^!: H_*(Y) \to H_*(X)$ the Gysin map induced by the diagonal embedding $d: M \to M \times M$.

Consider, for example, two closed subspaces $X, X' \subset M$ with $Y := X \times X'$ and $d^{-1}(X \times X') = X \cap X'$. For $\alpha \in F(X)$ and $\beta \in F(X')$ we have the constructible function $\alpha \times \beta \in F(X \times X')$, with

$$\alpha \times \beta((p, p')) := \alpha(p) \cdot \beta(p') \quad \text{for all } p \in X, p' \in X'.$$

Then $d^*(\alpha \times \beta) = \alpha \cdot \beta$. Let α resp. β be constructible with respect to a Whitney b-regular stratification of X resp. X' with strata S resp. S'. Then $\alpha \times \beta$ is constructible with respect to the Whitney b-regular product stratification of $X \times X'$ with strata $S \times S'$. Assume now that these stratifications of X and X' are *transversal* in the sense that all strata S of X are transversal to all strata S' of X'. Then the diagonal embedding $d: M \to M \times M$ is transversal to the product stratification of $X \times X'$ so that d is *non-characteristic* with respect to $\alpha \times \beta$. So by Corollary 2.9 and the multiplicativity (2.7) of the MacPherson Chern class one gets in this way a proof of Theorem 1.2. with

$$\begin{aligned} d^!(c_*(\alpha) \times c_*(\beta)) &= d^!(c_*(\alpha \times \beta)) \\ &= c(TM) \cap c_*(\alpha \cdot \beta)) \in H_*(X \cap X')\,. \end{aligned} \tag{2.15}$$

Theorem 2.3 states the same result under the assumption that α and β are *splayed*, which should be seen as another transversality assumption, similar but different from the *stratified transversality* used above. In fact, splayedness implies *locally* in the analytic topology this stratified transversality: Assume $M = V_1 \times V_2$ is a product of complex analytic manifolds, with $\alpha = \pi_1^*(\alpha')$ and $\beta = \pi_2^*(\beta')$ for some $\alpha' \in F(V_1)$ and $\beta' \in F(V_2)$, with $\pi_i: V_1 \times V_2 \to V_i$ the projection $(i = 1, 2)$. Take a Whitney b-regular stratification of V_1 resp. V_2 with strata S resp. S', so that α' resp. β' are constructible with respect to them. Then α resp. β is constructible with respect to the two Whitney b-regular stratifications of M with strata $S \times V_2$ resp. $V_1 \times S'$, which (trivially) intersect transversaly. And this implies again the global non-characteristic property:

splayed $\Rightarrow$ local stratified transversality $\Rightarrow$ non-characteristic.

Then the proof of Theorem 2.3 follows (as before) from Corollary 2.9 and the multiplicativity (2.7) of the MacPherson Chern class. In some sense this implication together with Corollary 2.9 give a "mechanism obtaining intersection-theoretic identities from local analytic data" as asked for in [2][Rem. 3.5].

In the following we give an even simpler proof of the implication

$$\text{splayed} \quad \Rightarrow \quad \text{non-characteristic},$$

which also applies to other interesting situations. To shorten the notation, and also to better emphasize the underlying principle of proof, let us introduce the *closed conic* subset

$$co(\alpha) := \operatorname{supp}(CC(\alpha)) \subset T^*M|X \subset T^*M$$

for $\alpha \in F(X)$ with X a closed subspace of the complex (algebraic) manifold. The three formal properties we need are:

(co1) *co* is locally defined in the sense that it commutes with restriction to open submanifolds of M.

(co2) We have the multiplicativity

$$co(\alpha \times \beta) \subset co(\alpha) \times co(\beta) \subset T^*M \times T^*M = T^*(M \times M).$$

(co3) For the projection $\pi : M \times N \to M$ from the product of two manifolds one has

$$co(\pi^*(\alpha)) \subset co(\alpha) \times T^*_N N \subset T^*M \times T^*N = T^*(M \times N),$$

with $T^*_N N$ the zero section of T^*N.

Note that in the case of $co(\alpha) := \operatorname{supp}(CC(\alpha))$ the support of the characteristic cycle $CC(\alpha)$ of a constructible function, these properties follow for example from (2.8) together with the multiplicativity (which holds in the algebraic context over a base field of characteristic zero):

$$Eu_{Z \times Z'} = Eu_Z \times Eu_{Z'} \,. \tag{2.16}$$

Lemma 2.10. *Let X, Y be two closed subspaces of the complex (algebraic) manifold M with given* splayed *constructible functions $\alpha \in F(X)$ and $\beta \in F(Y)$. Then the diagonal embedding $d : M \to M \times M$ is* non-characteristic *with respect to $co(\alpha \times \beta)$.*

Proof. Note that even if we are working in a complex algebraic context, the non-characteristic property in this context follows already from the corresponding non-characteristic property of the associated complex analytically constructible functions. Moreover, by (co1) this property can be locally checked in the analytic topology. By the splayedness condition, we can assume $M = V_1 \times V_2$ is a product of analytic manifolds, with $\alpha = \pi_1^*(\alpha')$ and $\beta = \pi_2^*(\beta')$ for some $\alpha' \in F(V_1)$ and $\beta' \in F(V_2)$, with $\pi_i : V_1 \times V_2 \to V_i$ the projection $(i = 1, 2)$. But then one gets by (co3):

$$co(\alpha) \subset co(\alpha') \times T^*_{V_2} V_2 \subset T^*V_1 \times T^*_{V_2} V_2$$

and

$$co(\beta) \subset T^*_{V_1} V_1 \times co(\beta')) \subset T^*_{V_1} V_1 \times T^*V_2 \,.$$

Assume now that $t((p,\omega),(p,\omega')) = (p,\omega+\omega') = 0 \in T^*M|p$ for some

$$((p,\omega),(p,\omega')) \in co(\alpha\times\beta) \subset co(\alpha)\times co(\beta)\,.$$

Then $\omega' = -\omega$ and

$$(p,\omega) \in co(\alpha) \cap a_*(co(\beta))\,,$$

with $a : T^*M \to T^*M$ the antipodal map. Therefore (p,ω) belongs to

$$(T^*V_1 \times T^*_{V_2}V_2) \cap (T^*_{V_1}V_1 \times T^*V_2) = T^*_{V_1}V_1 \times T^*_{V_2}V_2\,,$$

the zero section of T^*M. □

Let us finish this section with some other interesting examples, where a similar notion of splayedness can be introduced.

Example 2.11. Let us consider the MacPherson Chern class transformation $c_* : F(X) \to CH_*(X)$ in the embedded algebraic context over a base field of characteristic zero (see, e.g., [20, 31]). Then one can introduce the notion of splayedness for two constructible functions as before, but asking the corresponding "splitting" locally in the Zariski- or étale topology (which in the complex algebraic context is of course much stronger than working locally in the analytic topolgy). Using the corresponding characteristic cycle map CC characterized by (2.8), all of the results and their proofs of this and the following chapter apply. For the important comparison of the non-characteristic pullback of characteristic cycles with the corresponding pullback for constructible function as in Theorem 3.3, one can use, for example, the Lefschetz principle to reduce this to the complex algebraic context studied here.

Example 2.12. Consider the embedded real subanalytic or semi-algebraic context, with X a closed subanalytic (or semi-algebraic) subset in an ambient real analytic (Nash-) manifold M, and $F(X,\mathbb{Z}_2)$ the corresponding group of $\mathbb{Z}_2$-valued constructible functions. Then one can consider (as in [12, 31]) the corresponding characteristic cycle map

$$CC_2 : F(X,\mathbb{Z}_2) \xrightarrow{\sim} L(X,M)\otimes\mathbb{Z}_2$$

to conic subanalytic (or semi-algebraic) Lagrangian cycles with $\mathbb{Z}_2$ coefficients in $T^*M|X$. Since we are working with $\mathbb{Z}_2$-coefficients, no orientability of M is needed. But here conic just means $\mathbb{R}^+$-invariant. For the Lagrangian approch to Stiefel-Whitney classes of such constructible functions, it is important to work only with those constructible functions, for which $CC_2(\alpha)$ is also $\mathbb{R}^*$-invariant so that it can be projectivised as in the complex context. This just means $a_*CC_2(\alpha) = CC_2(\alpha)$ for the antipodal map $a : T^*M \to T^*M$, and corresponds by a beautiful observation of Fu and McCrory ([12]) to the classical *local Euler condition* of Sullivan (a sort of self duality). Introducing the local splayedness condition as before, one gets first the important property that for α and β *splayed* and both satisfying the *local Euler condition*, also $\alpha\cdot\beta$ satisfies the local Euler condition. For example, if X, Y are two splayed closed subanalytic (or semi-algebraic) *Euler* *subspaces* of M (i.e., such that $1_X, 1_Y$ satisfy the local Euler condition), then also the intersection $X \cap Y$ is an Euler space.

Let $F^{Eu}(X;\mathbb{Z}_2) \subset F(X,\mathbb{Z}_2)$ denote the corresponding subgroup of constructible functions satisfying the "local Euler condition". Then one can introduce the Stiefel-Whitney class transformation

$$w_* : F^{Eu}(X;\mathbb{Z}_2) \to H_*^{BM}(X,\mathbb{Z}_2)$$

with the help of the corresponding $\mathbb{R}^*$-invariant Lagrangian cycles very similar to the approach to the MacPherson Chern class c_* in the complex context discussed in the next chapter (see [12, 31]). Then all our results and proofs also apply to this context (as will be explained somewhere else).

If, for example, X, Y are two splayed closed subanalytic (or semi-algebraic) *Euler subspaces* of M, then

$$w_*(X) \cdot w_*(Y) = w(TM) \cap w_*(X \cap Y) \in H_*^{BM}(M,\mathbb{Z}_2)\,. \tag{2.17}$$

For a similar result about the Stiefel-Whitney class of the transversal intersection of two Euler spaces in the *pl*-context compare with [24].

Example 2.13. We consider the micro-local view on sheaf theory developed by Kashiwara and Schapira [19]. Here M is a real analytic manifold, and $F, G \in D^b(M)$ are bounded complexes of sheaves. Let us call them *splayed*, if there is near any given point $p \in M$ a local analytic isomorphism $M = V_1 \times V_2$ of analytic manifolds so that $F = \pi_1^*(F')$ and $G = \pi_2^*(G')$ for some $F' \in D^b(V_1)$ and $G' \in D^b(V_2)$, with $\pi_i : V_1 \times V_2 \to V_i$ the projection ($i = 1, 2$). Here the pullback just means the usual sheaf theoretic pullback.

Let, for example, $F \in D^b(M)$ be a sheaf complex, all of whose cohomology sheaves are locally constant. Then F and G are splayed for all $G \in D^b(M)$.

Back to the general case, assume only that F and G are splayed. Then the diagonal embedding $d : M \to M \times M$ is *non-characteristic* with respect to the *micro-support*

$$SS(F \boxtimes^L G) \subset T^*(M \times M)$$

of $F \boxtimes^L G$. In fact our proof of Lemma 2.10 applies to the closed $\mathbb{R}^+$-conic subset $co(F) := SS(F) \subset T^*M$, which satisfies the three properties (co1-3) by [19][Prop. 5.4.1, Prop. 5.4.5].

Assume that we are in the complex analytic context, with $F, G \in D_c^b(X)$ constructible sheaf complexes (with rational coefficients), which are perverse (up to a shift) for the *middle perversity* (see [19][Sec. 10.3] or [29][Chapter 6]). If F and G are splayed, then also

$$F \otimes^L G = d^*(F \boxtimes^L G)$$

is a perverse sheaf up to a shift (compare also with [19][Lem. 10.3.9]), since

$$F \otimes^L G = d^*(F \boxtimes^L G) \simeq d^!(F \boxtimes^L G)[2 \cdot \dim_{\mathbb{C}}(M)] \tag{2.18}$$

due to the non-characteristic property (see [19][Cor. 5.4.11]). Let, for example, X, Y be two *splayed* pure dimensional complex analytic subsets. Then also the corresponding middle perversity *intersection cohomology complexes* IC_X, IC_Y are splayed, since these commute with smooth pullback (see [29][Sec.6.0.2, Lem.6.0.3]. Here we are using the convention that $IC_X = \mathbb{Q}_X$ for X smooth). Then also $X \cap Y$ is pure dimensional of dimension $\dim_C(X) + \dim_C(Y) - \dim_C(M)$, with

$$IC_{X\cap Y} \simeq d^*(IC_X \boxtimes^L IC_Y)\,.$$

If X, Y are also compact, then one can consider the corresponding Goresky-MacPherson L-classes [17], and it becomes natural to ask for the following intersection formula:

$$L_*(X) \cdot L_*(Y) \stackrel{?}{=} L(TM) \cap L_*(X \cap Y) \in H_*(M, \mathbb{Q})\,, \tag{2.19}$$

with $L(TM)$ the Hirzebruch L-class. The Lagrangian approach taken up in this paper doesn't apply to the theory of L-classes of singular spaces. But the method of proof developed in [28] based on *Verdier specialization* might work, since the non-characteristic property of the diagonal embedding d with respect to the micro-support $SS(F \boxtimes^L G)$ implies by [19][Cor. 5.4.10(i)] that the support

$$\mathrm{supp}(\mu_M(F \boxtimes^L G))$$

of the *microlocalization* $\mu_M(F \boxtimes^L G)$ of two splayed sheaf complexes is contained in the zero-section of $T^*_M(M \times M)$.

For a result similar to (2.19) about the Goresky-MacPherson L-class of a transversal intersection in the *pl*-context compare with [25].

Example 2.14. We are working with coherent left $\mathcal{D}$-modules on the complex analytic manifold M (see, e.g., [19][Chapter XI]). Let us call two such modules $\mathcal{F}, \mathcal{G}$ *splayed*, if there is near any given point $p \in M$ a local analytic isomorphism $M = V_1 \times V_2$ of analytic manifolds so that $\mathcal{F} = \pi_1^{-1}(\mathcal{F}')$ and $\mathcal{G} = \pi_2^{-1}(\mathcal{G}')$ for some $\mathcal{D}$-modules $\mathcal{F}'$ on V_1 and $\mathcal{G}'$ on V_2, with $\pi_i : V_1 \times V_2 \to V_i$ the projection $(i = 1, 2)$. Here the pullback means this time the pullback of left $\mathcal{D}$-modules. Then the diagonal embedding $d : M \to M \times M$ is *non-characteristic* with respect to the *characteristic variety*

$$\mathrm{char}(\mathcal{F} \underline{\boxtimes} \mathcal{G}) \subset T^*(M \times M)$$

of the $\mathcal{D}$-module cross product $\mathcal{F} \underline{\boxtimes} \mathcal{G}$. In fact our proof of Lemma 2.10 applies to the closed complex analytic $\mathbb{C}^*$-conic subset

$$co(\mathcal{G}) := \mathrm{char}(\mathcal{G}) \subset T^*M\,,$$

which satisfies the properties (co1-3) by [19][(11.2.22), Prop.11.2.12]. Then also

$$\mathcal{F} \underline{\otimes} \mathcal{G} := d^{-1}(\mathcal{F} \underline{\boxtimes} \mathcal{G})$$

is a *coherent* $\mathcal{D}$-module by the non-characteristic property (see [19][Prop.11.2.12]). If we are working with *holonomic* $\mathcal{D}$-modules, then their characterictic varieties are also closed conic *Lagrangian* subsets of T^*M. Moreover, one can also directly define for a holonomic $\mathcal{D}$-module $\mathcal{G}$ its *characteristic cycle* $CC(\mathcal{G})$ supported on $\mathrm{char}(\mathcal{G})$. And it was Ginsburg [16], who first used from this $\mathcal{D}$-module view point the "non-characteristic" condition in his study of the behavior of the MacPherson

Chern class transformation with respect to a suitable convolution product (and compare also with [15] for his $\mathcal{D}$-module approach to the MacPherson Chern class transformation).

Finally one can similarly introduce the notion of splayedness for $F \in D^b(M)$ a bounded sheaf complex (with complex coefficients) and $\mathcal{G}$ a coherent left $\mathcal{D}$-module, using the usual (resp. $\mathcal{D}$-module) pullback for F resp. $\mathcal{G}$. Then the closed conic subsets $co(F) = SS(F) \subset T^*M$ and $co(\mathcal{G}) = \mathrm{char}(\mathcal{G}) \subset T^*M$ both satisfy the properties (co1) and (co3). Therefore one gets for F and $\mathcal{G}$ *splayed* as in the proof of Lemma 2.10 the estimate

$$SS(F) \cap \mathrm{char}(\mathcal{G}) \subset T^*_M M. \tag{2.20}$$

So if F is also a subanalytically constructible sheaf complex, one gets that $(F, \mathcal{G})$ is an *elliptic pair* in the sense of [27].

3. Pullback of Lagrangian cycles

In this section we work again in the embedded complex analytic or algebraic context, with Y a closed subspace in the complex manifold N. Let us first recall the main ingredients in MacPherson's definition of his (dual) Chern classes of a constructible function and the well known by now relation to the *theory of characteristic cycles* (cf. [15, 20, 26, 31]). Then the main characters of this story can be best visualized in the commutative diagram

$$\begin{array}{ccccc} F(Y) & \xleftarrow[\sim]{\check{E}u} & Z_*(Y) & \xrightarrow{\check{c}_*^{Ma}} & H_*(Y) \\ \| & & cn \downarrow \wr & & \| \\ F(Y) & \xrightarrow[\sim]{CC} & L(Y,M) & \xrightarrow{c(T^*M|Y) \cap s_*} & H_*(Y). \end{array} \tag{3.1}$$

Here $F(Y)$ and $Z_*(Y)$ are the groups of constructible functions and cycles in the corresponding complex analytic or algebraic context (where we allow locally finite sums in the analytic context). Similarly $H_*(Y)$ denotes the Borel-Moore homology group in even degrees $H^{BM}_{2*}(Y, \mathbb{Z})$ or in the algebraic context the Chow group $CH_*(Y)$.

The transformation $\check{E}u$ associates to a closed irreducible subset Z of Y the constructible function given by the *dual Euler obstruction*

$$\check{E}u_Z := (-1)^{\dim(Z)} \cdot Eu_Z$$

of Z, and is linearly extended to cycles. Then $\check{E}u$ is an isomorphism of groups, since $Eu_Z|Z_{\mathrm{reg}}$ is constant of value 1. The transformation $\check{c}_*^{Ma}$ is similarly defined by associating to an irreducible Z the total dual Chern-Mather class $\check{c}_*^{Ma}(Z)$ of Z viewed in $H_*(Y)$. One has the following description of the dual Chern-Mather class of Z in terms of the Segre class of the *conormal space* $T^*_Z N := \overline{T^*_{Z_{reg}} N}$ of Z in N, which is a conic Lagrangian subspace in $T^*N|X$:

$$\check{c}^{Ma}_*(Z) = c(T^*N|Z) \cap s_*(T^*_Z N) \tag{3.2}$$

after, e.g., [26, Lemme (1.2.1)] or [20, Lemma 1]. The Segre class is defined by (compare [13, Sec.4.1]):

$$\begin{aligned} s_*(T^*_Z N) &:= \hat{\pi}_*(c(\mathcal{O}(-1))^{-1} \cap [\mathbb{P}(T^*_Z N \oplus \mathbf{1})]) \\ &= \sum_{i \geq 0} \pi_*(c^1(\mathcal{O}(1))^i \cap [\mathbb{P}(T^*_Z N \oplus \mathbf{1})]). \end{aligned} \tag{3.3}$$

Here $\mathcal{O}(-1)$ denotes the tautological line subbundle on the projective completion $\hat{\pi} : \mathbb{P}(T^*N|_Z \oplus \mathbf{1}) \to Z$ with $\mathcal{O}(1)$ as its dual. Note that we work with the projective completion to loose no information contained in the zero section.

Example 3.1. Let Z be a closed complex submanifold of N, and consider the Lagrangian cycle $[T^*_Z N]$. Then $s_*([T^*_Z N]) = (c(T^*_Z N))^{-1} \cap [Z]$ and

$$c(T^*N|Z) \cap s_*([T^*_Z N]) = c(T^*N|Z) \cap (c(T^*_Z N))^{-1} \cap [Z] = c(T^*Z) \cap [Z].$$

Here we use the Whitney formula for the total Chern class c and the exact sequence of vector bundles:

$$0 \to T^*_Z N \to T^*N|Z \to T^*Z \to 0.$$

By definition, $L(Y, N)$ is the group of all cycles generated by the conormal spaces $T^*_Z N$ for $Z \subset Y$ closed and irreducible. The vertical map cn in diagram (3.1) is the correspondence $Z \mapsto T^*_Z N$. Then (3.2) obviously implies the commutativity of the right square in (3.1).

The *dual MacPherson Chern class transformation* is defined by

$$\check{c}_* := \check{c}^{Ma}_* \circ \check{E}u^{-1} : \; F(Y) \to H_*(Y). \tag{3.4}$$

This agrees up to a sign with MacPherson's original definition [23] of his Chern class transformation c_*, namely

$$\check{c}_i(\alpha) = (-1)^i \cdot c_i(\alpha) \in H_i(Y)\,. \tag{3.5}$$

The commutativity of the left square in (3.1) follows either by definition, if the characteristic cycle map CC is defined (as done in [20, 26]) by

$$CC(\check{E}u_Z) = [T^*_Z N] \tag{3.6}$$

for $Z \subset Y$ a closed irreducible subspace. Working in the complex algebraic or analytic context in the classical topology, one can also take another more refined definition based on *stratified Morse theory for constructible functions* (as done in [31, 33] and [29][Sec. 5.0.3]):

$$CC(\alpha) := \sum_S (-1)^{\dim(S)} \cdot \chi((NMD(S), \alpha)) \cdot [T^*_{\bar{S}} N] \tag{3.7}$$

for $\alpha \in F(Y)$ constructible with respect to a given Whitney b-regular stratification of Y with connected strata S. Here $\chi((NMD(S), \alpha))$ is the Euler characteristic of a suitable *normal Morse datum* $NMD(S)$ of Y weighted by α, which only depends on the stratum S (but the details are not needed in this paper). Then the equality (3.6) follows from [29][(6.35), p. 293 and p. 323–324].

Let us consider a morphism of manifolds $f : M \to N$ as in the diagram (2.10), with $X := f^{-1}(Y)$ and the same notations as in (2.10), e.g., with the induced map $f' : f^*(T^*N|Y) \to T^*N|Y$. Then we get a similar diagram for the *projective completions*:

$$\begin{array}{ccccc} \mathbb{P}(T^*M|X \oplus \mathbf{1}) & \xleftarrow{\hat{t}} & U \subset \mathbb{P}(f^*(T^*N|Y) \oplus \mathbf{1}) & \xrightarrow{\hat{f}} & \mathbb{P}(T^*N|Y \oplus \mathbf{1}) \\ \downarrow{\scriptstyle \hat{\pi}_X} & & \downarrow{\scriptstyle \hat{\pi}} & & \downarrow{\scriptstyle \hat{\pi}_Y} \\ X & = & X & \xrightarrow{f} & Y. \end{array} \tag{3.8}$$

The right square is cartesian, but the map $\hat{t}$ is only defined on the complement U of $\mathbb{P}(\mathrm{Ker}(t) \oplus \{0\})$. For the application to Segre-classes it is important to note, that

$$\hat{t}^*(\mathcal{O}_X(1)) \simeq (\hat{f}^*(\mathcal{O}_Y(1)))|U\,. \tag{3.9}$$

One has the following characterization (compare with [29][Lem. 4.3.1] for a counterpart in the real algebraic resp. analytic context):

Lemma 3.2. *Let $C \subset T^*N|Y$ be a closed conic complex algebraic resp. analytic subset. Then the following conditions are equivalent:*

1. $f : M \to N$ *is* non-characteristic *for C, i.e., $f'^{-1}(C) \cap \mathrm{Ker}(t)$ is contained in the zero section $f^*(T^*_N N|Y)$ of the vector bundle $f^*(T^*N|Y)$.*
2. *The map $t : f'^{-1}(C) \to T^*M$ is* proper *and therefore finite (since its fibers are subspaces of an affine vector space).*

Proof. Note that (2) $\Rightarrow$ (1) is obvious, since C is conic. So lets discuss the other implication. Looking at the projective completions, one gets a commutative diagram

$$\begin{array}{ccccc} T^*M|X & \xleftarrow{t} & f'^{-1}(C) & \xrightarrow{f'} & C \\ \downarrow & & \downarrow & & \downarrow \\ \mathbb{P}(T^*M|X \oplus \mathbf{1}) & \xleftarrow{\hat{t}} & \hat{f}^{-1}(\hat{C}) & \xrightarrow{\hat{f}} & \hat{C} \\ \downarrow{\scriptstyle \hat{\pi}_X} & & \downarrow{\scriptstyle \hat{\pi}} & & \downarrow{\scriptstyle \hat{\pi}_Y} \\ X & = & X & \xrightarrow{f} & Y. \end{array} \tag{3.10}$$

The upper vertical maps are the natural inclusions, and all squares except the lower left square are cartesian. Note that $\hat{t}$ is defined on $\hat{f}^{-1}(\hat{C})$, since $\hat{f}^{-1}(\hat{C}) \subset U$ by the assumption (1). Then $\hat{t}|\hat{f}^{-1}(\hat{C})$ is proper, since $\hat{\pi}_X \circ \hat{t} = \hat{\pi}$ and $\hat{\pi}, \hat{\pi}_X$ are proper. But then $t|f'^{-1}(C)$ is proper by base change. □

So if $f : M \to N$ is non-characterstic with respect to the closed conic complex algebraic resp. analytic subset $C \subset T^*N|Y$, we can define for the closed conic complex algebraic resp. analytic subset $C' := t(f'^{-1}(C)) \subset T^*M|X$ the induced group homomorphism

$$t_* \circ f'^! : H_*(C) \to H_*(C') . \tag{3.11}$$

Here we use the map $f' : T^*N \to f^*T^*N$ of ambient complex (algebraic) manifolds for the refined Gysin map $f'^! : H_*(C) \to H_*(f^{-1}(C))$. Note that t_* is degree preserving, whereas

$$f'^! : H_i(C) \to H_{i+m-n}(f^{-1}(C)) ,$$

with $n = \dim(N), m = \dim(M)$ and $m - n = \dim(f^*T^*N) - \dim(T^*N)$. Assume that C resp. C' are pure d- resp. d'-dimensional, with $d' := d + m - n$ (as will be the case for C Lagrangian with $d = n$ and $d' = m$). Then we get an induced pullback map of cycles:

$$t_* \circ f'^! : Z_d(C) \simeq H_d(C) \to H_{d'}(C') \simeq Z_{d'}(C')$$

behaving nicely with respect to *Segre classes*:

$$f^!(s_*([C])) = s_*(t_* f'^![C]) \in H_*(X) . \tag{3.12}$$

In fact, one has the following sequence of equalities (with the maps of projective completions as in (3.8)):

$$\begin{aligned} f^! s_*([C]) &= \sum_{i\geq 0} f^!\left(\hat{\pi}_{Y*}(c^1(\mathcal{O}_Y(1))^i \cap [\hat{C}])\right) \\ &= \sum_{i\geq 0} \hat{\pi}_*\left(\hat{f}^!(c^1(\mathcal{O}_Y(1))^i \cap [\hat{C}])\right) \\ &= \sum_{i\geq 0} \hat{\pi}_*\left(c^1(\hat{f}^*\mathcal{O}_Y(1))^i \cap (\hat{f}^![\hat{C}])\right) \\ &= \sum_{i\geq 0} \hat{\pi}_{X*}\hat{t}_*\left(c^1(\hat{t}^*\mathcal{O}_X(1))^i \cap (\hat{f}^![\hat{C}])\right) \\ &= \sum_{i\geq 0} \hat{\pi}_{X*}\left(c^1(\mathcal{O}_X(1))^i \cap (\hat{t}_*\hat{f}^![\hat{C}])\right) = s_*(t_* f^![C]) . \end{aligned}$$

Here we are using:

1. the base change isomorphism $f^!\hat{\pi}_{Y*} = \hat{\pi}_*\hat{f}^!$,
2. the compability $\hat{f}^!(c^1(\cdot) \cap (\cdot)) = c^1(\hat{f}^*(\cdot)) \cap \hat{f}^!(\cdot)$,
3. the isomorphism (3.9) together with $\hat{\pi}_* = \hat{\pi}_{X*}\hat{t}_*$ by the functoriality of push-down,
4. the projection formula $\hat{t}_*(c^1(\hat{t}^*\cdot) \cap (\cdot)) = c^1(\cdot) \cap \hat{t}_*(\cdot)$.

The non-characteristic pullback map (3.11) is also functorial in f. Let $g : V \to M$ be another morphism of complex (algebraic) manifolds, with $Z := g^{-1}(X) =$

$(f \circ g)^{-1}(Y)$. Consider the cartesian diagram

$$\begin{array}{ccccc} g^*f^*T^*N & \xrightarrow{g''} & f^*T^*N & \xrightarrow{f'} & T^*N \\ {\scriptstyle t'_f}\downarrow & & \downarrow{\scriptstyle t_f} & & \\ g^*T^*M & \xrightarrow[g']{} & T^*M & & \\ {\scriptstyle t_g}\downarrow & & & & \\ T^*V\,, & & & & \end{array} \tag{3.13}$$

with

$$f' \circ g'' = (f \circ g)' : g^*f^*T^*N = (f \circ g)^*T^*N \to T^*N$$

and

$$t_g \circ t'_f = t_{f \circ g} : g^*f^*T^*N = (f \circ g)^*T^*N \to T^*V\,.$$

Using Lemma 3.2, one easily gets that $f \circ g$ is non-characteristic with respect to the closed conic complex algebraic resp. analytic subset $C \subset T^*N|Y$, if and only if f is non-characteristic with respect to C and g is non-characteristic with respect to $C' := t(f'^{-1}(C)) \subset T^*M|X$. Moreover, in this case one gets

$$(t_{g*}g'^!) \circ (t_{f*}f'^!) = t_{g*}t'_{f*}g''^!f'^! = t_{f \circ g*}(f \circ g)^! : H_*(C) \to H_*(C'')\,, \tag{3.14}$$

with $C'' := t_g(g'^{-1}(C')) = t_{f \circ g}(f \circ g)'^{-1}(C) \subset T^*V|Z$. Here we are using the functorialities $t_{g*}t'_{f*} = t_{f \circ g*}$ and $g''^!f'^! = (f \circ g)'^!$ together with the base change isomorphism $g'^!t_{f*} = t'_{f*}g''^!$.

Now we can come to the main result of this chapter.

Theorem 3.3. *Let $f : M \to N$ be a morphism of complex (algebraic) manifolds of (complex) dimensions $m = \dim(M), n = \dim(N)$, and $Y \subset N$ be a closed subspace, with $X := f^{-1}(Y) \subset M$. Assume that f is non-characteristic with respect to the support $C := \mathrm{supp}(CC(\gamma)) \subset T^*N|Y$ of the characteristic cycle $CC(\gamma)$ of a constructible function $\gamma \in F(Y)$. Then $C' := t_f(f'^{-1}(C))$ is pure m-dimensional, with*

$$t_{f*}f'^!(CC(\gamma)) = (-1)^{m-n} \cdot CC(f^*(\gamma))\,. \tag{3.15}$$

In particular, the left hand side is a Lagrangian cycle in $T^*M|X$.

So by (3.2), (3.4) and (3.12) we get

$$\begin{aligned} f^!(c(T^*N)^{-1} \cap \check{c}_*(\gamma)) &= f^!s_*(CC(\gamma)) \\ &= (-1)^{m-n} \cdot s_*(CC(f^*(\gamma))) \\ &= (-1)^{m-n} \cdot c(T^*M)^{-1} \cap \check{c}_*(f^*(\gamma))\,. \end{aligned} \tag{3.16}$$

And this is just a reformulation of our main Theorem 2.5 in terms of the dual Chern MacPherson class, since $c^i(V^*) = (-1)^i \cdot c^i(V)$ for a complex vector bundle V. Recall that also $\check{c}_i(\gamma) = (-1)^i \cdot c_i(\gamma)$. Finally the sign $(-1)^{m-n}$ is disapearing in Theorem 2.5, since $t_{f*}f'^! : H_i(C) \to H_{i+m-n}(C')$.

Remark 3.4. Theorem 3.3 is similar to the result of [19][Prop.9.4.3] for constructible sheaf complexes in the context of real geometry, but formulated under the stronger assumption of the non-characteristic property with respect to the *micro-support* instead of the support of the characteristic cycles. Moreover, if one applies [19][Prop.9.4.3] to our complex analytic (or algebraic) context, then also the sign $(-1)^{m-n}$ is not appearing, because a different orientation convention is used for the ambient manifolds T^*N, T^*M (needed for the definition of the Gysin map $f'^!$. Compare [29][Rem.5.0.3]).

Let us now discuss the proof of Theorem 3.3. By using the graph embedding and the functoriality (3.14) of the non-characteristic pullback, one can consider separately the case of a submersion $f : M \to N$ of complex (algebraic) manifolds, and that of a closed embedding $i : M \to N$. Moreover it can be (étale) locally checked on $X \subset M$. So for the case of a submersion we can assume $f : M \times N \to N$ is a projection of complex manifolds. Then the claim follows from

$$t_{f*}f'^{-1}([T^*_Z N]) = [T^*_M M] \times [T^*_Z N]$$

and the mutiplicativity of the Euler obstruction

$$Eu_{M\times Z} = Eu_M \times Eu_Z = 1_M \times Eu_Z = f^*(Eu_Z)$$

for $Z \subset Y$ a closed irreducible subspace. Then

$$CC(Eu_Z) = (-1)^{\dim(Z)} \cdot [T^*_Z N]$$

and

$$CC(Eu_{M\times Z}) = (-1)^{\dim(Z)+\dim(M)} \cdot [T^*_M M] \times [T^*_Z N]\,.$$

This also explains the sign $(-1)^{\dim(M)}$ appearing, with $dim(M)$ the fiber dimension of the projection $f : M \times N \to N$. Here the multiplicativity of the Euler obstruction is also equivalent to the multiplicativity of the characteristic cycle map CC. If one uses the refined Morse theoretical definition (3.7), then this multiplicativity follows from [29][(5.6) and (5.24)], and the sign $(-1)^{\dim(M)}$ comes from the use of $(-1)^{\dim(S)}$ in the definition (3.7).

The local study of a closed embedding $i : M \to N$ can be reduced by induction (and the functoriality (3.14) of the non-characteristic pullback) to the case $i : M = \{f = 0\} \hookrightarrow N$ of the inclusion of a smooth hypersurface, with $f : N \to \mathbb{C}$ a submersion. Assume i is non-characteristic for $T^*_Z N$, with $Z \subset Y \subset N$ a closed irreducible subspace. This just means $df(M) \cap T^*_Z N = \emptyset$. By shrinking N, we can also assume $df(N) \cap T^*_Z N = \emptyset$. Then $f : Z_{\text{reg}} \to \mathbb{C}$ is also a submersion, in particular $Z \not\subset M = \{f = 0\}$. Consider the exact sequence of vector bundles on N:

$$0 \to < df >= \mathrm{Ker}(p) \to T^*N \to T^*_f \to 0\,,$$

with the projection $p : T^*N \to T^*_f$ dual to the inclusion $T_f \to TN$ of the subvector bundle of tangents to the fibers of f. Then $T^*_Z N \cap \mathrm{Ker}(p)$ is contained in the zero-section $T^*_N N$ of T^*N. Therefore $p : T^*_Z \to T^*_f$ is proper and finite (as in the proof of Lemma 3.2), with

$$p(T^*_Z N) = T^*_Z f := \overline{T^*_{Z_{\text{reg}}} f}$$

the *relative conormal space* of $f|Z$ (i.e., the closure of the fiberwise conormal bundle of $f : Z_{\text{reg}} \subset N \to \mathbb{C}$). In particular, $T^*_Z f$ is irreducible of dimension $n = \dim(N) = \dim(T^*_Z N)$. Moreover $p|T^*_{Z_{reg}} N$ is also injective so that

$$p_*([T^*_Z N]) = [T^*_Z f] \, .$$

Consider the cartesian diagram

$$\begin{array}{ccc} i^*T^*N & \xrightarrow{i'} & T^*N \\ {\scriptstyle t}\downarrow & & \downarrow{\scriptstyle p} \\ T^*M & \xrightarrow[k]{} & T^*_f \, , \end{array} \tag{3.17}$$

witk $k : T^*M = T^*_f|M \to T^*_f$ the inclusion of the fiber over $\{f = 0\} = M$. Then

$$t(i'^{-1}(T^*_Z N)) = k^{-1}(p(T^*_Z N)) \subset T^*M$$

is of pure dimension $n - 1 = \dim(M)$. Moreover, by base change we get

$$\begin{aligned} t_*(i'^!(CC(\check{E}u_Z)]) &= t_*(i'^!([T^*_Z N])) \\ &= k^!(p_*([T^*_Z N])) \\ &= k^!([T^*_Z f]) \, . \end{aligned} \tag{3.18}$$

And by [26][Thm.4.3] one has

$$k^!([T^*_Z f]) = CC(-\psi_f(\check{E}u_Z)) \, , \tag{3.19}$$

with $\psi_f : F(Y) \to F(X)$ the *nearby cycles* for constructible functions, which are calculated as weighted Euler characteristics of local Milnor fibers (compare [31][2.4.7]). Then the sign in (3.19) is again due to the use of $(-1)^{\dim(S)}$ in the definition (3.7), because this local Milnor fiber is transversal to the strata S of an adapted Whitney stratification, cutting down the complex dimension by one.

Finally we only have to show $\psi_f(\check{E}u_Z) = i^*(\check{E}u_Z)$. But the difference is just the *vanishing cycles* (see [31][2.4.8]):

$$\phi_f(\check{E}u_Z) := \psi_f(\check{E}u_Z) - i^*(\check{E}u_Z) \, ,$$

with $\phi_f(\check{E}u_Z)(x) = 0$ for all $x \in X$ by [6][Thm.3.1]. Or one can use here [30][(15)] as an application of the *micro-local intersection formula* (see [30][(13),(14)]):

$$-\phi_f(\check{E}u_Z)(x) = \sharp_{df_x}(df(N) \cap CC(\check{E}u_Z)) \, , \tag{3.20}$$

since $CC(\check{E}u_Z) = [T^*_Z N]$ and $df(N) \cap T^*_Z N = \emptyset$ by the non-characteristic assumption.

Let us finish this paper with another nice application of Theorem 3.3.

Corollary 3.5. *Let M a complex (algebraic) manifold of dimension $m = \dim(M)$, with $\alpha, \beta \in F(M)$ given constructible functions. Assume that the diagonal embedding $d : M \to M \times M$ is non-characteristic with respect to* $\mathrm{supp}(CC(\alpha \times \beta))$ *(e.g., $\alpha, \beta \in F(M)$ are splayed or stratified transversal), with* $\mathrm{supp}(\alpha \cdot \beta)$ *compact.*

Then also $\mathrm{supp}(CC(\alpha) \cap CC(\beta)) \subset T^*M$ *is compact, with*

$$\chi(M; \alpha \cdot \beta) = (-1)^m \cdot \deg(CC(\alpha) \cap CC(\beta)) \,. \tag{3.21}$$

Proof. Consider the cartesian diagram

$$\begin{array}{ccccc} T^*M & \xrightarrow{(\mathrm{id},a)} & T^*M \times_M T^*M & \xrightarrow{d'} & T^*(M \times M) \\ {\scriptstyle \pi_M}\downarrow & & \downarrow{\scriptstyle t} & & \\ M & \xrightarrow[s]{} & T^*M \,, & & \end{array} \tag{3.22}$$

with $s : M \to T^*M$ the zero-section and $a : T^*M \to T^*M$ the antipodal map. Then one gets by the *global index formula* for characteristic cycles (see, e.g., [30][Cor. 0.1], which also covers the corresponding context in real geometry):

$$\chi(M; \alpha \cdot \beta) = k_* s^!(CC(\alpha \cdot \beta)) \,,$$

with $k : M \to pt$ a constant map. Here in our complex geometric context, this is also a very special case of the functoriality of the (dual) Chern MacPherson class with respect to the constant proper map $k : \mathrm{supp}(\alpha \cdot \beta) \to pt$, since

$$s^!(CC(\alpha \cdot \beta)) = \check{c}_0(\alpha \cdot \beta) = c_0(\alpha \cdot \beta) \in H_0(\mathrm{supp}(\alpha \cdot \beta)) \,. \tag{3.23}$$

And this equality follows from [13][Example 4.1.8]:

$$\{c(T^*M) \cap s_*(CC(\alpha \cdot \beta))\}_0 \,,$$

with $\{-\}_0$ the degree zero part, calculates for the m-dimensional conic cycle $CC(\alpha \cdot \beta)$ in the vector bundle T^*M of rank m the *intersection with the zero-section*. But by the commutative diagram (3.1) we know that

$$c(T^*M) \cap s_*(CC(\alpha \cdot \beta)) = \check{c}_*(\alpha \cdot \beta) \,.$$

By Theorem 3.3 and the multiplicativity of the characteristic cycle map CC we also have

$$CC(\alpha \cdot \beta) = (-1)^m \cdot t_* d'^!(CC(\alpha) \times CC(\beta)) \,.$$

And by the non-characteristic (or splayedness) assumption,

$$t : d'^{-1}(\mathrm{supp}(CC(\alpha) \times CC(\beta))) \to T^*M$$

is proper. But then by base change, also π_M as a map

$$(\mathrm{id}, a)^{-1} d'^{-1}(\mathrm{supp}(CC(\alpha) \times CC(\beta))) = \mathrm{supp}(CC(\alpha) \cap a_* CC(\beta)) \to M$$

is proper, with image contained in the compact subset $\mathrm{supp}(\alpha \cdot \beta) \subset M$. Note that $a^! = a_* : H_*(T^*M) \to H_*(T^*M)$, since $a^2 = \mathrm{id} : T^*M \to T^*M$. Finally by the base change $s^! t_* = \pi_{M*}(\mathrm{id}, a)^!$ one gets

$$\begin{aligned}\chi(M; \alpha \cdot \beta) &= (-1)^m \cdot k_* \pi_{M*}(CC(\alpha) \cap a_*(CC(\beta)) \\ &= (-1)^m \cdot \deg(CC(\alpha) \cap a_* CC(\beta)) .\end{aligned}$$

And in our complex context we also have $a_* CC(\beta) = CC(\beta)$. □

Remark 3.6. In [18] a counterpart of Corollary 3.5 in the context of real geometry is discussed under a "stratified tranversality" assumption. See also [27][(1.4), Part II] for a far reaching generalization to *elliptic pairs*. Note that both references use a different orientation convention, so that the sign $(-1)^m$ in (3.21) disappears.

Acknowledgment

This paper is an extended version of a talk given at a conference in Merida (Mexico 2014) for the celebration of the 60th birthday of Pepe Seade. Here I would like to thank the organizers for the invitation to this wonderful conference. It is a pleasure to thank Pepe Seade for so many discussions over the years on the theory of Chern classes of singular spaces. The author is also greatful to Paolo Aluffi und Eleonore Faber for their inspiring papers as well as for some communications regarding the subject of this paper. The author was supported by the SFB 878 groups, geometry and actions.

References

[1] P. Aluffi, and J.P. Brasselet: *Une nouvelle preuve de la concordance des classes définies par M.H. Schwartz et par R. MacPherson.* Bull. Soc. Math. France **136** (2008), 159–166.

[2] P. Aluffi, and E. Faber: *Splayed divisors and their Chern classes.* J. Lond. Math. Soc. **88** (2013), 563-579.

[3] P. Aluffi, and E. Faber: *Chern classes of splayed intersections.* Canadian Journal of Math. **67** (2015), 1201–1218.

[4] K. Behrend: *Donaldson-Thomas type invariants via microlocal geometry.* Ann. of Math. **170** (2009), 1307–1338.

[5] J.P. Brasselet: *Existence des classes de Chern en théorie bivariante.* Astérisque, **101–102** (1981), 7–22.

[6] J.P: Brasselet, Lê Dũng Tráng and J. Seade: *Euler obstruction and indices of vector fields.* Topology **39** (2000), 1193–1208.

[7] J.P. Brasselet, J. Schürmann and S. Yokura: *Hirzebruch classes and motivic Chern classes for singular spaces.* Journal of Topology and Analysis, **2**, (2010), 1–55.

[8] J.P. Brasselet, J. Schürmann and S. Yokura: *On the uniqueness of bivariant Chern classes and bivariant Riemann–Roch transformations.* Advances in Math., **210** (2007), 797–812.

[9] J.P. Brasselet and M.-H. Schwartz: *Sur les classes de Chern d'une ensemble analytique complexe.* Astérisque **82–83** (1981), 93–148.

[10] R. Callejas-Bedregal, M.F.Z. Morgado and J. Seade: *On the Milnor classes of local complete intersections.* arXiv:1208.5084

[11] E. Faber: *Towards transversality of singular varieties: splayed divisors.* Publ. Res. Inst. Math. Sci. **49** (2013), 393–412.

[12] J. Fu, and C. McCrory: *Stiefel-Whitney classes and the conormal cycle of a singular variety.* Trans. Amer. Math. Soc. **349** (1997), 809–835.

[13] W. Fulton: *Intersection theory*, Springer Verlag, 1984.

[14] W. Fulton and R. MacPherson: *Categorical frameworks for the study of singular spaces.* Memoirs of Amer. Math. Soc. **243**, 1981.

[15] V. Ginsburg: *Characteristic cycles and vanishing cycles.* Inv. Math. **84** (1986), 327–402.

[16] V. Ginsburg: *g-Modules, Springer's representations and bivariant Chern classes.* Adv. in Math. **61** (1986), 1–48.

[17] M. Goresky and R. MacPherson: *Intersection homology theory.* Topology **149** (1980), 155–162.

[18] M. Grinberg, and R. MacPherson: *Euler characteristics and Lagrangian intersections.* In: Symplectic geometry and topology, IAS/Park City Math. Ser. **7**, Amer. Math. Soc. Providence (1999), 265–293.

[19] M. Kashiwara, and P. Schapira: *Sheaves on Manifolds.* Springer, Berlin Heidelberg, 1990.

[20] G. Kennedy: *MacPherson's Chern classes of singular varieties.* Com. Algebra. **9** (1990), 2821–2839.

[21] M. Kwiecinski: *Formule de produit pour les classes caractéristiques de Chern-Schwartz-MacPherson et homologie d'intersection.* C.R. Acad. Sci. Paris **314** (1992), 625–628.

[22] M. Kwiecinski, and S. Yokura: *Product formula for twisted MacPherson Classes.* Proc. Japan Acad. **68** (1992), 167–171.

[23] R. MacPherson: *C*hern classes for singular varieties. Ann. of Math. **100** (1974), 423–432.

[24] A. Matsui: *Intersection formula for Stiefel-Whitney homology classes.* Tohoku Math. J. **40** (1988), 315–322.

[25] A. Matsui: *Hirzebruch L-homology classes and the intersection formula.* Kodai Math. J. **12** (1989), 56–71

[26] C. Sabbah: *Quelques remarques sur la géométrie des espaces conormaux.* Astérisque **130** (1985), 161–192.

[27] P. Schapira and J.P. Schneiders: *Index theorems for elliptic pairs.* Astérisque **224** (1994).

[28] J. Schürmann: *A generalized Verdier-type Riemann-Roch theorem for Chern-Schwartz-MacPherson classes.* math.AG/0202175

[29] J. Schürmann: *Topology of singular spaces and constructible sheaves.* Mathematics Institute of the Polish Academy of Sciences, Mathematical Monographs (New Series), **63**, Birkhäuser Verlag, Basel, 2003.

[30] J. Schürmann: *A general intersction formula for Lagrangian cycles.* Comp. Math. **140** (2004), 1037–1052.

[31] J. Schürmann: *Lectures on characteristic classes of constructible functions.* Trends Math.: Topics in cohomological studies of algebraic varieties (Ed. P. Pragacz), 175–201, Birkhäuser, Basel, 2005.

[32] J. Schürmann: *Characteristic classes of mixed Hodge modules.* in "Topology of Stratified Spaces", MSRI Publications Vol. **58**, Cambridge University Press (2010) , 419–470.

[33] J. Schürmann, and M. Tibăr: *Index formula for MacPherson cycles of affine algebraic varieties.* Tohoku Math. J. **62** (2010), 29-44.

[34] M.H. Schwartz: *Classes caractéristiques définies par une stratification d'une variétés analytique complexe.* C. R. Acad. Sci.Paris **t. 260** (1965), 3262–3264, 3535–3537.

[35] M.H. Schwartz: *Classes et caractères de Chern des espaces linéaires.* Pub. Int. Univ. Lille, *2 Fasc. 3* (1980) and C. R. Acad. Sci. Paris Sér. I **295** (1982), 399–402.

[36] S. Yokura: *On a Verdier-type Riemann-Roch for Chern-Schwartz-MacPherson class.* Topology and its Appl. **94** (1999), 315–327.

Jörg Schürmann
Mathematische Institut
Universität Münster
Einsteinstr. 62
48149 Münster
Germany
e-mail: `jschuerm@uni-muenster.de`

FSC
www.fsc.org
MIX
Papier aus verantwortungsvollen Quellen
Paper from responsible sources
FSC® C105338

If you have any concerns about our products,
you can contact us on
ProductSafety@springernature.com

In case Publisher is established outside the EU,
the EU authorized representative is:
Springer Nature Customer Service Center GmbH
Europaplatz 3, 69115 Heidelberg, Germany

Printed by Libri Plureos GmbH
in Hamburg, Germany